Advances in Intelligent Systems and Computing

Volume 1237

AF870064

The series "Advances in Intelligent Systems and Computing" contains publications on theory, applications, and design methods of Intelligent Systems and Intelligent Computing. Virtually all disciplines such as engineering, natural sciences, computer and information science, ICT, economics, business, e-commerce, environment, healthcare, life science are covered. The list of topics spans all the areas of modern intelligent systems and computing such as: computational intelligence, soft computing including neural networks, fuzzy systems, evolutionary computing and the fusion of these paradigms, social intelligence, ambient intelligence, computational neuroscience, artificial life, virtual worlds and society, cognitive science and systems, Perception and Vision, DNA and immune based systems, self-organizing and adaptive systems, e-Learning and teaching, human-centered and human-centric computing, recommender systems, intelligent control, robotics and mechatronics including human-machine teaming, knowledge-based paradigms, learning paradigms, machine ethics, intelligent data analysis, knowledge management, intelligent agents, intelligent decision making and support, intelligent network security, trust management, interactive entertainment, Web intelligence and multimedia.

The publications within "Advances in Intelligent Systems and Computing" are primarily proceedings of important conferences, symposia and congresses. They cover significant recent developments in the field, both of a foundational and applicable character. An important characteristic feature of the series is the short publication time and world-wide distribution. This permits a rapid and broad dissemination of research results.

**** Indexing: The books of this series are submitted to ISI Proceedings, EI-Compendex, DBLP, SCOPUS, Google Scholar and Springerlink ****

More information about this series at http://www.springer.com/series/11156

Yucheng Dong · Enrique Herrera-Viedma ·
Kenji Matsui · Shigeru Omatsu ·
Alfonso González Briones ·
Sara Rodríguez González
Editors

Distributed Computing and Artificial Intelligence, 17th International Conference

Springer

Editors
Yucheng Dong
Business School
Sichuan University
Chengdu, China

Kenji Matsui
Dept. of System Design
Osaka Institute of Technology
Osaka, Japan

Alfonso González Briones
GRASIA Research Group
Facultad de Informática
Universidad Complutense de Madrid
Madrid, Spain

Enrique Herrera-Viedma
Andalusian Research Institute on
Data Science and Computational
Intelligence (DaSCI)
University of Granada
Granada, Spain

Shigeru Omatsu
Hiroshima University
Osaka, Japan

Sara Rodríguez González
IoT European Digital Innovation Hub
Bioinformatics Intelligent Systems
and Educational Technology Research Group
Department of Computer Science
Faculty of Science
University of Salamanca
Salamanca, Spain

ISSN 2194-5357 ISSN 2194-5365 (electronic)
Advances in Intelligent Systems and Computing
ISBN 978-3-030-53035-8 ISBN 978-3-030-53036-5 (eBook)
https://doi.org/10.1007/978-3-030-53036-5

This Springer imprint is published by the registered company Springer Nature Switzerland AG
The registered company address is: Gewerbestrasse 11, 6330 Cham, Switzerland

Preface

Research on Intelligent Distributed Systems has matured during the last decade, and many effective applications are now deployed. Nowadays, technologies such as Internet of Things (IoT), Industrial Internet of Things (IIoT), Big Data, Blockchain, distributed computing, in general, are changing constantly as a result of the large research and technical effort being undertaken in both universities and businesses. Most computing systems from personal laptops to edge/fog/cloud computing systems are available for parallel and distributed computing. Distributed computing performs an increasingly important role in modern signal/data processing, information fusion, and electronics engineering (e.g. electronic commerce, mobile communications and wireless devices). Particularly, applying artificial intelligence in distributed environments is becoming an element of high added value and economic potential.

The 17th International Symposium on Distributed Computing and Artificial Intelligence 2020 (DCAI 2020) is a forum to present applications of innovative techniques for solving complex problems in these areas. The exchange of ideas between scientists and technicians from both academic and business areas is essential to facilitate the development of systems that meet the demands of today's society. The technology transfer in this field is still a challenge, and for that reason, this type of contributions will be specially considered in this symposium. This conference is the forum in which to present application of innovative techniques to complex problems. This year's technical programme will present both high quality and diversity, with contributions in well-established and evolving areas of research. Specifically, 83 papers were submitted to main track and special sessions, by authors from 26 different countries (Algeria, Angola, Brazil, Bulgaria, China, Colombia, Croatia, Denmark, Ecuador, France, Greece, India, Iran, Italy, Japan, Mexico, Nigeria, Perú, Poland, Portugal, Russia, Saudi Arabia, Spain, Taiwan, Tunisia, Venezuela), representing a truly "wide area network" of research activity. The DCAI'20 technical programme has selected 35 papers and, as in past editions, it will be special issues in ranked journals such as Information Fusion, Neurocomputing, Electronics, IEEE Open Journal of the Communications, Smart Cities, ADCAIJ. These special issues will cover extended versions of the most highly regarded works. Moreover, DCAI'20 Special Sessions have been a very useful tool in order to complement the regular programme with new or emerging topics of particular interest to the participating community.

This symposium is organized by the University of L'Aquila (Italy). We would like to thank all the contributing authors, the members of the Programme Committee, the sponsors (IBM, Armundia Group, EurAI, AEPIA, APPIA, CINI, OIT, UGR, HU, SCU, USAL, AIR Institute and UNIVAQ) and the Organizing Committee of the University of Salamanca for their hard and highly valuable work; and the funding supporting of the project "Intelligent and sustainable mobility supported by multi-agent systems and edge computing (InEDGEMobility): Towards Sustainable Intelligent Mobility: Blockchain-based framework for IoT Security", Reference: RTI2018-095390-B-C32, financed by the Spanish Ministry of Science, Innovation and Universities (MCIU), the State Research Agency (AEI) and the European Regional Development Fund (FEDER), and finally, the Local Organization members and the Programme Committee members for their hard work, which was essential for the success of DCAI'20.

June 2020

Yucheng Dong
Enrique Herrera-Viedma
Kenji Matsui
Shigeru Omatsu
Alfonso González Briones
Sara Rodríguez González

Organization

Honorary Chairman

Masataka Inoue	President of Osaka Institute of Technology, Japan

Program Committee Chairs

Yuncheng Dong	Sichuan University, China
Enrique Herrera Viedma	University of Granada, Spain
Sara Rodríguez	University of Salamanca, Spain

Workshop Chair

Alfonso González Briones	Universidad Complutense de Madrid, Spain

Advisory Board

Sigeru Omatu	Hiroshima University, Japan
Francisco Herrera	University of Granada, Spain
Kenji Matsui	Osaka Institute of Technology, Japan

Scientific Committee

Ana Almeida	ISEP-IPP, Portugal
Gustavo Almeida	Instituto Federal do Espírito Santo, Brazil
Giner Alor Hernandez	Instituto Tecnologico de Orizaba, Mexico
Fidel Aznar	Universidad de Alicante, Spain
Zbigniew Banaszak	Warsaw University of Technology, Faculty of Management, Dept. of Business Informatics, Poland
Olfa Belkahla Driss	University of Manouba, Tunisia

Carmen Benavides	.
Holger Billhardt	Universidad Rey Juan Carlos, Spain
Amel Borgi	ISI/LIPAH, Université de Tunis El Manar, Tunisia
Lourdes Borrajo	.
Adel Boukhadra	National high School of Computer Science, Algeria
Edgardo Bucciarelli	University of Chieti-Pescara, Italy
Juan Carlos Burguillo	University of Vigo, Spain
Francisco Javier Calle	Departamento de Informática. Universidad Carlos III de Madrid, Spain
Rui Camacho	University of Porto, Portugal
Davide Carneiro	University of Minho, Portugal
Ana Carolina	.
Carlos Carrascosa	GTI-IA DSIC Universidad Politecnica de Valencia, Spain
Luis Castillo	Autonomous University of Manizales, Colombia
Rafael Corchuelo	University of Seville, Spain
Paulo Cortez	University of Minho, Portugal
Ângelo Costa	University of Minho, Portugal
Stefania Costantini	Dipartimento di Ingegneria e Scienze dell'Informazione e Matematica, Univ. dell'Aquila, Italy
Giovanni De Gasperis	Dipartimento di Ingegneria e Scienze dell'Informazione e Matematica, Italy
Fernando De La Prieta	University of Salamanca, Spain
Carlos Alejandro De Luna-Ortega	Universidad Politecnica de Aguascalientes, Mexico
Raffaele Dell'Aversana	Università "D'Annunzio" di Chieti-Pescara, Italy
Fernando Diaz	University of Valladolid, Spain
Worawan Diaz Carballo	Thammasat University, Thailand
Youcef Djenouri	LRIA_USTHB, Algeria
António Jorge Do Nascimento Morais	Universidade Aberta, Portugal
Ramon Fabregat	Universitat de Girona, Spain
Ana Faria	ISEP, Portugal
Pedro Faria	Polytechnic of Porto, Portugal
Florentino Fdez-Riverola	University of Vigo, Spain
Alberto Fernandez	CETINIA. University Rey Juan Carlos, Spain
Felix Freitag	Universitat Politècnica de Catalunya, Spain
Toru Fujinaka	Hiroshima University, Japan
Francisco Garcia-Sanchez	University of Murcia, Spain
Marisol García Valls	Universitat Politècnica de València, Spain
Irina Georgescu	Academy of Economic Studies, Romania

Abdallah Ghourabi	Higher School of Telecommunications SupCom, Tunisia
Ana Belén Gil González	University of Salamanca, Spain
Arkadiusz Gola	Lublin University of Technology, Poland
Juan Gomez Romero	University of Granada, Spain
Carina Gonzalez	Universidad de La Laguna, Spain
Angélica González Arrieta	Universidad de Salamanca, Spain
David Griol	Universidad Carlos III de Madrid, Spain
Aurélie Hurault	IRIT: ENSEEIHT, France
Elisa Huzita	State University of Maringa, Brazil
Gustavo Isaza	University of Caldas, Colombia
Patricia Jiménez	Universidad de Huelva, Spain
Bo Noerregaard Joergensen	University of Southern Denmark, Denmark
Vicente Julian	Universitat Politècnica de València, Spain
Geylani Kardas	Ege University International Computer Institute, Turkey
Amin Khan	UiT The Arctic University of Norway, Norway
Naoufel Khayati	COSMOS Laboratory: ENSI, Tunisia
Egons Lavendelis	Riga Technical University, Latvia
Rosalia Laza	Universidad de Vigo, Spain
Tiancheng Li	Northwestern Polytechnical University, China
Ivan Lopez-Arevalo	Cinvestav: Tamaulipas, Mexico
Daniel López-Sánchez	BISITE, Spain
Benedita Malheiro	Instituto Superior de Engenharia do Porto, Portugal
Eleni Mangina	UCD, Ireland
Fabio Marques	University of Aveiro, Portugal
Goreti Marreiros	ISEP/IPP-GECAD, Portugal
Angel Martin Del Rey	Department of Applied Mathematics, Universidad de Salamanca, Spain
Ester Martinez-Martin	Universidad de Alicante, Spain
Philippe Mathieu	University of Lille 1, France
Kenji Matsui	Osaka Institute of Technology, Japan
Shimpei Matsumoto	Hiroshima Institute of Technology, Japan
Rene Meier	Lucerne University of Applied Sciences, Switzerland
Mohd Saberi Mohamad	Universiti Malaysia Kelantan, Malaysia
Jose M. Molina	Universidad Carlos III de Madrid, Spain
Miguel Molina-Solana	Data Science Institute: Imperial College London, UK
Stefania Monica	Università degli Studi di Parma, Italy
Paulo Moura Oliveira	UTAD University, Portugal
Paulo Mourao	University of Minho, Portugal
Susana Muñoz Hernández	Universidad Politécnica de Madrid, Spain
Antonio J. R. Neves	University of Aveiro, Portugal

Jose Neves	University of Minho, Portugal
Julio Cesar Nievola	Pontifícia Universidade Católica do Paraná: PUCPR Programa de Pós Graduação em Informática Aplicada, Brazil
Nadia Nouali-Taboudjemat	CERIST, Algeria
Paulo Novais	University of Minho, Portugal
José Luis Oliveira	University of Aveiro, Portugal
Sigeru Omatu	Osaka Institute of Technology, Japan
Mauricio Orozco-Alzate	Universidad Nacional de Colombia, Colombia
Miguel Angel Patricio	Universidad Carlos III de Madrid, Spain
Juan Pavón	Universidad Complutense de Madrid, Spain
Reyes Pavón	University of Vigo, Spain
Stefan-Gheorghe Pentiuc	University Stefan cel Mare Suceava, Romania
Antonio Pereira	Escola Superior de Tecnologia e Gestão do IPLeiria, Portugal
Tiago Pinto	University of Salamanca, Spain
Julio Ponce	Universidad Autónoma de Aguascalientes, Mexico
Juan-Luis Posadas-Yague	Universitat Politècnica de València, Spain
Jose-Luis Poza-Luján	Universitat Politècnica de València, Spain
Isabel Praça	GECAD/ISEP, Portugal
Radu-Emil Precup	Politehnica University of Timisoara, Romania
Mar Pujol	Universidad de Alicante, Spain
Araceli Queiruga-Dios	Department of Applied Mathematics, Universidad de Salamanca, Spain
Mariano Raboso Mateos	Facultad de Informática: Universidad Pontificia de Salamanca, Spain
Manuel Resinas	University of Seville, Spain
Jaime A. Rincon	Universitat Politècnica de València, Spain
Ramon Rizo	Universidad de Alicante, Spain
Sara Rodríguez	University of Salamanca, Spain
Luiz Romao	Univille, Mexico
Gustavo Santos-Garcia	Universidad de Salamanca, Spain
Ichiro Satoh	National Institute of Informatics, Japan
Emilio Serrano	Universidad Politécnica de Madrid, Spain
Mina Sheikhalishahi	Consiglio Nazionale delle Ricerche, Italy
Amin Shokri Gazafroudi	Universidad de Salamanca, Spain
Fábio Silva	University of Minho, Portugal
Nuno Silva	DEI & GECAD: ISEP: IPP, Portugal
Pedro Sousa	University of Minho, Portugal
Masaru Teranishi	Hiroshima Institute of Technology, Japan
Adrià Torrens Urrutia	Universitat Rovira i Virgili, Spain
Volodymyr Turchenko	Research Institute for Intelligent Computing Systems, Ternopil National Economic University, Ukraine

Zita Vale	GECAD: ISEP/IPP, Portugal
Miguel A. Vega-Rodríguez	University of Extremadura, Spain
Maria João Viamonte	Instituto Superior de Engenharia do Porto, Portugal
Paulo Vieira	Insituto Politécnico da Guarda, Portugal
José Ramón Villar	University of Oviedo, Spain
Friederike Wall	Alpen-Adria-Universitaet Klagenfurt, Austria
Zhu Wang	XINGTANG Telecommunications Technology Co., Ltd., China
Li Weigang	University of Brasilia, Brazil
Bozena Wozna-Szczesniak	Institute of Mathematics and Computer Science, Jan Dlugosz University in Czestochowa
Takuya Yoshihiro	Faculty of Systems Engineering, Wakayama University, Japan
Michifumi Yoshioka	Osaka Pref. Univ., Japan
Andre Zúquete	University of Aveiro, Portugal
Zhaoxia Guo	Sichuan University, China
Zhen Zhang	Dalian University of Technology, China
Hengjie Zhang	Hohai University, China

Organizing Committee

Juan M. Corchado Rodríguez	University of Salamanca, Spain AIR Institute, Spain
Fernando De la Prieta	University of Salamanca, Spain
Sara Rodríguez González	University of Salamanca, Spain
Javier Prieto Tejedor	University of Salamanca, Spain AIR Institute, Spain
Pablo Chamoso Santos	University of Salamanca, Spain
Belén Pérez Lancho	University of Salamanca, Spain
Ana Belén Gil González	University of Salamanca, Spain
Ana De Luis Reboredo	University of Salamanca, Spain
Angélica González Arrieta	University of Salamanca, Spain
Emilio S. Corchado Rodríguez	University of Salamanca, Spain
Angel Luis Sánchez Lázaro	University of Salamanca, Spain
Alfonso González Briones	University Complutense of Madrid, Spain
Yeray Mezquita Martín	University of Salamanca, Spain
Enrique Goyenechea	University of Salamanca, Spain AIR Institute, Spain
Javier J. Martín Limorti	University of Salamanca, Spain
Alberto Rivas Camacho	University of Salamanca, Spain
Ines Sitton Candanedo	University of Salamanca, Spain
Elena Hernández Nieves	University of Salamanca, Spain

Beatriz Bellido	University of Salamanca, Spain
María Alonso	University of Salamanca, Spain
Diego Valdeolmillos	AIR Institute, Spain
Roberto Casado Vara	University of Salamanca, Spain
Sergio Marquez	University of Salamanca, Spain
Jorge Herrera	University of Salamanca, Spain
Marta Plaza Hernández	University of Salamanca, Spain
Guillermo Hernández González	AIR Institute, Spain
Luis Carlos Martínez de Iturrate	University of Salamanca, Spain AIR Institute, Spain
Ricardo S. Alonso Rincón	University of Salamanca, Spain
Javier Parra	University of Salamanca, Spain
Niloufar Shoeibi	University of Salamanca, Spain
Zakieh Alizadeh-Sani	University of Salamanca, Spain

Local Organizing Committee

Pierpaolo Vittorini	University of L'Aquila, Italy
Tania Di Mascio	University of L'Aquila, Italy
Giovanni De Gasperis	University of L'Aquila, Italy
Federica Caruso	University of L'Aquila, Italy
Alessandra Galassi	University of L'Aquila, Italy

DCAI 2020 Sponsors

Contents

A Risk-Driven Model for Traffic Simulation 1
Philippe Mathieu and Antoine Nongaillard

SMARTSEC4COP: Smart Cyber-Grooming Detection Using Natural Language Processing and Convolutional Neural Networks 11
Fabián Muñoz, Gustavo Isaza, and Luis Castillo

Improving BERT with Focal Loss for Paragraph Segmentation of Novels 21
Riku Iikura, Makoto Okada, and Naoki Mori

Parallel Implementation of Nearest Feature Line and Rectified Nearest Feature Line Segment Classifiers Using OpenMP 31
Ana-Lorena Uribe-Hurtado, Eduardo-José Villegas-Jaramillo, and Mauricio Orozco-Alzate

Let's Accept a Mission Impossible with Formal Argumentation, or Not 41
Ryuta Arisaka and Takayuki Ito

Analysis of Partial Semantic Segmentation for Images of Four-Scene Comics 51
Akira Terauchi, Naoki Mori, and Miki Ueno

Two Agent-Oriented Programming Approaches Checked Against a Coordination Problem 60
Eleonora Iotti, Giuseppe Petrosino, Stefania Monica, and Federico Bergenti

Context-Aware Information for Smart Retailers 71
Ichiro Satoh

Morphometric Characteristics in Discrete Domain for Brain Tumor Recognition 81
Jesús Silva, Jack Zilberman, Narledis Núñez Bravo, Noel Varela, and Omar Bonerge Pineda Lezama

Filtering Distributed Information to Build a Plausible Scene for Autonomous and Connected Vehicles 89
Guillaume Hutzler, Hanna Klaudel, and Abderrahmane Sali

A Lightweight Pedestrian Detection Model for Edge Computing Systems 102
Wen-Hui Chen, Han-Yang Kuo, Yu-Chen Lin, and Cheng-Han Tsai

Collaborative Recommendations in Online Judges Using Autoencoder Neural Networks 113
Paolo Fantozzi and Luigi Laura

Natural Language Inference in Ordinary and Support Verb Constructions 124
Ignazio Mauro Mirto

Comparative Analysis Between Different Automatic Learning Environments for Sentiment Analysis 134
Amelec Viloria, Noel Varela, Jesús Vargas, and Omar Bonerge Pineda Lezama

Context-Aware Music Recommender System Based on Automatic Detection of the User's Physical Activity 142
Alejandra Ospina-Bohórquez, Ana B. Gil-González, María N. Moreno-García, and Ana de Luis-Reboredo

Classification of Chest Diseases Using Deep Learning 152
Jesús Silva, Jack Zilberman, Yisel Pinillos Patiño, Noel Varela, and Omar Bonerge Pineda Lezama

Mobile Device-Based Speech Enhancement System Using Lip-Reading 159
Tomonori Nakahara, Kohei Fukuyama, Mitsuru Hamada, Kenji Matsui, Yoshihisa Nakatoh, Yumiko O. Kato, Alberto Rivas, and Juan Manuel Corchado

Mobile Networks and Internet of Things: Contributions to Smart Human Mobility 168
Luís Rosa, Fábio Silva, and Cesar Analide

Design and Implementation of a System to Determine Property Tax Through the Processing and Analysis of Satellite Images 179
Jesús Silva, Darwin Solano, Roberto Jimenez, and Omar Bonerge Pineda Lezama

Multi-step Ultraviolet Index Forecasting Using Long Short-Term Memory Networks 187
Pedro Oliveira, Bruno Fernandes, Cesar Analide, and Paulo Novais

Multispectral Image Analysis for the Detection of Diseases in Coffee Production 198
Jesús Silva, Noel Varela, and Omar Bonerge Pineda Lezama

Photograph Classification Based on Main Theme and Multiple Values by Deep Neural Networks 206
Toshinori Aoki and Miki Ueno

In-Vehicle Violence Detection in Carpooling: A Brief Survey Towards a General Surveillance System 211
Francisco S. Marcondes, Dalila Durães, Filipe Gonçalves, Joaquim Fonseca, José Machado, and Paulo Novais

Multiagent Systems and Role-Playing Games Applied to Natural Resources Management 221
Vinícius Borges Martins and Diana Francisca Adamatti

A Systematic Review to Multiagent Systems and Regulatory Networks 231
Nilzair Barreto Agostinho, Adriano Velasque Wherhli, and Diana Francisca Adamatti

The Reversibility of Cellular Automata on Trees with Loops 241
A. Martín del Rey, E. Frutos Bernal, D. Hernández Serrano, and R. Casado Vara

Virtual Reality Tool for Learning Sign Language in Spanish 251
Amelec Viloria, Isabel Llerena, and Omar Bonerge Pineda Lezama

Data Augmentation Using Gaussian Mixture Model on CSV Files 258
Ashish Arora, Niloufar Shoeibi, Vishwani Sati, Alfonso González-Briones, Pablo Chamoso, and Emilio Corchado

Sentiment Analysis in Twitter: Impact of Morphological Characteristics 266
Jesús Silva, Juan Manuel Cera, Jesús Vargas, and Omar Bonerge Pineda Lezama

The Use of Artificial Intelligence for Clinical Coding Automation: A Bibliometric Analysis 274
A. Ramalho, J. Souza, and A. Freitas

A Feature Based Approach on Behavior Analysis of the Users on Twitter: A Case Study of AusOpen Tennis Championship 284
Niloufar Shoeibi, Alberto Martín Mateos, Alberto Rivas Camacho, and Juan M. Corchado

S-COGIT: A Natural Language Processing Tool for Linguistic Analysis of the Social Interaction Between Individuals with Attention-Deficit Disorder 295
Jairo I. Vélez, Luis Fernando Castillo, and Manuel González Bedia

A Machine Learning Platform for Stock Investment Recommendation Systems 303
Elena Hernández-Nieves, Álvaro Bartolomé del Canto, Pablo Chamoso-Santos, Fernando de la Prieta-Pintado, and Juan M. Corchado-Rodríguez

Applying Machine Learning Classifiers in Argumentation Context 314
Luís Conceição, João Carneiro, Goreti Marreiros, and Paulo Novais

Type-Theory of Parametric Algorithms with Restricted Computations 321
Roussanka Loukanova

Author Index 333

A Risk-Driven Model for Traffic Simulation

Philippe Mathieu and Antoine Nongaillard(✉)

Univ. Lille, CNRS, Centrale Lille, UMR 9189 - CRIStAL, 59000 Lille, France
{Philippe.Mathieu,Antoine.Nongaillard}@univ-lille.fr

Abstract. With the advent of the autonomous vehicle and the transformation appearing in the automobile sector in the next decade, road traffic simulation has taken off again. In particular, it is one of the few ways to test an autonomous vehicle *in silico* [6]. To achieve this, current traffic generators must increase their realism. We argue here that one of the major points of this realism concerns the consideration of risk in driving models. We propose here an individual and self-organizing driving model based on customisable risk-taking factors. In this model, interactions create accidents. Each driver, individually, does not generate any accident, but the collectivity does. Accidents here are unpredictable emerging phenomena resulting from individual deterministic behaviours. Thanks to this model, the risk-taking factor of vehicles improves the realism of the simulations.

Keyword: Trafic simulation

1 Introduction

In traffic simulation as in any other field, a simulation is an abstraction of reality. This abstraction may be more or less detailed. With regard to road traffic, at a low level of detail, an undifferentiated flow of vehicles is sufficient: this is the case, for example, when vehicles simply have a "decorative" role. At a high level of detail, depending on the objective of the study, one can consider, for example, the condition of the pavement, the braking capacity of the vehicle or the drunkenness of the driver.

We are positioning, as well as Scaner [4], Sumo [8], Traficgen [3], Movesim [11] or MATISSE [1]... at an intermediate level of detail where we consider individualised vehicles with their own behaviour that drive on a flat and flawless road infrastructure (e.g. whose topology is provided by a geographical information system such as *OpenStreetMap* or *GoogleMaps*).

In recent years, many traffic generators and driving simulators have been developed at this granularity, both from the research point of view and the industrial point of view, with very specific objectives different from ours. These different tools were mainly focused on testing road infrastructure [5], coordination between vehicles [3,9] or the simulation of realistic flows [7].

Y. Dong et al. (Eds.): DCAI 2020, AISC 1237, pp. 1–10, 2021.
https://doi.org/10.1007/978-3-030-53036-5_1

In these tools, vehicles must simply be able to move. May the environment be a perfect world where everyone can brake in time with complete information, or even instantly, is not a problem. Mostly, these behaviours are based on the IDM (*Intelligent Driver Model*) [10] for the longitudinal behaviour, associated with the MOBIL model [7] for lane switching. These driving models are mainly based on an unlimited knowledge of the environment. Each vehicle always perceives its predecessor no matter how far they are and the behaviour is designed to avoid any accident. The risk factor is not explicitly taken into account in these models.

Only MITSIMLab [2] and MATISSE [1] handle collisions. However, in MITSIMLab, an accident is defined before the execution of the simulation and is characterised by a specific time and position. In MATISSE, accidents are simply due to drivers distraction (by disrupting the perception of a driver). None of these tools propose an accident management model with parametric risk-taking based on inter-vehicle interactions.

Our objective is to build efficient test environments. In this purpose, we aim at identifying a model that can simulate traffic where accidents emerge from interactions between different vehicles based on the risk that each is willing to take. To do this, we propose a deterministic behavioural model in the same line as IDM but calibrated thanks to a risk-taking vector. Risk-taking is considered through four aspects: in the cruising speed, in the perception radius, in the desired inter-vehicular distance and in the distance required to switch from a lane to another one. In Sect. 2, we present our risk-based driving model and the calibration of different parameters such as the inter-vehicle distance or the radius of the perception area. Section 3 empirically validates this model and Sect. 4 provides an overview of the work and presents future perspectives.

2 Driving Model

A model is an abstraction from reality: it never reflects perfectly a real situation and can always be complicated to be a little closer to the real world. Our objective is not to complicate the model for pleasure but rather to identify a minimal accident generation model, to capture the essence of the phenomenon that can be reused by all simulators of this type. The traffic simulator implemented in this paper is based on a multi-agent system where each vehicle is an agent enjoying an individual parametric behaviour. All agents perceive, decide and act simultaneously in this multi-agent system. We propose a behavioural model that can be set differently, leading a various range of driving behaviour ranging from safety-first to a road-raging. In our driving model, vehicles are able to accelerate, to brake or to switch lane. Vehicles are also able to regulate their speed according to an inter-vehicular distance desired or a cruising speed. No inconsistent behaviour is considered, like making a U-turn where it is prohibited. We assume that, as it will be the case in a near future, vehicles are able to know the position, the acceleration and braking capacity, as well as computing the speed of neighbouring vehicles (thanks to LIDAR and radar). If the information are not available, vehicles replace missing information by their own as an approximation.

The presented model does not separate between the driver and vehicle models. Our model is regulated using a cruising speed when no other vehicle is perceived or using an inter-vehicular distance when the vehicle is among others. We argue that driving necessarily involves a certain amount of "*risk-taking*" specific to each driver. This risk-taking is considered in this model through the risk-taking vector $k = (k_0, k_1, k_2, k_3)$ in which each of the components is a percentage corresponding to a different decision-making aspect. For instance: k_0 represents compliance with the speed limit required by the environment; k_1 represents the cognitive distance, i.e. the distance below which a vehicle uses the information it perceives in its reasoning; k_2 corresponds to the respect of safety distances; k_3 represents the minimum space that the vehicle is willing to have with the other vehicles in front and behind on the lane it wants to switch on.

This vector can easily be extended depending on the granularity of the model. It is limited to these four aspects in this paper. Each vehicle is characterised by its own risk-taking vector. Let us note that even if a driver does not take any risk, it does not mean that such a driver will not have an accident. He will not cause an accident due to its behaviour, but he can still suffer from another driver behaviour leading both into a crash. However, if no vehicle takes risk, there will be no accident. Let us precise that even if several other vehicles are within the perception radius of a given vehicle, only the closest is considered.

2.1 Formalisation of the Model

As often in literature, the $\mathcal{E}$ environment is a set of w_n lanes: $\mathcal{W} = \{w_1, w_2, w_n\}$. The longitudinal component of the environment is continuous while the lateral component is discrete.

A vehicle A is characterised by four status parameters and three behavioural parameters. The status parameters are factual and describe the vehicle within its environment by:

- its position, consisting of an instantaneous position $x_A(t)$ and the lane the vehicle is on at t: $w_A(t)$.
- a perception function φ_A which defines a cognitive area (considered constant and uniform here) of radius h_A in which the vehicle A collects relevant information for its decision making;
- an instantaneous speed: $v_A(t)$ bounded such that: $v_A(t) \in [0, v_A^{\max}]$, where $v_A^{\max}$ represents the maximum speed of the vehicle A;
- an instantaneous acceleration: $\gamma_A(t)$. We assume here that acceleration and braking are constant functions: $\gamma_A(t) \in \{-f_A, 0, a_A\}$. There is no gradation but only a choice in the action: either a vehicle A maintains its speed ($\gamma_A(t+1) = 0$), or it brakes ($\gamma_A(t+1) = -f_A$), or it accelerates ($\gamma_A(t+1) = a_A$).

Behavioural parameters characterise the actions of a vehicle by:

- a desired speed $v_A^*(t)$ that the vehicle is willing to achieve;
- an inter-vehicle distance $\delta_A^*(t)$ that the driver wishes to maintain;

- a risk-taking vector $k = (k_0, k_1, k_2, k_3)$ in which each component corresponds to the risk-taking factor towards a different decision-making aspect.

The φ_A perception function of a vehicle A allows the definition of the environment perceived by this vehicle at time t, denoted by $\mathcal{E}_A(t)$. $\varphi_A : \mathcal{E} \times t \rightarrow \mathcal{E} \Rightarrow \varphi_A(\mathcal{E}, t) = \mathcal{E}_A(t)$.

Any vehicle is associated with an instantaneous position. Thus, the distance between a vehicle A and any vehicle e it perceives can then be calculated: $\forall e \in \mathcal{E}_A(t), \delta_A(e,t) = \sqrt{|x_A(t) - x_e(t)|^2}$. In other words, the environment perceived by a vehicle A at time t is defined as all vehicles from the environment separated from vehicle A by a distance less than the radius of its perception area: $\mathcal{E}_A(t) = \{e \in \mathcal{E} \mid \delta_A(e,t) < h_A\}$. Any vehicle outside this "*cognitive*" perception zone is not part of the vehicle's reasoning and can be considered unknown by the vehicle.

2.2 Movement Rules and Regulation Mechanism

The status parameters control the movement of a vehicle according to well-known physical laws. It is easy to calculate the future position of a vehicle from its current position and instantaneous speed: $x_A(t+\Delta t) = x_A(t) + v(t+\Delta t)\Delta t$ where Δt represents the discrete time sampling. To simplify, the approximation of the new position is computed through this equation but it could also be estimated using the average between the initial speed and the final speed over Δt instead, without altering the results described Sect. 3. Similarly, it is easy to determine its future speed from its current speed and instantaneous acceleration: $v_A(t+\Delta t) = v_A(t) + \gamma_A(t+\Delta t)\Delta t$. The speed variation of a vehicle A results from an acceleration choice. Behavioural parameters are involved in the mechanisms of acceleration or speed regulation.

The acceleration of a vehicle A is determined according to the inter-vehicle distance $\delta^*_A(t)$ that this vehicle is willing to maintain if a vehicle B is perceived at front. If A does not perceive other vehicles or if other vehicles are outside the cognitive perception area of A, the acceleration depends on the cruising speed to achieve.

Let's start with the simpler case where the mechanism is based on the observation of the instantaneous speed. In this case, a vehicle aims at driving at the maximum possible speed: $v^*_A(t) = v^{\max}_A$.

$$\gamma_A(t+1) = \begin{cases} a_A & \text{if } v_A(t) < v^*_A(t) \\ 0 & \text{if } v_A(t) = v^*_A(t) \\ -f_A & \text{otherwise} \end{cases}$$

If another vehicle $B \in \mathcal{E}_A(t)$ is in a close neighbourhood, the acceleration regulation mechanism is based on the inter-vehicle distance and is written:

$$\gamma_A(t+1) = \begin{cases} -f_A & \text{if } \delta_A(B,t) < \delta^*_A(t) \\ 0 & \text{if } \delta_A(B,t) = \delta^*_A(t) \end{cases}$$

2.3 Calibration of Parameters

Some parameters like the desired inter-vehicular distance or the perception radius can be calibrated according to a threshold value that ensures a vehicle safety. We have chosen to calibrate our parameters in relation to the stopping distance. The stopping distance of a vehicle A (SD_A) depends on its instantaneous speed and braking capacity as follows: $SD_A(t) = \frac{-v_A(t)^2}{-2f_A}$. This distance SD_A therefore depends on the speed of a vehicle. The faster it drives, the greater the distance. This distance represents the distance required to stop the vehicle completely.

Maximum Speed: the k_0 Risk Factor. The risk factor k_0 is used to determine the maximum speed a vehicle is willing to drive. It can be used to simulate respectful drivers as well as unconstrained ones. The maximum speed $v_A^{\max}$ can be calibrated from the physical speed limit that the vehicle can reach $v_\varphi^{\lim}$ and the driver's interpretation of the limit imposed by the environment $v_\mathcal{E}^{\lim}$ (which he can respect or not).

We can thus calibrate $v_A^{\max}$ according to the following relationship: $v_A^{\max} = \min(k_0\, v_\mathcal{E}^{\lim}, v_\varphi^{\lim})$, where $k_0 \in \mathbb{R}_+^*$ represents the risk-taking with regard to the compliance with the speed limit imposed by the environment.

Perception Area: the k_1 Risk Factor. The risk factor k_1 impacts the perception radius and can be used to simulate old and young drivers for instance. Young drivers are more aware of the environment while old ones may have difficulties to perceive as many information as younger drivers do.

The perception function defines an area of radius h_A. The size of this perception area could depend on external factors such as visibility but that is not considered in this study. This radius can be indexed to the stopping distance SD when the vehicle is driving at its maximum speed. Indeed, in the case of a perception area narrower than this stopping distance, there is always a risk of a front-accident, not perceived soon enough. The driver drives too fast with regard to his or her perception and braking abilities.

Let us remind that the perception radius is constant and cannot change during a simulation. The radius of the perception area can be calibrated directly to the stopping distance required to brake when the vehicle is travelling at full speed ($SD^{\max}$). This value is used to give a meaningful reference point when calibrating the behaviour: $h_A = k_1 SD_A^{\max}(t)$.

Inter-Vehicle Distance: the k_2 Risk Factor. The risk factor k_2 impacts the stopping distance to define the inter-vehicular distance desired by a driver. It can be used to simulate drivers that urge others either to free the lane or to accelerate, as well as peaceful drivers that fear other vehicles proximity.

Let us now assume that the vehicle has a wider perception area than its stopping distance. The inter-vehicle distance desired by a vehicle can be calibrated in relation to its stopping distance: $\delta_A^*(t) = k_2\, SD_A(t)$ where k_2 is the risk factor for compliance with the safety distance. If $k_2 > 1$, then the vehicle will respect more than the required inter-vehicle distance.

In real life, no driver behaves in this way (otherwise, there would be no such traffic density on the roads). A driver distances himself from another one depending on the situation, not based on the position of the preceding vehicle at time t, but from the position he will occupy at time $t+1$. In other words, you have to be able to stop before the vehicle you are following stops himself, otherwise there will be a collision.

$$RD_A(t) = \frac{v_A(t)^2}{2f_A} - \frac{v_B(t)^2}{2f_B}$$

An important difference between the two thresholds is that the stopping distance depends only on the vehicle settings while the reasoned distance also depends on the characteristics of the other vehicle.

We assume that vehicles are able to know the position, speed and acceleration of neighbouring vehicles. If the information are not available, vehicles replace missing information by their own as an approximation.

In order to calibrate the desired inter-vehicular distance, several cases are possible:

- if $\delta^*_A(t) > SD_A(t)$: no risk of collision. The safety distances are maintained such that even if the vehicle in front were to stop instantly, A would still have time to brake without causing a collision.
- if $RD_A(t) < \delta^*_A(t) < SD_A(t)$: moderate risk. If the vehicle in front brakes normally, no collision occurs. However, if the vehicle in front collides with another, he would brake abnormally and cause a new collision at the rear.
- if $\delta^*_A(t) < RD_A(t)$: very significant risk. Collision is not guaranteed, but depends completely on the behaviour of other vehicles, and in particular on the duration of the braking they may perform. The longer they brake, the higher the risk of collision is.

An accident is a complex phenomenon that depends on distances, capacities of drivers, duration of braking... If a vehicle slams on the brakes, it may not have/cause an accident, while several vehicles later on the same lane, a chain of interactions may lead to a crash. An accident emerges from various interactions among vehicles and cannot be foreseen.

Lane Switching: the k_3 Risk Factor. Let us remind that, according to the environment described, the longitudinal component is continuous while the lateral component is discrete. A vehicle switches from a lane to another as soon as the space available satisfies its criterion. This space is calibrated compared to the stopping of the vehicle considered once again. In order to make a decision, the vehicle that is willing to switch lane considers the inter-vehicular distance that would exists with the vehicles already on this lane.

A risk factor k_3 can be assigned in a similar way than the calibration performed for the safety distance based on k_2. In this work, the distance required in front and at back are based on the same coefficient k_3. However, the proposed model can be extended to differentiate these distance to simulate a behaviour that can switch lane at the rear of a vehicle, only considering safety related with the following vehicle.

Thanks to the risk-taking vector, we can express a wide variety of behaviour, taking more or less risk in different aspects.

3 Validation of the Behavioural Model

We might wish to validate this model from real data, but accidentology databases very rarely have the causes of accidents: we do not know in the general case if an accident has occurred due to inattention or abrupt lane switching. These databases generally contain the dates, the locations and number of accidents, which does not allow the validation of a behavioural model (e.g., databases from the French Inter-Ministerial Road Safety Organisation (ONISR)). Our evaluation protocol therefore aims at finding the stylised facts based on simulations, in order to empirically validate the correct functioning of the model that we propose.

All simulations presented in this paper are carried out using the TrafficGen simulator [3]. The simulation environment used is a 20 km road with a number of lanes ranging from 1 and 4 according to the simulations.

We determined the ranges of variation from real data. From the technical data sheet of the best-selling vehicle in 2018 in France (*Clio4*), we were able to establish an interval of plausible variation for acceleration between 2.3 m/s^2 and 3.1 m/s^2 (during a passage of 0 km/h and 100 km/h, by varying the engines and finishes). Braking varies between -8.5 m/s^2 and -11 m/s^2, which corresponds to the average braking respectively on wet (with grip reduced) and dry environment.

3.1 Perception Distance

The perception distance of a vehicle represents the area in which it perceives information. The information thus retrieved modifies the vehicle's actions according to its behaviour: for example, slowing down or switching to another lane on a too abrupt approach at another vehicle. Vehicles with a reduced perception distance will logically cause many more accidents than vehicles with a greater perception distance, allowing them to obtain information "earlier". The purpose of this experience is therefore to illustrate the influence of this perception distance on the number of accidents.

Vehicles are generated uniformly with a maximum possible speed between 50 km/h and 130 km/h. The acceleration is set at 2.7 m/s^2 and the deceleration at -9.5 m/s^2. We vary the risk-taking parameter k_1 which allows the perception distance to be calibrated as a function of the stopping distance. We vary k_1 between 0 and 1.5 in order to obtain a perception distance ranging from 0 to 100 m.

The Fig. 1 shows that the increase of perception distance reduces the accident rate. When only one lane is available, the accident rate becomes very low as soon as the perception distance exceeds a certain threshold, allowing vehicles to brake.

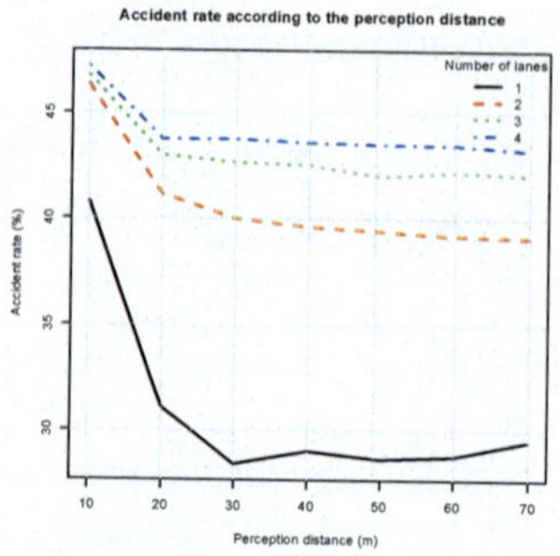

Fig. 1. Effect of perception distance on accident rate according to the number of lanes available.

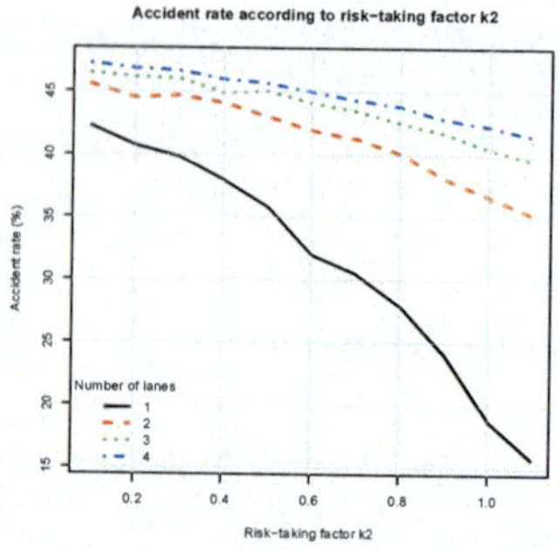

Fig. 2. Effect of inter-vehicle distance on the accident rate, according to the number of lanes.

All vehicles then focus on the speed of the slowest vehicle. On the other hand, if the number of lanes available increase, the accident rate is immediately higher. It is due to the nature of accident that is changing. Accidents then come from lane switching and not from lack of perception associated with a too-late braking. Indeed, the vehicles change lane first before braking. Although the vehicles only move out if they believe they have enough space (corresponding to their own stopping distance) of accidents occur due to the speed difference between vehicles sometimes important.

3.2 Inter-vehicle Distance

Vehicles choose the inter-vehicle space they wish to maintain with the vehicle in front of them. This distance is calibrated according to the vehicle's stopping distance and the risk-taking factor k_2. We remind you that with a risk factor of $k_2 = 0.8$, a vehicle maintains 80% of its stopping distance with the front-vehicle. A factor of $k_2 = 1.0$ means that the vehicle scrupulously respects the safety distances, which nevertheless do not guarantee the absence of an accident.

The experimental conditions are the same as before (20 km road, from 1 to 4 lanes, 10% chance of a new vehicle appearing per lane). The average speed difference between the vehicles is 50 km/h. The perception halo covers an area of 50 m, equivalent to a risk taking factor of $k_1 = 0.4$.

Figure 2 shows that with k_2 close to 0, the safety distances are not respected at all, leading to as a result of a very high accident rate. The value is limited to 40% due to the traffic density chosen during the experiments. The more k_2 is close to 1.0, the fewer accidents there are because each vehicle has the necessary perception and braking capacity to stop as soon as an obstacle is detected.

3.3 Lane Switching

When a vehicle switches lanes, it may have to be inserted behind a first vehicle and in front of a second. The space it considers necessary to be inserted is calibrated via the risk factor k_3 which weights its own stopping distance. If $k_3 = 0.75$, the driver is ready to change lanes as soon as 75% of his stopping distance is available on the destination lane. The experimental conditions are the same as before. The average speed difference between the vehicles is 50 km/h. The perception halo covers an area of 50 meters, equivalent to a risk taking factor $k_1 = 0.4$. The only risks that vehicles take occur when switching lanes where they can cut off the road from another vehicle, which may not have time to brake.

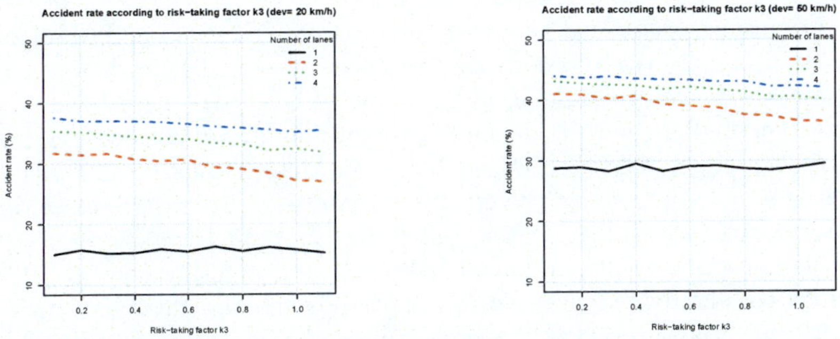

Fig. 3. Impact of the inter-vehicle distance during lane switching on the accident rate, according the number of lanes, when the average speed deviation between vehicles is 20 km/h (left side) and 50 km/h (right side).

We can see that the more important the factor is ($k_3 = 1.0$), the less there is an accident. One of the determining factors is the speed deviation of vehicles. Our model is based on the assumption that a vehicle does not know the speed of the other vehicles, but uses its own speed to calculate distances. If $k_3 = 1.0$, a vehicle makes sure it has enough stopping distance to stop. However, if the vehicle in front of which you want to be driven is travelling at a much higher speed than yours, then it may result in a collision. The comparison of Figs. 3 shows that for a smaller speed difference, the number of accidents observed is much lower for a value of k_3 identical. The slower the vehicles travel, the more it is easy to manage vehicles that suddenly interleave. The slope of these curves strongly depends on the traffic density. The more dense the traffic is, the higher the slope. Since, the density is quite weak here, the slope is also quite weak, only leading to a variation of several percent.

4 Conclusion

With the advent of the autonomous vehicle and following safety standards that are imposed on him, the test in silico of the autonomous vehicle has become an essential complement to the test in vivo. This is the technique used right now by major companies in the field like Waymo or Tesla. To carry out these tests successfully, the simulators must allow the vehicle to be put under stress in order to test its reactions. We argue that one of the key points concerns the creation of unplanned accident situations. Many traffic generators already exist, but most of them either do not allow accidents to occur (Vissim, Sumo: aiming at an ideal behaviour, without accident) or allow them to occur but within the context of stochastic individual behaviour (Matisse: accidents are independent from the behaviour of other vehicles). Our objective is then to improve the autonomous vehicle test by generating accidents in traffic simulators, which we achieve here thanks to the multi-agent approach and its parametric deterministic behaviours that offer remarkable expressiveness.

We propose a risk-based model for expressing realistic and configurable behaviours that collectively constitute a complex system that leads to the emergence of accidents (not random accident). Here, it is the interactions that create accidents. The higher the risk, the higher the probability of accidents occurring, but we do not know where or when. We show that thanks to this risk-based model, it is possible to exhibit the stylised facts classically recognised in driving simulation and accidentology. After describing the model, we show the results of the different experiments we have carried out and describe how the different phenomena commonly encountered in situations such as brutal lane switching or vehicles that follow too closely are identifiable in these experiments.

References

1. Al-Zinati, M., Wenkstern, R.: MATISSE 2.0: a large-scale multi-agent simulation system for agent-based its. In: IAT 2015, pp. 328–335 (2015)
2. Ben-Akiva, M., Koutsopoulos, H.N., Toledo, T., Yang, Q., Choudhury, C.F., Antoniou, C., Balakrishna, R.: Traffic simulation with MITSIMLab. In: Fundamentals of Traffic Simulation, pp. 233–268 (2010)
3. Bonhomme, A., Mathieu, P., Picault, S.: A versatile multi-agent traffic simulator framework based on real data. IJAIT **25**(1), 20 (2016)
4. Champion, A., Mandiau, R., Kolski, C., Heidet, A., Kemeny, A.: Traffic generation with the SCANeR^{TM} simulator: towards a multi-agent architecture. In: DSC, pp. 311–324 (1999)
5. Fellendorf, M.: VISSIM: a microscopic simulation tool to evaluate actuated signal control including bus priority. In: 64th ITEAM, pp. 1–9 (1994)
6. Kalra, N., Paddock, S.M.: Driving to safety (2016)
7. Kesting, A., Treiber, M., Helbing, D.: General lane-changing model mobil for car-following models. Transp. Res. Board **1**(1999), 86–94 (2007)
8. Krajzewicz, D., Erdmann, J., Behrisch, M., Bieker, L.: Recent development and applications of SUMO - Simulation of Urban MObility. IJASM **5**(3–4), 128–138 (2012)
9. Tlig, M., Buffet, O., Simonin, O.: Cooperative behaviors for the self-regulation of autonomous vehicles in space sharing conflicts. In: ICTAI, pp. 1126–1132 (2012)
10. Treiber, M., Hennecke, A., Helbing, D.: Congested traffic states in empirical observations and microscopic simulations. Phys. Rev. E **62**(2), 1805 (2000)
11. Treiber, M., Kesting, A.: An open-source microscopic traffic simulator. IEEE ITSM **2**(3), 6–13 (2010)

SMARTSEC4COP: Smart Cyber-Grooming Detection Using Natural Language Processing and Convolutional Neural Networks

Fabián Muñoz[1], Gustavo Isaza[2(✉)], and Luis Castillo[2]

[1] Universidad Tecnológica de Pereira, V. Julita, Pereira, Colombia
falmo@utp.edu.co

[2] Universidad de Caldas, C. 65 26-10, Manizales, Colombia
{gustavo.isaza,luis.castillo}@ucaldas.edu.co

Abstract. This paper aims to present the design and implementation of a prototype that recognizes grooming attacks in the context of COP (child online protection) using Natural Language Processing and Machine Learning hybrid model, via Convolutional Neural Networks (CNN). The solution uses a vector representation of words as the semantic model and the implementation of the model was made using TensorFlow, evaluating the classification of grooming for a text (dialogue) prepared asynchronously in a controlled environment according to methodologies, techniques, frameworks and multiple proposed techniques with his development described. The model predicts a high number of false positives, therefore low precision and F-score, but a high 88.4% accuracy and 0.81 AUROC (Area under the Receiver Operating Characteristic).

Keywords: Natural language processing · Convolutional neural networks · Grooming detection

1 Introduction

The fight against Cybergrooming have done different efforts, as in the United States from 1998 to 2009 where the Copa 2000 (Commission on Child Online Protection) commission was created, a law in the United States of America, with the declared purpose of restricting access by minors to any material defined as harmful, later was disarticulated by the supreme court for violating the first and fifth amendment of this country [1]. Later, Initiatives to promote the correct use of information technologies emerged as an active work of training from parents as promoters of responsible and safe use. In the United Kingdom, the CEOP (Child Exploitation and Online Protection Centre), was created in 2006 as part of the NCA, basing its methodology on reports from different entities to seek out child sex offenders and bring them to justice. The COP initiative [2], which was formed as a specialized program on the cybersecurity agenda, was presented by the United Nations secretariat, heads of state, ministers and leaders of international organizations. The initiative was launched at the end of 2008

Y. Dong et al. (Eds.): DCAI 2020, AISC 1237, pp. 11–20, 2021.
https://doi.org/10.1007/978-3-030-53036-5_2

and is an international collaboration promoting the protection of minors online around the world. These efforts, join with many others [3] and recent studies as [4] shows important data from results of the research sexting and risk on Children in Cyberspace shape the global efforts on the subject and due to this obvious and felt need to develop components that include methods of detection and prevention of attacks in the context of the protection of online minors. This paper shows a machine learning model using neural networks and NLP (Natural Language Processing), allowing to approximate a representation of knowledge for semantic analysis, contribute to the discovery of pseudo-intelligent information in these environments and diminishing human intervention for the characterization of implicit anomalous behaviors and the Detection of messages that potentially represent these attacks. In line with this, the development and implementation of the prototype of a system for detecting attacks to Childs, called SMARTSEC4COP is proposed.

2 Materials and Methods

As technological resources in the development of this research, the hardware used for characteristic processing systems 4 Cores $\times$ 8 Threads Memory Corsair Dominator DDR34 $\times$ 8 GB 1600 MHz, 1 GPU NVIDIA GeForce GTX 9804GBGPU2NVIDIA GeForce GTX TITAN X12 GB. The software and operating systems, on which image-based containers were created with Tensorflow using Docker, allowed an efficient change and version management, Jupyter was used with the Python language and Markdown was used to document online. Jupyter notebook server 4.2.1 Python, TensorFlow 0.12.1, nltk 3.2.1, scikit-learn 0.17.1 Container Image Debian 8 JessieDocker 1.12.5, Host system Ubuntu 14.04.1; the details of the implementation are described in the github repository https://github.com/gisazae/Tensorflow-Examples/blob/master/IntegracionCorpus_checkpoint.ipynb.

For the design of the prototype, the PAN corpus [5] was adapted, Fig. 1 present a prototype solution in the detection of this type of attacks.

The process starts with a preprocessing stage where the messages and tokens from the chosen corpus are analyzed, this is aimed at filtering information that is not relevant and that generates inconveniences when used, the techniques of tokenization, the dimensions of the data sets, the lengths of messages and words are studied to form the filters and validations necessary to have a set of data of sufficient value. As a second step is the training of a matrix for vector representation of words in the thematic domain of the problem giving value to its semantic context, this matrix is known for various problems as matrix of embeddings, being this as the key transformation technique chosen to work with an unstructured set of data for classification and semantic representation tasks.

Finally, the training and prediction processes are based on the architectural specification of the classification model with convolutional neural networks, where the trained embeddings matrix is retrieved to transform each token of the messages to their vectoral representation and hyperparameters for his structure are tested.

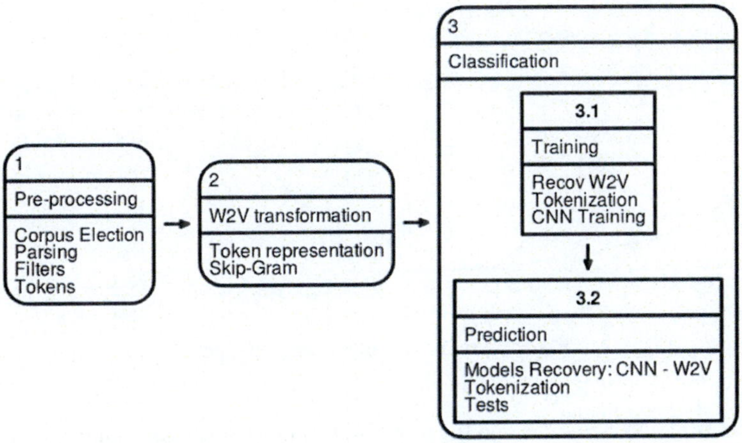

Fig. 1. Process diagram

2.1 Preprocessing

The corpus used to train the prototype was the manufactured for the PAN CLEF 2012 competition, given its characteristics and sources of information, constitutes it as a point of reference for different research projects [5]. The dimensions of the dataset in Table 1 is a very important aspect, serve to close gaps between research and development of applied technologies, for this reason the corpus was created with properties close to reality, with a low number of true positives and a high number of possible false positives and true negatives. In realistic scenarios and as mentioned in the OECD document, the percentage of conversations with attacks in relation to regular conversations is very low (Fig. 2).

Table 1. Division of the PAN CLEF training and testing dataset

Dataset	Training	Testing
Conversations	2.016	3.737
Messages	79.331	125.702
Attackers	40.978	65.239
Victims msgs	38.353	60.463
Other conversations	824.276	1.933.079

The desired characteristics when messages are tokenized were similar to Tweets, which adapts very well to the tokenizer *Tweet_Tokenizer* of the library NLTK [6], with lexicon handling used in short messages, emoticons, hashtags and other elements present in these types of messages, allowing for instance the reduction of length for cases of

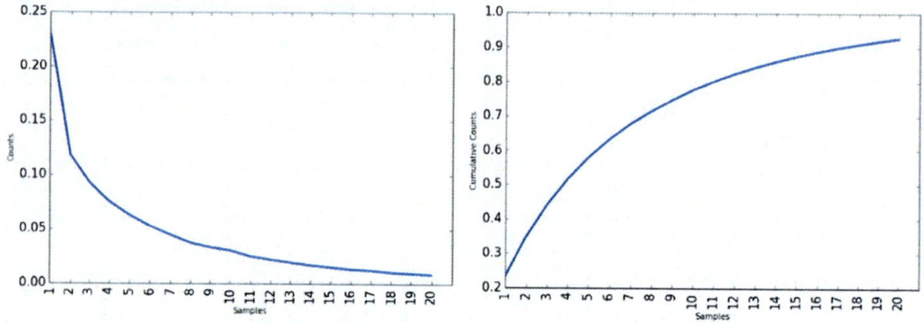

Fig. 2. Number of tokens on messages

repeated continuous characters and the transformation to lowercase, such characteristics used on the messages processing. After making the tokenization, more than 90% of the messages have a shorter length to 20 tokens. The frequency of the 50 most common tokens in the collection of training after going through the tokenizer, which under the described characteristics leaves a collection with 6'594.132 tokens of a vocabulary with 173,188 elements. From this information we review the Hapaxes which are 103,425, elements that only appear once in the collection, this prevents obtaining a clear meaning and the context in which they are used, therefore are eliminated as independent elements and grouped under a same term. The length of the tokens had 97,565 different words, mentioned 743,842,922,321 times in the Google books Ngrams [7], which for 2011 was 4% of all books printed in this language. In this work, the authors found that for unique words according to their length, 0.008% of them were greater than 20 letters, 0.016% had 20 letters and 80% were between 4 and 10 letters. For the number of mentions, found that 0.001% of the words used in the texts had length 20, above this length appeared 6 times less and 80% have between 2 and 7 letters. For these reasons it was decided to eliminate the tokens with length greater than 20, which covers 99% of the words to be used in this research SMARTSEC4COP and delete the token whose use is less than 0.001%, that is, 6.175 unique tokens that are repeated in the data collection a total of 21,994 times. The output of this stages allowed a dimensional reduction and validation of data, leaving 60,255 unique elements in 7'068,462 tokens of 922,399 messages from the training collection.

2.2 W2V Transformation

To perform classification and clustering tasks, a structured representation of the information is required, such as Bag of words [8], which is a simple and efficient model to represent a text as the set of words used on it by using this model the order of the words is lost, different sentences can contain the same words and therefore lose meaning. Algorithms that convert words into vector representations, such Word2Vec [9]; the use of predictive models like Word2Vec should be used. The Word2Vec model is divided in

2, continuous bag of words (CBOW) and skip-gram (SG), these models represent vectorially the words in the contexts that are trained unsupervised, and by calculating cosine similarity allows to give a semantic similarity between words. The implementation of Skip Gram was performed using TensorFlow [10]. Entries are built with a vocabulary processor, using the pre-processing pseudo code that tokenizes the entire corpus with the Tweet Tokenizer, then extracts a vocabulary of elements and assigns an identifier in a vector of correspondence for each item repeated more than once, otherwise, assigns the "0" identifier corresponding to an unknown token (UNK), leaving 60,256 unique items. Model performs a false task of training, since you do not obtain the full model itself, what you want to obtain are finally the optimized values for each row of the matrix of embedding's that represents the location of each token in a vector space of 128 dimensions in this case. The training process performs a search of the vector representation of the input tokens according to the batch size with the lookup function. To optimize, estimated error was reduced by Noise Contrastive Estimation (*nceloss*), which samples the negative data comparing the expected data in the prediction to perform the reduction of error, so it becomes possible to perform this type of training that by its dimensions can be unfeasible. Quality of the matrix after the training process can be evaluated by measuring similarity between words or reasoning by analogy and the loss calculation value, in this case during the training process cosine similarity was used to visualize the result of nearby tokens; The model was trained with a configuration of 2 skip windows getting 2 skips per window, which gives a batch of size 100 with 50 tokens each. 3 training sessions of 150,000 steps in approximately 4 h each were run on the 7,068,462 tokens of the corpus, recovering and continuing with the status of the previous process.

2.3 Classification Model

The representative matrix of the sentence feeds the convolutions, which have a number of filters of size 2, 3, 4 or 5 to have logic of n-grams, as defined in the architecture, in this way a set of phrases with related senses, by their structure of tokens could form similar semantics and can be filtered correctly in a similar way. From this result, samples of maximum values are made on all filters with *MaxPooling*, leaving for each filter size a reduced matrix, these matrices are concatenated going through a dropout to reduce the Overfitting preventing the co-adaptation of neurons. Finally, in the output layer, the prediction is made by a multiplication with the number of classes to be identified: sentence with meaning related to grooming (positive attack) or not (negative attack), the highest value of the two classes is the prediction of the network. The calculation of loss or error is done with the measure of cross-entropy, specifically the function softmax cross entropy with logits. The network trains using 100 messages of 50 tokens at a time, coming from a list with messages of topics mixed randomly (grooming and other topics). Since these two groups are unbalanced in an approximate ratio of 10:1, with a smaller number of samples for the class of attacks, a technique was needed that allowed to combat this inconvenience in a simple way by over-sampling the group with low number of data. Finally, after the training process, the model is recovered and fed only with the test data to evaluate its quality. In this step the confusion matrix is constructed and precision, recall, specificity, accuracy, F-score, ROC curve and area under the ROC curve are calculated, all of them to evaluate and analyze with sufficient information how good is the model built for solve the problem, this sequence is explained in pseudocode in Algorithm 1.

Algorithm 1 Pseudocode to train and test neural network

```
linesMsgsAtt ← READFILE(attacksFile)
linesMsgsOth ← READFILE(othersFile)
linesTokensMsgsAtt ← PREPROCESS(linesMsgsAtt)
linesTokensMsgsOth ← PREPROCESS(linesMsgsOth)
linesTokensMsgsAtt ← OVERSAMPLING(linesTokensMsgsAtt)

Msgs ← linesTokensMsgsOth + linesTokensMsgsAtt

vocabulary ← RESTOREVOCABULARY()

msgsIds ← TRANSFORM(msgs; vocabulary)
msgsIds ← MIX(msgsIds)
    batch(training; evaluation) ← SPLIT(msgsIds)
  graph ← CREATEGRAPHTENSORFLOW()
    graph(embeddings) ← RESTOREMODEL() → Embeddings and weights
  for all epochs do
   if evaluate then
     result(loss; accuracy) ← EVALMODEL(model; batch)
   else if save model then
     SAVEMODEL(model)
    else
    model ← TRAINMODEL(graph; batch)
   end if
 end for
```

3 Results

In the Table 2 the characteristics configured for the training of the Skip-gram vector representation model are specified, this produce the matrix that later is imported in the classification stage and the loss curves during training phase.

The classification model was trained in 8 different configurations that are observed in the Table 3, which among others, includes changes in number and size of filters, percentage of evaluation during training, number of epochs and the origin of the embedding's matrix.

The data observed in the Table 3 and the ROC curves, with the configuration 4 (Fig. 3a), allows to know the quality of the model given its hyper-parameters in relation to the quantity of steps of training or epochs (Fig. 3b shows the error curve), in them, it is found that the classifier has a good recall and a high value for the area under the ROC curve, which means a very useful model, even from settings with little training.

Table 2. Training of the embedding's matrix

Configuration	1	2	3
ID	1491765517	1491808882	1493773958
Embedded dimensions	128	128	128
Skip window	2	2	2
Skips	2	2	2
Batch size	100	100	100
Vocabulary	60.256	60.256	60.256
Steps	150.000	150.000	150.000
Optimizer	ADAM	ADAM	ADAM
Restored from	N/A	1491765517	1491808882
Corpus tokens	7.068.462	7.068.462	7.068.462
NCE sampling	64	64	64
Time	4 h 10 m	4 h 11 m	4 h 7 m
Loss avg	6.4	6	6

Table 3. Configurations and quality results of the classifier

Config.	1	2	3	4	6	7
Batch size	100	100	100	100	100	100
% Evaluation	5%	7%	5%	4%	3%	4%
Filters size	2, 3	2, 3	2, 3, 4	2, 3	2, 3, 4, 5	2, 3, 4
Epochs	1	20	5	5	50	50
Filters by size	128	128	128	192	192	192
Time	9 m	2 h 59 m	1 h 12 m	58 m	1 d 16 m	16 h 29 m
Final loss eval.	0.388	0.289	0.317	0.323	0.228	0.236
Final accuracy	**0.822**	**0.874**	**0.858**	**0.852**	**0.900**	**0.900**
Recall	0.779	0.772	0.773	0.726	0.736	0.702
Specificity	0.840	0.869	0.868	0.888	0.879	0.896
Accuracy	0.837	0.863	0.863	0.879	0.871	0.884
F-score	0.362	0.402	0.401	0.417	0.404	0.420

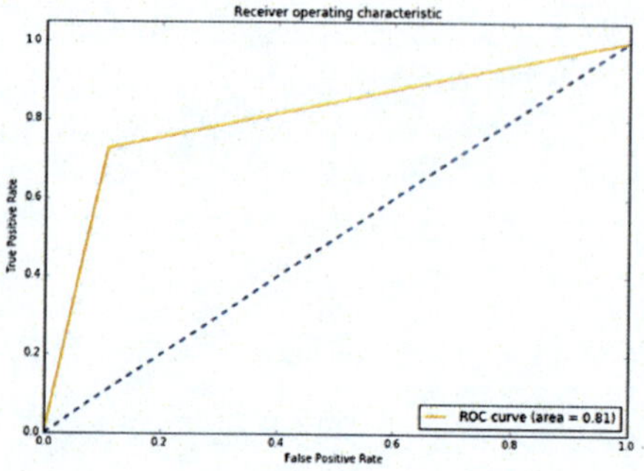

Fig. 3a. ROC curve model 4 (1493592613)

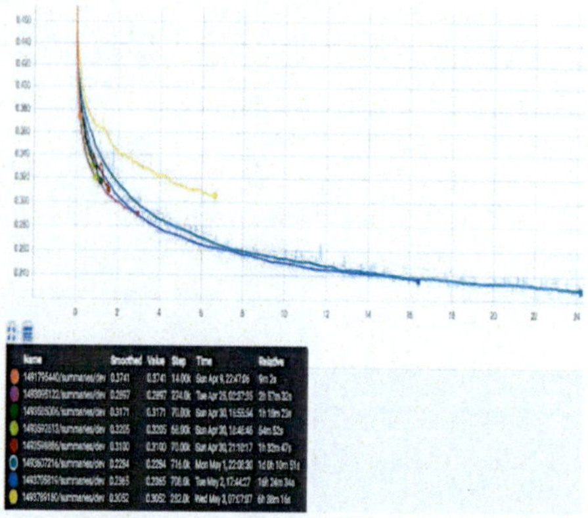

Fig. 3b. Evaluation error curve for all models in training time

4 Conclusions and Future Work

This computational prototype, SMARTSEC4COP, focuses on prevent Cybergrooming attacks using natural language processing and machine learning, implemented in Tensorflow and NLTK, using a model that predicts a high number of false positives, good accuracy and F-score. Direct classification was made a transformation of the dictionary of words used in conversations written over the Internet to vectors (W2V). This transformation was carried out with the implementation of a skip-gram model that performs a training "false task" to obtain the vector representation. From this vector representation, a cross-network graph was constructed with information from Empath (Stanford's Lexicon), This graph can be used and adjusted in related investigations, such as in the analysis of stages in attacks on minors. Finally, the general architecture of the computational model was constructed. The results were analyzed by presenting tables with different hyperparameter configurations for the Skip Gram model and classifier, related to the detection metric values. Loss curves during training were presented, and ROC curves as an additional metric to determine the sensitivity of the classifier. The results of the metrics show a quite useful model considering that it performs message-oriented classification, it has values of accuracy (accuracy) and area under the ROC curve over 80%. However, in this problem where the number of messages in the context of grooming is so low compared to the number of conversations and messages from other contexts, it can be concluded that it is a consistent and useful result since it captures a high number of true positives.

Future works could be scaling the semantic representation of the analysis of interactions from an ontology language, OWL OWL2 SWRL, with the purpose of extending an ontology with inference and reasoning capabilities, thus achieving to extend new information discovery techniques. In the future, it is expected to have the support of experts in the areas of psychology, sociology, and the like, to validate the relevance of the interactions built by the designed models. To evaluate the use of vector representation of sentences which serves as a characteristic to carry out classification. Similarly, this type of representation can be used to anonymize the messages of study subjects or suspects in follow-up. Among other characteristics for the classifier, POStagging, time data between messages, models such as BOW (bag of words) and by relating lexicons (Vocabularies) categories can be obtained that serve as characteristics, these can be empath or LWIC. The methodology for detection can be improved with a recurring process of evaluation of the conversation, where each sentence is constantly classified and characteristics are extracted, such as For the partial classification, by messages or sentences, recurrent neural networks can be used to improve precision since natural language is a sequential process, also as mentioned in the previous point for the flow of conversation is also possible with recurrent networks However, depending on the messages sent at each moment to give a probabilistic value of the course of the conversation, as alternative options to neural networks, it is also possible to use a tree classifier to make predictions about the entire conversation, since the matrix of embeddings It relates the semantic meaning of the words, using all the data sets, training, testing and others that can be integrated, allows having a much more extensive domain and a semantic sense in multiple contexts for each possible word. Classification models have been carried out so far in CPU, but the use of TensorFlow allows training and classification with GPUs almost immediately, it is only dependent on a configuration and hardware availability, as well as some minimum adjustments already mentioned in the implementation. Perform knowledge extraction and discovery processes or semantic on the networks or graphs constructed, validating this information with other investigations and proposals, for example, stage flow diagrams in the problem of grooming. Working with languages other than English, although it presents a challenge with the -data set, it is possible to train a model based on the same architecture described and implemented in this research project. Perform tests with different regularization techniques such as L2 since only dropout was used

Acknowledgment. This work was supported by Universidad de Caldas and Universidad Tecnológica de Pereira in their research groups GITIR and SIRIUS.

References

1. Smith, M.S.: Internet: Status report on legislative attempts to protect children from unsuitable material on the Web (2008). http://ovidsp.ovid.com/ovidweb.cgi?T=JS&PAGE=reference&D=psyc6&NEWS=N&AN=2008-10767-006
2. ITU-D 2010. Child Online Protection: Statistical Framework and Indicators 2010. https://www.itu.int/dmspub/itu-d/opb/ind/D-IND-COP.01-11-2010-PDF-E.pdf
3. Webster, S., Davidson, J., Bifulco, A., Gottschalk, P., Caretti, V., Pham, T., Grove-Hills, J., Turley, C., Tompkins, C., Ciulla, S., Milazzo, V., Schimmenti, A., Craparo, G.: Final Report European Online Grooming Project, European Online Grooming Project, p. 152, March 2012
4. Kopecký, K., René, S.: Sexting in the population of children and its risks (quantitative research). Int. J. Cyber Criminol. **12**, 376–391 (2019). https://doi.org/10.5281/zenodo.3365620
5. Inches, G., Crestani, F.: Overview of the International Sexual Predator Identification Competition at PAN-2012 (2012)
6. Bird, S., Klein, E., Beijing, E.L.: Natural Language Processing with Python, 1st edn. O'Reilly Media Inc, Sebastopol (2009). ISBN 9780596803346

7. Norvig, P.: English Letter Frequency Counts: Mayzner Revisited or ETAOIN SRHLDCU (2013). http://norvig.com/mayzner.html, http://norvig.com/mayzner.html, Achieved at: http://www.webcitation.org/6b56XqsfK
8. Harris, Z.S.: Distributional structure. Word **10**(2–3), 146–162 (1954)
9. Le, Q., Mikolov, T.: Distributed representations of sentences and documents. In: 31st International Conference on Machine Learning, ICML 2014, 4 (2014)
10. Abadi, M., Agarwal, A., Barham, P., Brevdo, E., Chen, Z., Citro, C., Corrado, G.S., Davis, A., Dean, J., Devin, M.: Tensorflow: Large-scale machine learning on heterogeneous distributed systems (2016). https://arxiv.org/abs/1603.04467

Improving BERT with Focal Loss for Paragraph Segmentation of Novels

Riku Iikura(✉), Makoto Okada, and Naoki Mori

Osaka Prefecture University, 1-1 Gakuen-cho, Naka-ku, Sakai, Osaka, Japan
iikura@ss.cs.osakafu-u.ac.jp

Abstract. In this study, we address the problem of paragraph segmentation from the perspective of understanding the content of a novel. Estimating the paragraph of a text can be considered a binary classification problem regarding whether two given sentences belong to the same paragraph. When the number of paragraphs is small relative to the number of sentences, it is necessary to consider the imbalance in the number of data. We applied the bidirectional encoder representations from transformer (BERT), which has shown high accuracy in various natural language processing tasks, to paragraph segmentation. We improved the performance of the model using the focal loss as the loss function of the classifier. As a result, the effectiveness of the proposed model was confirmed on multiple datasets with different ratios of data in each class.

Keywords: Natural language processing · Text segmentation · Imbalanced classification · BERT · Focal loss

1 Introduction

For a computer to generate a novel automatically, it is essential that the computer understand the structure of the text specific to the novel. One of the important techniques for improving the readability of the sentences in a novel is to divide the sentences into paragraphs. Placing new paragraphs in the appropriate positions based on the transitions of scenes and topics will help readers fully understand the story. Therefore, the position where the new paragraph starts contains important sensory information for people who write or read novels. Based on this assumption, in this study, as a stepwise approach with the automatic creation of a novel as the ultimate objective, we estimate paragraph boundaries from the perspective of a computer's understanding of the story in the novel.

When estimating where the new paragraph starts, it is possible to consider it as a binary classification problem regarding whether two target sentences belong to the same paragraph, in other words, whether a boundary appears as a paragraph between the two sentences. In this case, the number of paragraphs is usually small relative to the number of sentences, and therefore it is necessary to consider the imbalance in the number of data.

Y. Dong et al. (Eds.): DCAI 2020, AISC 1237, pp. 21–30, 2021.
https://doi.org/10.1007/978-3-030-53036-5_3

In this study, we apply the bidirectional encoder representations from transformer (BERT) [1], which has been shown to be highly accurate in various natural language processing tasks, to the paragraph segmentation problem. Furthermore, we attempt to improve the performance of the model using the focal loss (FL) [2] as the loss function of the classifier.

The contributions of this study are as follows.

- We applied BERT to the paragraph segmentation of novels as imbalanced classification, and demonstrated that the model performance was improved by using FL as the loss function.
- We proved that the proposed method is effective for multiple data with different class size ratios.

2 Related Works

2.1 Imbalanced Classification

Imbalanced classification problems can be generally classified into two approaches: re-sampling methods and cost-sensitive learning. A re-sampling method over-samples minority classes or under-samples majority classes to eliminate imbalances in the number of data. The simplest over-sampling method randomly duplicates instances of a minority class. However, this can cause an over-learning because the data distribution becomes monotonous. To solve this problem, SMOTE [3], an algorithm for generating new minority class data points through interpolation, was proposed. In contrast, cost-sensitive learning is a method for improving the classifier itself by applying a loss function with different weights to each class instead of changing the distribution of the training data. The simplest and most common value used as the weight is the value corresponding to the number of data in each class. FL has been proposed as a loss function that considers not only the importance according to the size of each class but also the difficulty of identifying each sample.

2.2 Text Segmentation

Text segmentation plays an important role in multiple tasks in natural language processing, such as document summarization and answering questions. To divide a sentence into multiple segments, it is necessary to understand the semantic structure of the sentence, which is closely related to natural language understanding by a computer.

Several unsupervised and supervised learning methods have been proposed for this task. Glavaš et al. [4] proposed an unsupervised algorithm for constructing a semantic relevance graph of a document using word embeddings and a measure of semantic relevance of short sentences. In addition, a method using supervised learning includes a model applying long short-term memory (LSTM), which is a type of recurrent neural network [5]. Such methods can efficiently model input sequences by controlling the flow of information over time.

Badjatiya et al. [6] proposed an attention-based convolutional neural network bidirectional LSTM model that introduces an attention mechanism and learns the relative importance of each sentence in the text to achieve a segmentation.

3 Datasets

For the experiments, we created a dataset from Charles Dickens's novels managed by Project Gutenberg[1], an electronic library. We used Oliver Twist, David Copperfield, and Great Expectations for the training data, and A Christmas Carol and A Tale of Two Cities for the development and test data, respectively. The data were divided into sentences using the PUNKT tokenizer from NLTK [7]. We defined a set of sentences from an indentation at the beginning of a sentence to the line breaks as a paragraph. A label of 0 was assigned if two consecutive sentences belonged to the same paragraph, and a label of 1 was assigned if they did not belong to the same paragraph.

Based on the above definition, a conversational sentence was defined as a single independent paragraph. Generally speaking, a conversation sentence starts with a quotation mark, and it is therefore easier to discriminate based on surface or symbolic grounds compared to a paragraph among descriptive sentences. Therefore, we generated Dataset A in which conversational sentences were counted as one independent paragraph, and Dataset B where conversational sentences were not counted as a single paragraph. Note that Dataset B was more imbalanced with respect to the ratio of each label number than Dataset A. Tables 1 and 2 show the statistics of each dataset and examples of the data included in Dataset A, respectively. In Dataset B, a label of 0 was given to the sample, as shown in example 4 in Table 1.

Table 1. Statistics of the datasets generated.

Data	#Labels (0 : 1)		#Words		
	Dataset A	Dataset B	Mean (std.)	Min	Max
Train	23972 : 14778	32912 : 5838	42.93 (29.73)	3	383
Dev	1242 : 707	1656 : 293	35.44 (27.88)	3	260
Test	4586 : 3258	6625 : 1219	41.33 (29.13)	3	349

4 Methodology

Here, we describe BERT in detail, which is the base language model used for paragraph segmentation in this study, along with the loss function used in classification.

[1] https://www.gutenberg.org/.

Table 2. Excerpts of example from Dataset A where conversational sentences are counted as a single independent paragraph.

#	Label	Sentence 1	Sentence 2
1	0	Marley was dead, to begin with	There is no doubt whatever about that
2	0	There is no doubt whatever about that	The register of his burial was signed by the clergyman, the clerk, the undertaker, and the chief mourner
3	1	You will, therefore, permit me to repeat, emphatically, that Marley was as dead as a door-nail	Scrooge knew he was dead?
4	1	"You don't mean that, I am sure?"	"I do," said Scrooge

4.1 BERT

BERT is a general-purpose language model based on a multiple bidirectional transformer [8] that outputs a distributed representation of an input sequence and words included in the sequence. BERT improves the performance of a language model by pre-training a large-scale corpus. For pre-learning, a masked word prediction was applied to predict the original word of a sentence, in which a portion of the input sentence had been replaced with a token "[MASK]," and next sentence prediction was used to correctly identify the continuity of the two sentences as the input.

To apply BERT in classification tasks such as polarity determination and document classification, the vector output for the token "[CLS]" added to the head of the input sentence was input to the classifier. At this time, fine-tuning using a pre-trained model can be applied to the tasks to be solved. As a result, a high accuracy was achieved in multiple tasks, such as the General Language Understanding Evaluation benchmark problem [9], as well as a benchmark problem for machine reading using the Stanford Question Answering Dataset [10].

4.2 Loss Function

Cross Entropy Loss. The loss function applied by BERT to solve the classification problem was generally softmax cross entropy loss. In this study, we used binary cross entropy (BCE) loss expressed using the following Eq. 1 for handling the binary classification problem.

$$\mathrm{BCE}(p, y) = -y \log(p) - (1-y)\log(1-p), \tag{1}$$

where y is the true label and $p \in [0, 1]$ is the model's estimated probability for the class with label $y = 1$. For notational convenience, we define p_t:

$$p_t = \begin{cases} p & \text{if } y = 1 \\ 1-p & \text{if } y = 0, \end{cases} \tag{2}$$

and rewrite Eq. 1 as Eq. 3.

$$\mathrm{BCE}(p_t) = -\log(p_t). \tag{3}$$

In general, for classification problems that target imbalanced data, weights $\alpha \in [0, 1]$ are introduced into the BCE loss to adjust the balance, as shown in Eq. 4, and consider the importance based on the size of each class. In many cases, the reciprocal of the number of data included in each class is adopted as a practical value of α.

$$\alpha\mathrm{BCE}(p_t) = -\alpha_t \log(p_t), \tag{4}$$

$$\alpha_t = \begin{cases} \alpha & \text{if } y = 1 \\ 1 - \alpha & \text{if } y = 0. \end{cases} \tag{5}$$

Madabushi et al. [11] showed the effectiveness of changing the loss function in a fully connected layer, which is the final layer of BERT, to an α-balanced BCE loss for the classification problem of imbalanced data for the identification of propaganda.

Focal Loss. FL is an effective loss function for the problem of object detection in the field of image processing. In the object detection problem, the background occupies a large part of the image, and therefore it is difficult to identify a specific object of the minority class. The abovementioned α-balanced BCE loss makes it possible to consider the importance based on the size of each class, but cannot distinguish the difficulty of identification for each class. However, FL introduces a modulation factor that attenuates the contribution of errors from easily identifiable examples and prevents overwhelming loss functions. This allows the model to effectively focus on examples that are difficult to identify. Specifically, a term $(1 - p_t)^\gamma$ containing $\gamma \geq 0$ is introduced into the BCE loss, which can be tuned as shown in Eq. 6.

$$\mathrm{FL}(p_t) = -\alpha_t (1 - p_t)^\gamma \log(p_t). \tag{6}$$

5 Experiment

Here, we describe the experiments and results for the paragraph segmentation of novels using BERT.

5.1 Experimental Settings

In this experiment, we used $\mathrm{BERT_{BASE}}$ (number of transformer blocks $L = 12$, hidden size $H = 768$, and number of self-attention heads $A = 12$), which is a pre-trained and publicly available English language model of BERT. We compared the performance of the classifier using each loss function when performing fine-tuning for the paragraph segmentation problem described above. The loss function used is as follows.

- **BCE loss.** Baseline method used in this experiment.
- **αBCE loss.** Method that introduces weight α to the BCE loss.
- **FL.** The values of the hyper-parameters α and γ are determined using a grid search with respect to the values of the evaluation metrics described below obtained for the development data.

5.2 Evaluation Metrics

When dealing with classification problems for imbalanced data, it is necessary to pay attention to the setting of the model evaluation metrics. In this study, we adopted the $F1$-score, Matthews correlation coefficient (MCC), and balanced accuracy as evaluation metrics for comparing models with different loss functions.

5.3 Experimental Results

Table 3 shows the average and standard deviation of the values of each evaluation metric obtained as a result of 10 experiments. Regarding the values of each evaluation metric for Datasets A and B, higher values were obtained when using the αBCE loss as the loss function in comparison to the BCE loss. This shows that it is effective to adjust the weight α according to the difference in class size. However, when FL was used, higher accuracy was obtained than when either loss function of the BCE loss or the αBCE loss was used. The results obtained for Dataset B showed a significant improvement in the performance of the model for paragraph segmentation using FL instead of the BCE loss or the αBCE loss as compared with Dataset A. Therefore, the greater the size difference between the two classes, the more effective is the usage of FL as a loss function.

In addition, it can be confirmed that the results for both Datasets A and B improved the performance of the model for the development data more than the test data. This is because the model parameters were adjusted based on the evaluation metrics obtained for the development data during this experiment. Regarding the test data using FL, each evaluation metric was higher than when using the BCE loss. Therefore, it can be stated that the model is versatile. However, to divide sentences into paragraphs more accurately, it is necessary to adjust the appropriate parameters for each novel.

One of the problems in solving paragraph boundary estimation as a classification problem was that it was difficult to understand the characteristics of paragraph boundaries due to their imbalance. In this experiment, the value of each metric was improved, therefore the relationship between each sentence of the novel is considered more accurately, and it became to be possible to more accurately determine whether there is a paragraph boundary between sentences.

In Fig. 1, the output of BERT for the samples included in Datasets A and B (specifically, a distributed representation of the "[CLS]" token added to the beginning of the sentence when two sentences are input) is visualized using t-distributed stochastic neighbor embedding [12]. Although there are parts where labels of 0 and 1 form a cluster, a part where both label samples are mixed

Table 3. Values of evaluation metrics.

Dataset		Method	$F1$-score		MCC		Balanced accuracy	
			Mean	Std.	Mean	Std.	Mean	Std.
A	Dev	BCE	0.8523	0.00419	0.7724	0.00666	0.8814	0.00324
		αBCE	0.8603	0.00439	0.7818	0.00700	0.8896	0.00364
		FL	**0.8630**	0.00360	**0.7863**	0.00547	**0.8914**	0.00304
	Test	BCE	0.8774	0.00196	0.7933	0.00291	0.8944	0.00173
		αBCE	0.8786	0.00208	0.7932	0.00378	0.8959	0.00177
		FL	**0.8797**	0.00230	**0.7949**	0.00398	**0.8970**	0.00194
B	Dev	BCE	0.7286	0.00597	0.6859	0.00699	0.8254	0.00409
		αBCE	0.7446	0.00515	0.6989	0.00635	0.8535	0.00526
		FL	**0.7583**	0.00721	**0.7144**	0.00856	**0.8671**	0.00533
	Test	BCE	0.7416	0.00576	0.6976	0.00633	0.8360	0.00480
		αBCE	0.7532	0.00663	0.7072	0.00766	0.8578	0.00594
		FL	**0.7548**	0.00460	**0.7086**	0.00535	**0.8631**	0.00522

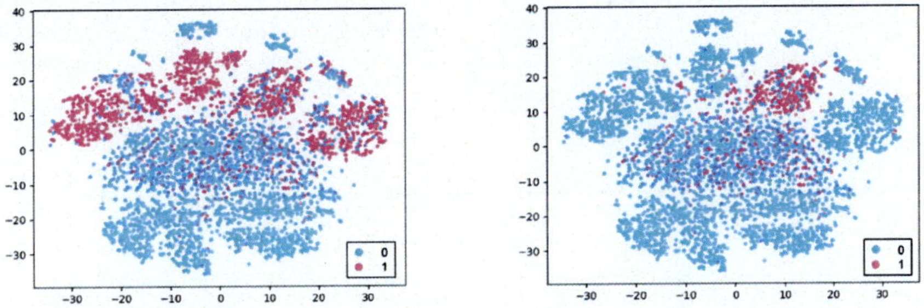

(a) Dataset A (b) Dataset B

Fig. 1. Output of BERT for the samples

can be seen near the center of Figs. 1(a) and 1(b). Comparing Figs. 1(a) and 1(b), we can see that Dataset A, where the conversation is counted as one independent paragraph, has a larger number of clusters composed of a single label than Dataset B. Samples in which each label forms a unique cluster are easy to estimate, whereas samples in which each label is mixed are difficult to estimate.

Table 4 shows examples of the label of 1 sample, which could be correctly estimated in all 10 trials using FL, but not when using the BCE loss. Additionally, we visualized the distributed representation of the final output of BERT for those samples, as shown in Fig. 2. Figure 2 shows that many of these samples belong to the part where the samples of each label are mixed in Fig. 1(b). From this result, it is considered that FL allows those samples that are difficult to estimate to contribute to learning more significantly.

Table 4. Examples correctly estimated as label 1 in Dataset B using FL.

Sentence 1	Sentence 2
In both countries it was clearer than crystal to the lords of the State preserves of loaves and fishes, that things in general were settled for ever	It was the year of Our Lord one thousand seven hundred and seventy-five
Madame Defarge knitted with nimble fingers and steady eyebrows, and saw nothing	Mr Jarvis Lorry and Miss Manette, emerging from the wine-shop thus, joined Monsieur Defarge in the doorway to which he had directed his own company just before
Before this rumour, the crowd gradually melted away, and perhaps the Guards came, and perhaps they never came, and this was the usual progress of a mob	Mr Cruncher did not assist at the closing sports, but had remained behind in the churchyard, to confer and condole with the undertakers
He sat her down just within the door, and held her, clinging to him	Defarge drew out the key, closed the door, locked it on the inside, took out the key again, and held it in his hand

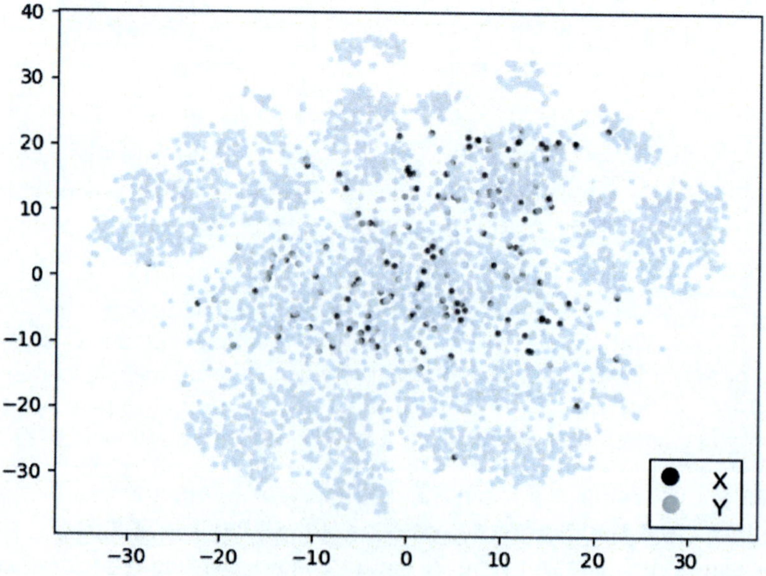

Fig. 2. Output of BERT for the samples included in Dataset B (X, label 1 correctly estimated using only FL; Y, other than X.)

6 Conclusions and Future Works

In this study, we considered the paragraph segmentation of a novel as a binary classification problem regarding whether two sentences belong to the same paragraph. Devising a way to handle an imbalance of the data, we introduced FL as a loss function in the BERT-based classifier. As a result, the classification performance for estimating the paragraph boundaries was improved as compared to the general case using the BCE loss.

The following are planned as future tasks.

- **Quantitative analysis of results.** In this paper, we examined the specific identification results of the proposed method by visualizing each distributed representation with T-SNE. Future studies will need to consider this result quantitatively.
- **Creating a more general model.** In the experiments described in this paper, although we used datasets created based on novels written by a single author, there was a difference in the accuracy of the paragraph segmentation between each novel. This result suggests the diversity of sentence structures used in different novels. Therefore, a more general-purpose model can be created using a dataset with novels by multiple authors.
- **Expanding the range of input sentences.** In the experiments conducted in this study, only two sentences were given as input sentences, and it was estimated whether these sentences belong to the same paragraph. However, considering the features of the sentence form used in a novel, it is expected that more sentence information will enable a more appropriate estimation of the paragraph boundaries.

Acknowledgement. This work was supported by JSPS KAKENHI Grant, Grant-in-Aid for Scientific Research(B), 19H04184.

References

1. Devlin, J., Chang, M.-W., Lee, K., Toutanova, K.: BERT: pre-training of deep bidirectional transformers for language understanding. arXiv preprint arXiv:1810.04805 (2018)
2. Lin, T.-Y., Goyal, P., Girshick, R.B., He, K., Dollár, P.: Focal loss for dense object detection. CoRR, abs/1708.02002 (2017)
3. Chawla, N.V., Bowyer, K.W., Hall, L.O., Philip Kegelmeyer, W.: SMOTE: synthetic minority over-sampling technique. J. Artif. Intell. Res. **16**(1), 321–357 (2002)
4. Glavaš, G., Nanni, F., Ponzetto, S.P.: Unsupervised text segmentation using semantic relatedness graphs. In: Proceedings of the Fifth Joint Conference on Lexical and Computational Semantics, Berlin, Germany, pp. 125–130. Association for Computational Linguistics, August 2016
5. Koshorek, O., Cohen, A., Mor, N., Rotman, M., Berant, J.: Text segmentation as a supervised learning task. In: Proceedings of the 2018 Conference of the North American Chapter of the Association for Computational Linguistics: Human Language Technologies, Volume 2 (Short Papers), New Orleans, Louisiana, pp. 469–473. Association for Computational Linguistics, June 2018

6. Badjatiya, P., Kurisinkel, L.J., Gupta, M., Varma, V.: Attention-based neural text segmentation. CoRR, abs/1808.09935 (2018)
7. Loper, E., Bird, S., Klein, E.: Natural Language Processing with Python. O'Reilly Media Inc., Sebastopol (2009)
8. Vaswani, A., Shazeer, N., Parmar, N., Uszkoreit, J., Jones, L., Gomez, A.N., Kaiser, L., Polosukhin, I.: Attention is all you need. In: Guyon, I., Luxburg, U.V., Bengio, S., Wallach, H., Fergus, R., Vishwanathan, S., Garnett, R. (eds.) Advances in Neural Information Processing Systems 30, pp. 5998–6008. Curran Associates, Inc. (2017)
9. Wang, A., Singh, A., Michael, J., Hill, F., Levy, O., Bowman, S.: GLUE: a multi-task benchmark and analysis platform for natural language understanding. In: Proceedings of the 2018 EMNLP Workshop BlackboxNLP: Analyzing and Interpreting Neural Networks for NLP, Brussels, Belgium, pp. 353–355. Association for Computational Linguistics, November 2018
10. Rajpurkar, P., Jia, R., Liang, P.: Know what you don't know: unanswerable questions for SQuAD. In: Proceedings of the 56th Annual Meeting of the Association for Computational Linguistics (Volume 2: Short Papers), Melbourne, Australia, pp. 784–789. Association for Computational Linguistics, July 2018
11. Madabushi, H.T., Kochkina, E., Castelle, M.: Cost-sensitive BERT for generalisable sentence classification on imbalanced data. In: Proceedings of the Second Workshop on Natural Language Processing for Internet Freedom: Censorship, Disinformation, and Propaganda, Hong Kong, China, pp. 125–134. Association for Computational Linguistics, November 2019
12. van der Maaten, L.J.P., Hinton, G.E.: Visualizing high-dimensional data using t-SNE. J. Mach. Learn. Res. **9**, 2579–2605 (2008)

Parallel Implementation of Nearest Feature Line and Rectified Nearest Feature Line Segment Classifiers Using OpenMP

Ana-Lorena Uribe-Hurtado(✉), Eduardo-José Villegas-Jaramillo, and Mauricio Orozco-Alzate

Departamento de Informática y Computación, Universidad Nacional de Colombia - Sede Manizales, Kilómetro 7 vía al Magdalena, Manizales 170003, Colombia
alhurtadou@unal.edu.co

Abstract. Parallelizing computationally expensive classification algorithms, such as the Rectified Nearest Feature Line Segment (RNFLS), remains a task for expert programmers due to the complexity involved in rewriting the application and the required knowledge of the available hardware and tools. A simple parallel implementation of the Nearest Feature Line (NFL) and the RNFLS algorithms using OpenMP over multi-core architectures is presented. Both non-parametric classifiers are used in the classification of datasets that contain few samples. The training and testing evaluation technique is used, with a 70–30 ratio respectively and seven datasets from the UCI repository, to verify the speedup and the accuracy of the classifier. Results and experiments derived from the parallel execution of both algorithms, calculating the average of 20 repetitions on a small architecture of 6 physical cores and 12 ones with multi-threading, show that accelerations of up to 21 times can be achieved with NFL and up to 13 times with RNFLS.

Keywords: Parallel computing · Nearest Feature Line · Rectified Nearest Feature Line · OpenMP

1 Introduction

Parallelizing algorithms is the art of distributing the tasks of a program such that it makes better use of the internal architecture of the computer [2,7]. Two ways to create parallel code are identified in [4]: automatic parallelization, vectorized instructions performed by processors from the compilation stage, and explicit programming, in which the programmer must identify the instructions to parallelize. Among the best-known APIs for implementing parallel applications are: Open Multi-Processing (OpenMP), Portable Operating System Interface (POSIX threads), StarSs API developed to extend OpenMP functionalities (OmpSs), Compute Unified Device Architecture (CUDA), Open Computing Language (OpenCL)

Y. Dong et al. (Eds.): DCAI 2020, AISC 1237, pp. 31–40, 2021.
https://doi.org/10.1007/978-3-030-53036-5_4

and Open Accelerators (OpenACC). The last three ones are designed to implement parallelism over heterogeneous architectures. OpenMP allows parallelization of portable and scalable multi-process applications and make use of the computational architecture through *pragma* directives, without requiring that the programmer has an advanced knowledge of it. Loops operations and vectorization are commonly used to parallelize the instructions of a program.

The Nearest Feature Line (NFL) and the Rectified Nearest Feature Line segment (RNFLS) are classification algorithms to deal with datasets having few samples: the so-called small sample size case. Implementations of both algorithms were written in [10], parallelizing the Leave One Out (LOO) test instead of paralellizing the algorithms themselves. Although the LOO test is computationally expensive, in this paper we want to show an alternative of parallelization of the algorithms themselves, using OpenMP, with a lighter evaluation technique as the repeated train and test evaluation technique presented in [3].

The NFL algorithm suffers from the problems of interpolation and extrapolation, this occurs because the algorithm does not verify whether the projection of the feature line violates the class territory of other samples when a feature line passes through objects that are far from each other [5]. Several proposals to improve NFL problems have been presented in [1] and [6]. The first one involves two additional processes: segmentation and rectification which are explained in Sect. 2.2. The second one proposes to transform the data of the training set of different classes, using spectral feature analysis into independent groups, before applying NFL. Finally, in [6], it is suggested to edit feature line segments using two strategies: generalization-based and representation-based eliminations.

We adopt the proposal developed in [5] to implement the parallelization; since that algorithm involves a high computational complexity. We present a parallelization approach of the NFL and the RNFLS algorithms using OpenMP to improve the elapsed times of both algorithms on multicore architectures. According to our review of the state-of-the-art, although other classifiers (see [11] and [9]) have been parallelized with OpenMP, we did not find other proposals for parallelization of the NFL and the RNFLS algorithms.

The rest of the paper is organized as follows: Sect. 2 shows the parallel normalization process, required for both NFL and RNFLS. Sections 2.1 and 2.2 briefly explain the algorithms, show their pseudo-codes and present the instructions used to parallelize them with OpenMP. Section 3 presents the experiments and results obtained with the parallelization. Finally, in Sect. 4, our conclusions are discussed.

2 OpenMP Parallelization of NFL and RNFLS Classifiers

The training and test sets, that are used by Algorithms 2 and 3, need to be normalized; see Algorithm 1. The threads concurrently execute the normalization, as follows: Instructions 1 and 7 create as many threads (N_c) as the architecture allows.

Instructions 2 and 8 are responsible for distributing Operations 4 and 10 between the threads. They parallelize the loops and distribute—evenly if possible when the proportion is not explicitly specified (static scheduling mode)—the normalization instructions associated with the indexes of each loop. For example, if the machine allows the launch of 12 threads and the loop normalizes 34 columns of the dataset, each thread concurrently processes two columns of the dataset. Thread 0 processes columns 0 and 1; thread 1 processes columns 2 and 3, thread 11 processes columns 22 and 23, until all the loop indexes are distributed among the threads. The thread that finishes the execution first, will process the subsequent columns 24 and 25, until the normalization is done.

Algorithm 1. Normalization phase for NFL and RNFLS

Inputs:
$\mathcal{T}$: training set, Γ:test set

Output:
$\mathcal{T}$: normalized training set and Γ: normalized test set,

Launch OpenMP N_c **threads**
Launch OpenMP for static schedule ▷ Each thread executes normalization, over a chunk of columns of $\mathcal{T}$
for each $\boldsymbol{t}_i \in \mathcal{T}$ **do**
  $\boldsymbol{t}_i = (\boldsymbol{t}_i - \boldsymbol{\mu}_{\mathcal{T}}) \oslash \boldsymbol{\sigma}_{\mathcal{T}}$ ▷ $\oslash$ denotes Hadamard product
end for
Close all the N_c **OpenMP threads**

Launch OpenMP N_c **threads**
Launch OpenMP for static schedule ▷ Each thread executes normalization over a chunk of columns of Γ
for each $\boldsymbol{\gamma}_i \in \Gamma$ **do**
  $\boldsymbol{\gamma}_i = (\boldsymbol{\gamma}_i - \boldsymbol{\mu}_{\mathcal{T}}) \oslash \boldsymbol{\sigma}_{\mathcal{T}}$
end for
Close all the N_c **OpenMP threads**

All operations involving addition, subtraction, division and multiplication between vectors are automatically vectorized, e.g. line 11 of Algorithm 2 and lines 10, 32 and 38 of Algorithm 3 among others, using the instruction `#pragma omp simd`.

The parallel classification algorithms are described below. They are based on the sequential ones originally presented in [10].

2.1 Nearest Feature Line and Its Parallelization

NFL is a non-parametric classifier. It was developed in [8] and attempts to improve the representation capability of a dataset—denoted as $\mathcal{T}$—, with a small sample size, using feature lines—denoted as $\mathcal{L}$—, which pass through each pair of samples belonging to the same class [5]. The lines are geometrically computed by a pair of connected points. Each line L_i is represented as a triplet $(\boldsymbol{t}_j, \boldsymbol{t}_k, \rho_i)$, where, $\boldsymbol{t}_i, \boldsymbol{t}_j$ are objects belonging to $\mathcal{T}$ and ρ_i is the label of the class of the feature line. In addition, $\rho_i = \theta_j = \theta_k$ because the lines only connect objects of the same class.

The parallel version of NFL using OpenMP is shown in Algorithm 2. Since this algorithm is closely related with the parallel version of RNFLS, an explanation is provided only after presenting Algorithm 3.

Algorithm 2. Parallel Nearest feature line classifier for one object γ of Γ

Inputs:

$\mathcal{T}$: training set, γ: test object, m:minimum, $\hat{\theta}$: empty label for one γ ▷ all input variables are shared

Output:

$\hat{\theta}$: estimated class label for one γ ▷ Calculate $\hat{\theta}$ for each object of Γ

Call Normalization phase ▷ see Algorithm 1
// Classification phase:
Launch OpenMP N_c threads ▷ Launch as many threads ▷ as the architecture allows
$minimum = \infty$ ▷ private for each thread
Launch OpenMP for static schedule
for $j = 1$ to $|\mathcal{T}|$-1 **do**
    Launch OpenMP simd
    for $k = j + 1$ to $|\mathcal{T}|$ **do**
        if $\theta_j == \theta_k$ **then**
            $\rho_i = \theta_j$
            $\tau = (\gamma - \boldsymbol{t}_j) \cdot (\boldsymbol{t}_k - \boldsymbol{t}_j)/(||(\boldsymbol{t}_k - \boldsymbol{t}_j)||)^2$
            $\boldsymbol{p} = \boldsymbol{t}_j + \tau(\boldsymbol{t}_k - \boldsymbol{t}_j)$
            $d_i = ||\gamma - \boldsymbol{p}||$ ▷ Distances to lines
            if $d_i < minimum$ **then**
                $minimum = d_i$
                $\rho = i$
            end if
        end if
    end for
end for
Launch OpenMP critical area
$m = \min_{N_C}(minimum)$ ▷ smallest distance
$\hat{\theta} = \rho$
Close OpenMP critical area
Close all the N_c OpenMP thread

2.2 Rectified Nearest Feature Line Segment and Its Parallelization

The authors of [5] diagnosed the interpolation and extrapolation problems of the NFL algorithm and proposed a rectification of it, called Rectified Nearest Feature Line Segment (RNFLS). The modification includes two additional processes: segmentation and rectification.

The rectification calculates the distance from $\boldsymbol{\gamma}$ as done in NFL if and only if the projection of $\boldsymbol{\gamma}$ is within the interpolating part of the line. If this does not happen, the distance from $\boldsymbol{\gamma}$ to $\boldsymbol{t}_j$ or $\boldsymbol{t}_k$ is calculated according to the extrapolating parts of the line where the orthogonal projection appears.

The rectification is implemented by checking whether the lines cross the territory of other classes. In the event that this occurs, such a line is excluded and is not taken into account to calculate the distances. A feature line violates the territory of another, if there is at least one object of the training set where the radius $(\boldsymbol{r})$ of its territory is less than its distance from the feature line.

The radius of each object in the training set is defined as the distance to the nearest training object within the same class.

Algorithm 3 performs three processes: normalization, calculation of training set radii and classification. The algorithm concurrently finds the radii $(\mathcal{R})$ between the pairs $(\boldsymbol{t}_i, \boldsymbol{t}_j)$ belonging to $\mathcal{T}$, evenly distributing among the threads the instructions contained in loop 5 and vectorizing, by command 7, the instructions in loop 8.

Each thread concurrently executes the code of the classifier and the check of the invasion of the lines, evenly distributing the operations, by means of instruction 22. All the threads find the local minimum value among the group of indices processed and, at the end of the loop, instruction 64 is in charge of finding, in a serial way, the global minimum among all the minimum values found by each chunk.

Algorithm 3. Parallel Rectified nearest feature line segment classifier for one object (γ) of $\boldsymbol{\Gamma}$

Inputs:
$\mathcal{T}$: training set, $\boldsymbol{\gamma}$: test object, m: minimum, $\hat{\theta}$: empty class label for $\boldsymbol{\gamma}$ ▷ all input variables are shared

Output:
$\hat{\theta}$: estimated class label for $\boldsymbol{\gamma}$ ▷ Calculate $\hat{\theta}$ for each object of Γ

1: Normalization phase ▷ see Algorithm 1
//Radii of the territories
2: $\mathcal{R} = \emptyset$
3: **Launch OpenMP** N_c **threads, shared** $(\mathcal{R})$**, private**$(\boldsymbol{t}_i, \boldsymbol{t}_j)$
4: **Launch OpenMP for static schedule**
5: **for each** $\boldsymbol{t}_i \in \mathcal{T}$ **do**
6: $r = \infty$ ▷ Initial guess for the radius
7: **Launch OpenMP simd** 1

Algorithm 3. Parallel Rectified nearest feature line segment classifier, Cont...

for all $\boldsymbol{t}_j \in \mathcal{T}$ **do**
  if $\theta_i \neq \theta_j$ **then** ▷ Chech whether class labels are different
    $d = ||\boldsymbol{t}_i - \boldsymbol{t}_j||$ ▷ Distance between objects of different classes
    if $d < r$ **then**
      $r = d$ ▷ Update the value of the radius
    end if
  end if
end for
$\mathcal{R} \cup \{r\}$ ▷ Include the current radius in the collection of radii
end for
Close all the N_c OpenMP thread
// Classification phase:
$minimum = \infty$ ▷ private for each thread
Launch OpenMP N_c threads shared($\mathcal{R}$) private($\boldsymbol{t}_i$, $\boldsymbol{t}_j$)
Launch OpenMP for static schedule
for $j = 1$ to $|\mathcal{T}|$-1 **do**
  Launch OpenMP simd
  for $k = j$ to $|\mathcal{T}|$ **do**
    if $\theta_j == \theta_k$ **then** ▷ Chech whether class labels are equal
      $\rho_i = \theta_j$ ▷ Label of the feature line segment
      // Verification of invasion to class territories
      $check = True$
      for $j = m$ to $|\mathcal{T}|$ **do** ▷ Check lines
        if $\theta_m \neq \rho_i$ **then**
          $\tau = (\gamma - \boldsymbol{t}_j) \cdot (\boldsymbol{t}_k - \boldsymbol{t}_j)/(||(\boldsymbol{t}_k - \boldsymbol{t}_j)||)^2$
          if $\tau < 0$ **then**
            $\boldsymbol{p} = \boldsymbol{t}_j$ ▷ First part of the segmentation process
          else if $\tau > 0$ **then**
            $\boldsymbol{p} = \boldsymbol{t}_k$ ▷ Second part of the segmentation process

3 Experimental Results

The experiments were calculated on an Intel® Xeon® CPU E5-2643 v3 @ 3.40 GHz, that was exclusively dedicated to run them. The training and test technique selected in each execution consisted of randomized permutations of 70% for the training set and 30% for the test set, see Table 1. The average of 20 executions of sequential and parallel implementations of NFL and RNFLS are shown in Tables 2 and 3, respectively. Seven datasets from the UCI repository were used. Table 1 shows the names of the datasets used, their number of objects, dimensions and the percentage of objects used for training and test in both algorithms.

The sequential algorithms are implemented in ANSI C. OpenMP is used for the parallelization, in order to compare accuracies, elapsed times and speedups

Algorithm 3. Parallel Rectified nearest feature line segment classifier, Cont...

```
                else
                    p = t_j + τ(t_k − t_j)          ▷ Projection onto the feature line
                end if
                d2line = ||t_m − p||                 ▷ Distance to the line
            end if
            if d2line < r_m then                    ▷ Check invasion to other territory
                check = False
            end if
        end for
        if check == True then
            τ = (γ − t_j) · (t_k − t_j)/(||(t_k − t_j)||)^2
            if τ < 0 then
                p = t_j
            else if τ > 0 then
                p = t_k
            else
                p = t_j + τ(t_k − t_j)
            end if
            d_i = ||γ − p||                          ▷ Distances to line segments
            if d_i < minimum then
                minimum = d_i
                ρ = i                                ▷ Label
            end if
        end if
    end if
  end for
end for
Launch OpenMP critical area
m = min_{N_C}(minimum)                               ▷ smallest distance
θ̂ = ρ                                                ▷ Assigned class label
Close OpenMP critical area
Close all N_c threads
```

on multicore architectures. Executions of both algorithms, using 2 and 12 cores, are shown in Tables 2 and 3. The speedup of the algorithms are calculated as: $SU = ETseq/ETpar$, the ratio between sequential and parallel elapsed times. The RNFLS algorithm, with a complexity of $\mathcal{O}(n^3)$, presents in its sequential version the highest elapsed times: 275.4 s and 392.27 s, see Table 3. It is important to parallelize the algorithm such that it makes better use of the machine resources. In this way the parallel version allows multiple repetitions, making it easier for the expert to visualize the data in pseudo real time. Although NFL is a computationally lighter algorithm, with complexity $\mathcal{O}(n^2)$, the speedup of the parallel implementation overpass, in all cases, the sequential version. Speedups of 1.7 and 3.7 times using only two cores are achieved, with the smallest and the largest datasets respectively; see Table 2.

Table 1. Dataset details, number of classes (#C), dimensions (Dim), rows of the dataset (Row), training (TR) and test (TE) set sizes.

Datasets	#C	Dim	Row	70%	30%
Iris	3	4	150	105	45
Wine	3	13	178	124	54
Glass	7	9	214	149	65
Ionosphere	2	34	351	245	106
Bupa	2	6	345	241	104
Wdbc	2	30	569	398	171
Pima	2	8	768	537	231

Table 2. Results of parallel NFL: sequential (SEQ), accuracy (ACC), standard deviation (STD), elapsed time (ET) in seconds and speedup (SU)

			SEQ	2 cores		12 cores	
Datasets	ACC	STD	ET	ET	SU	ET	SU
Iris	0.8356 ±	0.0480	0.0158	0.0090	1.7561	0.0040	3.9085
Wine	0.9586 ±	0.0293	0.0543	0.0090	6.0408	0.0105	5.1648
Glass	0.5967 ±	0.0488	0.0584	0.0179	3.2631	0.0113	5.1796
Ionosphere	0.6959 ±	0.0421	1.6828	0.2754	6.1096	0.0792	21.2354
Bupa	0.5862 ±	0.0445	0.3518	0.0823	4.2742	0.0314	11.1907
Wdbc	0.5865 ±	0.0305	5.2857	1.1500	4.5965	0.2847	18.5673
Pima	0.5689 ±	0.0277	4.1140	1.0991	3.7430	0.3354	12.2646

Figure 1 shows: the elapsed times in the left vertical axis and the speedups in the right one. These curves are drawn against the number of cores. Speedups are computed by considering the sequential elapsed times as references. Notice that, in almost all cases, speedups are monotonically increasing except for the largest numbers of cores with the iris dataset. Since iris is the smallest dataset, the computational cost of the distribution of the load, among many cores, may be comparable with the cost of the algorithm computation itself.

Table 3. Results of parallel RNFLS: sequential (SEQ), accuracy (ACC), standard deviation (STD), elapsed time (ET) in seconds and speedup (SU)

			SEQ	2 cores		12 cores	
Datasets	ACC	STD	ET	ET	SU	ET	SU
Iris	0.9256 ±	0.0299	0.5787	0.3117	1.8564	0.0732	7.9075
Wine	0.9574 ±	0.3290	2.6401	1.1038	2.3917	0.2938	8.9868
Glass	0.6610 ±	0.0533	2.4748	1.0907	2.2690	0.2644	9.3585
Ionosphere	0.7212 ±	0.2688	81.8352	24.4647	3.3450	6.1456	13.3161
Bupa	0.6276 ±	0.0453	11.5756	4.3549	2.6580	1.3382	8.6503
Wdbc	0.5911 ±	0.0279	275.4488	84.7739	3.2492	22.5450	12.2178
Pima	0.6000 ±	0.0259	392.2739	129.8544	3.0209	33.4431	11.7296

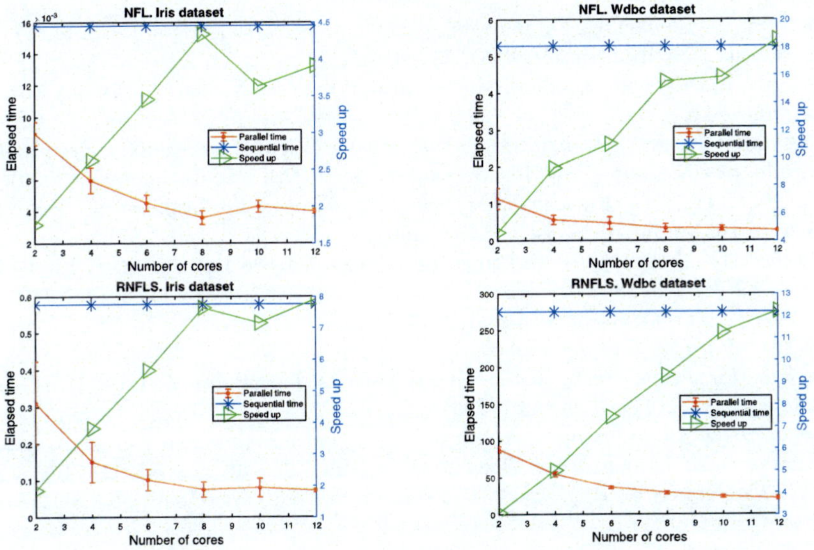

Fig. 1. Results using up to 12 cores for NFL (top) and RNFLS (bottom)

4 Conclusions

Developing applications using OpenMP facilitates their parallel implementation and allows a better use of the machine architecture, reducing elapsed times.

This parallelization is an interesting research topic in sectors such as industry, since it could improve the performance of the classification algorithms in production chains, where automatic object classification is typically required.

The results derived from the parallel execution of both algorithms, calculating the average of 20 repetitions on a small computational architecture of 6 physical cores and 12 with multi-threading, show that accelerations of up to 21 times can be achieved with NFL and up to 13 times with RNFLS, with a dataset (ionosphere) of 351 objects and 34 dimensions.

Future work may also consider implementing, under these semi-automatic parallelization technologies, the new versions of the NFL algorithm, cited in this paper, that improve the accuracy of the NFL classifier.

Acknowledgments. The authors acknowledge the support provided by Facultad de Administración, Universidad Nacional de Colombia - Sede Manizales (UNAL) and Grupo de Ambientes Inteligentes Adaptativos - GAIA to attend DCAI'20.

References

1. Altincay, H., Erenel, Z.: Avoiding the interpolation inaccuracy in nearest feature line classifier by spectral feature analysis. Pattern Recognit. Lett. **34**, 1372–1380 (2013)
2. Berman, F., Snyder, L.: On mapping parallel algorithms into parallel architectures. J. Parallel Distrib. Comput. **4**(5), 439–458 (1987)
3. Bramer, M.: Estimating the predictive accuracy of a classifier, pp. 79–92. Springer, London (2016)
4. Chapman, B., Jost, G., van der Pas, R.: Using OpenMP: portable shared memory parallel programming, vol. 46. Massachusetts Institute of Technology (2008)
5. Du, H., Chen, Y.Q.: Rectified nearest feature line segment for pattern classification. Pattern Recognit. **40**(5), 1486–1497 (2007)
6. Kamaei, K., Altincay, H.: Editing the nearest feature line classifier. Intell. Data Anal. **19**(3), 563–580 (2015)
7. Kung, H.: The structure of parallel algorithms. In: Advances in Computers, vol. 19, pp. 65 – 112. Elsevier (1980)
8. Li, S.Z., Lu, J.: Face recognition using the nearest feature line method. IEEE Trans. Neural Netw. **10**(2), 439–443 (1999)
9. Reyes-Ortiz, J.L., Oneto, L., Anguita, D.: Big data analytics in the cloud: spark on hadoop vs MPI/openMP on Beowulf. Procedia Comput. Sci. **53**, 121–130 (2015). iNNS Conference on Big Data 2015 Program San Francisco, CA, USA 8-10 August 2015
10. Uribe-Hurtado, A., Villegas-Jaramillo, E.J., Orozco-Alzate, M.: Leave-one-out evaluation of the nearest feature line and the rectified nearest feature line segment classifiers using multi-core architectures. Ingeniería y Ciencia **14**(27), 75–99 (2018)
11. Xiao, B., Biros, G.: Parallel algorithms for nearest neighbor search problems in high dimensions. SIAM J. Sci. Comput. **38**, S667–S699 (2016)

Let's Accept a Mission Impossible with Formal Argumentation, or Not

Ryuta Arisaka(✉) and Takayuki Ito

Nagoya Institute of Technology, Nagoya, Japan
ryutaarisaka@gmail.com, ito.takayuki@nitech.ac.jp

Abstract. We show with a plain example how causal dependency among arguments could lead to an infeasible acceptability judgement in abstract argumentation resulting in acceptance of some argument when no arguments that could engender it are acceptable. While the dependency is somewhat like necessary support relation in the literature, differences are still clearly noticeable, as we are to show. We present abstract argumentation theory with a causal relation, which can be termed causal abstract argumentation or as bipolar argumentation with causal support interpretation. We formulate causal acceptability semantics for this theory as the counterparts of acceptability semantics in abstract argumentation, also detailing the relation between them.

Keywords: Formal argumentation · Causality · Acceptability semantics · Bipolar argumentation

1 Introduction

Let us consider the following argumentation graph: $a_1 \qquad\quad a_2 \longleftrightarrow a_3$. It represents an argumentation of 3 arguments a_1, a_2 and a_3, as described below, and mutual attacks between a_2 and a_3.

a_1: We shall go to Montparnasse while in Paris.
a_2: Sure, we go by metro from here.
a_3: Too many people on the metro, let's just walk there, it's gonna be a 40 min walk.

Abstract argumentation [10] is a formal argumentation theory to infer from an argumentation graph which arguments may be acceptable. So it seems that we could make use of it for this example.

While there are a number of criteria for characterising the acceptability semantics, two concepts, (1) conflict-freeness: *no members of a set attack a member of the same set*, and (2) defence: *a set of arguments defnd an argument just when any argument attacking the argument is attacked by at least one member of the set*, are typically the core. Complete semantics, the basis of many other semantics, is such that each of its members is a set of arguments that is conflict-free and that includes every argument it defends. For the above example, it is $\{\{a_1\}, \{a_1, a_2\}, \{a_1, a_3\}\}$.

Y. Dong et al. (Eds.): DCAI 2020, AISC 1237, pp. 41–50, 2021.
https://doi.org/10.1007/978-3-030-53036-5_5

There is then the difference between credulous and skeptical acceptabilities. An argument is credulously (resp. skeptically) acceptable just when it is in some (resp. every) member of the semantics. An ordinary understanding is: acceptance of any skeptically acceptable argument holds in every circumstance, which is therefore more certainly acceptable than an argument that is only credulously acceptable.

1.1 Research Problem

That does not seem to apply to our example, however, where only a_1 is skeptically acceptable. For, what does it even mean to accept to go to Montparnasse if no means to get there could be agreed upon? Indeed, accepting a_1 alone seems less reasonable than accepting it together either with a_2 or with a_3. Thus, we have a good reason to expect $\{a_1, a_2\}$ and $\{a_1, a_3\}$ to be two credulously acceptable sets of arguments, while $\{\}$ (an empty set) should become the only one skeptically acceptable set of arguments. This expectation disparity arises from the means-purpose relation between a_2 (or a_3) and a_1, that it is strange to accept the purpose if every means is not accepted.

1.2 The Objective

To address the odd semantic prediction, in this work we consider causal abstract argumentation or alternatively bipolar abstract argumentation with causal support relation, and formulate causal acceptability semantics for this theory as the counterparts of acceptability semantics in abstract argumentation. We study in detail the relation among the semantics.

After the following small section for related work, we go through technical preliminaries (in Sect. 2), and then present causal abstract argumentation or bipolar argumentation with causal support interpretation with theoretical results (in Sect. 3), before drawing conclusions.

1.3 Related Work

Any argument-to-argument relation that conditions acceptability of arguments enforces a certain dependency relation. Bipolar argumentation (abstract argumentation with attack and support relations) with necessary interpretation of support [13,14] expresses the relation that any accepted argument forces acceptance of every argument that supports it. However, let us draw a necessary support relation from a_2 to a_1 as well as from a_3 to a_1. Then, acceptance of a_1 would imply acceptance of both a_2 and a_3. However, because a_2 and a_3 are attacking each other, the conflict-freeness prevents the acceptance, from which we obtain that a_1 is not an acceptable argument. This prediction does not match our expectation for this argumentation example (see the above discussion). With another interpretation termed evidential support [15], acceptance of a_1 would imply acceptance of a_2 or a_3 but only if a_2 or a_3 is unattacked. The evidential interpretation, by introducing such special arguments, restrict the applicability too narrowly for our purpose. Common to both of the interpretations, there is a circular dependency problem. Suppose two arguments a_x and a_y that do not attack each other, but suppose that a_x is a means to a_y which in turn is a means to a_x. Bipolar argumentation with necessary or evidential support interpretation would accept both a_x and a_y. Our interpretation on the contrary does not accept either of the arguments, since if acceptance of a_x requires that of a_y first and if acceptance of a_y requires that of a_x first, that is the chicken-and-egg problem, for which we do not conclude both and simultaneously that the chicken came before the egg and that the egg came before the chicken.

Dependencies can be expressed in abstract dialectical frameworks (ADF) [5] locally per argument, which determines every argument's acceptability status based on acceptability statuses of the arguments attacking it. However, the level of abstractness of a dependency relation in ADF is not as high as bipolar argumentations' [7–9], since ultimately we have to know the conclusion as to which set of arguments are acceptable first in order to obtain matching local conditions. More abstract an approach, called may-must argumentation [2], that still respects locality is known, Unlike ADF, each argument in may-must argumentation only specifies the number of attacking arguments that have to be rejected (resp. accepted) in order that the attacked argument can be or must be accepted (resp. rejected). However, so far may-must argumentation is studied only for the attack, offering not the type of dependency that suits our purpose.

Two other formalisms seem able to model the above causality more closely. First, the dynamic relations as found in particular in [1,3] can express causal relation. Specifically, the theory allows generation (as well as elimination and alteration) of an argument from acceptable arguments. A downside of the theory, however, is in its technical complexity in thinking in terms of a transition system which is unnecessary for this work targeting argumentations with no dynamic change. Second, there is the premise-conclusion relation of structured argumentation [11,12]. A work [4] in fact considers a causal relation within the formalism. However, with structured argumentation it is hard to identify intuitively the connection between abstract argumentation and abstract argumentation with a causal relation as we described, since a raw node in its argumentation graph expresses a conclusion or a premise which is not itself an argument. As per the philosophy of abstract argumentation in which the primary components to evaluate an argumentation with are just arguments and argument-to-argument relations, we too consider only those in this work for a general argumentation graph. Moreover, in [4], it is assumed that two arguments as a cause of the same argument are incompatible, which is, while the case in the particular simple example we saw, not generally a norm. Thus, it deserves attention how causality can be understood in the context of abstract argumentation, and what basis its acceptability semantics forms for a corresponding acceptability semantics in causal abstract argumentation or bipolar argumentation with causal support interpretation.

2 Technical Preliminaries

Abstract argumentation considers an argumentation as a graph of: nodes representing arguments; and edges representing an attack of the source argument on the target argument [10].

Let $\mathcal{A}$ denote the class of abstract entities that we understand as arguments, then a (finite) abstract argumentation is a pair (A, R) with $A \subseteq_{\text{fin}} \mathcal{A}$ and $R \subseteq A \times A$. We denote the class of all abstract argumentations by $\mathcal{F}^{\mathsf{D}}$. From here on, we denote: a member of $\mathcal{A}$ by a; a finite subset of $\mathcal{A}$ by A; and a member of $\mathcal{F}^{\mathsf{D}}$ by F^{D}, all with or without a subscript.

One of the main objectives of representing an argumentation formally as a graph is to infer from it which set(s) of arguments may be accepted. Acceptability of a set of arguments in any $F^{\mathsf{D}} \equiv (A, R)$ $(\in \mathcal{F}^{\mathsf{D}})$ is determined by whether it satisfies certain criteria (and there can be many reasonable criteria), and the set of all acceptable sets of arguments of F^{D} under given criteria is called (acceptability) semantics of F^{D} for the given argumentation with the criteria, which is obviously a subset of 2^{2^A}.

The following two concepts of conflict-freeness and defence are generally important for characterisation of an abstract argumentation semantics. For any $F^{\mathsf{D}} \equiv (A, R)$ $(\in \mathcal{F}^{\mathsf{D}})$, $A_1 \subseteq A$ is said to be conflict-free if and only if, or iff, there are no $a_x, a_y \in A_1$ with $(a_x, a_y) \in R$. $A_1 \subseteq A$ is said to defend $a \in A$ iff every $a_2 \in A$ attacking a is attacked by at least one member of A_1. With these, $A_1 \subseteq A$ is said to be:

- admissible iff A_1 is conflict-free and A_1 defends every $a \in A_1$.[1] We denote the set of all admissible sets of F^{D} by $\mathsf{D}(\mathsf{ad}, F^{\mathsf{D}})$.
- complete iff $A_1 \in \mathsf{D}(\mathsf{ad}, F^{\mathsf{D}})$ and every $a \in A$ A_1 defends is in A_1. We denote the set of all complete sets of F^{D} by $\mathsf{D}(\mathsf{co}, F^{\mathsf{D}})$.
- preferred iff $A_1 \in \mathsf{D}(\mathsf{co}, F^{\mathsf{D}})$ and also there is no $A_x \in \mathsf{D}(\mathsf{co}, F^{\mathsf{D}})$ such that $A_1 \subset A_x$. We denote the set of all preferred sets of F^{D} by $\mathsf{D}(\mathsf{pr}, F^{\mathsf{D}})$.
- stable iff $A_1 \in \mathsf{D}(\mathsf{co}, F^{\mathsf{D}})$ and also every $a_x \in (A \backslash A_1)$ is attacked by a member of A_1. We denote the set of all stable sets of F^{D} by $\mathsf{D}(\mathsf{st}, F^{\mathsf{D}})$.
- grounded iff $A_1 = \bigcap_{A_x \in \mathsf{D}(\mathsf{co}, F^{\mathsf{D}})} A_x$. We denote the set of all grounded sets of F^{D} by $\mathsf{D}(\mathsf{gr}, F^{\mathsf{D}})$.

$\mathsf{D}(\mathsf{co}, F^{\mathsf{D}})$, $\mathsf{D}(\mathsf{pr}, F^{\mathsf{D}})$, $\mathsf{D}(\mathsf{st}, F^{\mathsf{D}})$ and resp. $\mathsf{D}(\mathsf{gr}, F^{\mathsf{D}})$ are called in particular the complete, the preferred, the stable, and resp. the grounded semantics of F^{D}.

3 Causal Abstract Argumentation or Bipolar Argumentation with Causal Support Interpretation

In this section, we present Causal Abstract Argumentation, which can be equally sensibly named Bipolar Argumentation with Causal Support Interpretation, and characterise its semantics.

[1] Here and everywhere, 'and' instead of 'and' is used when the context in which it appears strongly indicate a truth value comparison. It has the semantics of classical logic conjunction.

Definition 1 (Causal abstract argumentation or bipolar argumentation with causal support interpretation). *We define a (finite) causal abstract argumentation to be a tuple* (A, R, R^{dep}) *with:* $A \subseteq_{fin} \mathcal{A}$*; and* $R, R^{dep} \subseteq A \times A$.

We denote the class of all (finite) causal argumentations by $\mathcal{F}$ *and refer to its member by* F *with or without a subscript.*

In addition to conflict-freeness and the self-defence, we consider another core notion of causal satisfaction.

Definition 2 (Causal satisfaction). *For any* $F \equiv (A, R, R^{dep})(\in \mathcal{F})$ *and any* $A_1 \subseteq A$*, we say that* A_1 *satisfies causality in* F *iff either of the following conditions holds for every* $a \in A_1$.

- *There is no* $a_x \in A$ *with* $(a_x, a) \in R^{dep}$.
- *There is some* $a_1 \in A_1$ *such that* a_1 *satisfies causality and that* $(a_1, a) \in R^{dep}$.

Example 1 (Illustration of causal satisfaction). Consider $a_1 \dashleftarrow a_2 \longleftrightarrow a_3$ (with a dashed arrow from a_3 to a_1), where we assume $a_x \dashrightarrow a_y$ or $a_y \dashleftarrow a_x$ represents $(a_x, a_y) \in R^{dep}$. Of all members of $2^{\{a_1, a_2, a_3\}}$, firstly $\{\}$ (an empty set) satisfies causality since it contains no arguments a with $(a, a) \in {R^{dep}}^*$ (reflexive-transitive closure of R^{dep}, meaning in short that a occurs in R^{dep}). $\{a_2\}, \{a_3\}$ and $\{a_2, a_3\}$ all satisfy causality due to the first condition of Definition 2. In the meantime, we have $(a_2, a_1), (a_3, a_1) \in R^{dep}$. Thus, $\{a_1\}$ does not satisfy the first condition. It also does not satisfy the second condition, since it contains neither a_2 nor a_3. The remaining $\{a_1, a_2\}, \{a_1, a_3\}$ and $\{a_1, a_2, a_3\}$ satisfy causality. ♣

As we briefly discussed in Sect. 1, a causality loop may lead to the situation where no subset of the set of arguments in the loop satisfies causality. To formalise this observation (Proposition 1 below), we recall the concept of a strongly connected component (SCC) with also the depth of a SCC for R^{dep}, and define the condition for an argument to be in such a situation (Definition 4).

Definition 3 (Causal SCC and causal SCC depth). *For any* $F \equiv (A, R, R^{dep})(\in \mathcal{F})$*, we say that* $(A_1, {R^{dep}}_1)$ *with* $A_1 \subseteq A$ *and* ${R^{dep}}_1 \equiv (R^{dep} \cap (A_1 \times A_1))$ *is a causal SCC iff, for every* $a_x \in A$ *and every* $a_y \in A_1$*, we have:* $\{(a_x, a_y), (a_y, a_x)\} \subseteq {R^{dep}}^*$ *iff* $a_x \in A_1$*. Again,* ${R^{dep}}^*$ *is the reflexive-transitive closure of* R^{dep}.

Let $\Delta : \mathcal{F} \times \mathcal{A} \rightarrow 2^{\mathcal{A}}$ *be such that, for any* $F \equiv (A, R, R^{dep})$ $(\in \mathcal{F})$ *and any* $a \in A$*,* $\Delta(F, a)$ *is the set of all arguments in a causal SCC that includes* a*. Let* $\delta : \mathcal{F} \times \mathcal{A} \rightarrow \mathbb{N}$ *be such that, for any* $F \equiv (A, R, R^{dep})$ $(\in \mathcal{F})$ *and any* $a \in A$*,* $\delta(F, a)$ *is:*

- 0 *iff there is no* $a_x \in \Delta(F, a)$ *and* $a_y \in (A \backslash \Delta(F, a))$ *such that* $(a_y, a_x) \in R^{dep}$.
- $1 + \max_{a_z \in A'} \delta(F, a_z)$ *with:* $A' = \{a_w \in (A \backslash \Delta(F, a)) \mid \exists a_u \in \Delta(F, a).(a_w, a_u) \in R^{dep}\}$.

Example 2 (Illustration of causal SCC and causal SCC depth). Consider $a_1 \dashleftarrow a_2 \longleftrightarrow a_3$ again. Let us denote this argumentation by F, we have: $\Delta(F, a_i) = \{a_i\}$ for each $i \in \{1, 2, 3\}$; $\delta(F, a_1) = 1$; and $\delta(F, a_2) = \delta(F, a_3) = 0$.

For another example, consider $a_1 \dashleftarrow a_2 \longleftrightarrow a_3$ Let us denote this argumentation by F', then we have: $\Delta(F', a_1) = \{a_1\}$; $\Delta(F', a_2) = \Delta(F', a_3) = \{a_2, a_3\}$; $\delta(F', a_1) = 1$; and $\delta(F', a_2) = \delta(F', a_3) = 0$. ♣

Definition 4 (Causality loop). *For any $F \equiv (A, R, R^{dep}) (\in \mathcal{F})$ and any $a \in A$, we say that a is in a causality loop iff either of the following conditions holds.*

- $\delta(F, a) = 0$ *and for every $a_x \in \Delta(F, a)$, there is some $a_y \in \Delta(F, a)$ with $(a_y, a_x) \in R^{dep}$.*
- $\delta(F, a) = n$ *and for every $i < n$ and every $a_x \in A$, if $\delta(F, a_x) = i$, then a_x is in a causality loop.*

Example 3 (Illustration of causality loop (Continued)). With F of Example 2, neither a_2 nor a_3 is in a causality loop. Thus, a_1 is also not in a causality loop. With F' of Example 2, both a_2 and a_3 are in a causality loop, by the first condition of Definition 4. By the second condition of Definition 4, a_1 is also in a causality loop. ♣

A causality loop has the significance that no argument in such a loop could ever satisfy causality, which is formally:

Proposition 1 (Causality loop and causality satisfaction). *For any $F \equiv (A, R, R^{dep}) (\in \mathcal{F})$ and any $a \in A$, if a is in a causality loop, then there is no $(\{a\} \cup A_1) \subseteq A$ such that $(\{a\} \cup A_1)$ satisfies causality.*

3.1 Acceptability Semantics

We now formulate acceptability semantics of $F \in \mathcal{F}$. Instead of inventing new terms, we just attach causal- in front of admissible, complete, preferred and grounded sets and semantics that are obtainable from (A, R).

We first define a causal-admissible set as one that satisfies conflict-freeness, self-defence, and also causality.

Definition 5 (Causal-admissible sets). *For any $F \equiv (A, R, R^{dep}) (\in \mathcal{F})$, we say that $A_1 \subseteq A$ is causal-admissible iff A_1 is admissible in (A, R) and A_1 satisfies causality in F. By $\Phi(\mathsf{ad}, F)$ we denote the set of all causal-admissible sets in F.*

Characterisation of the other causal-sets, in particular the causal-complete set on which the others rely, requires a little more elaboration. Since the role of R^{dep} is to filter out arguments that do not respect causality, it never acts to adjust a member of a causal-complete (and other) set to include more arguments than there originally were with $R^{dep} = \emptyset$. To therefore find causal-complete set(s) of (A, R, R^{dep}) from a complete set of (A, R), algorithmically we need the following procedure.

1. If the complete set is conflict-free, satisfies causality and includes every argument it defends, then it is its causal-complete set.
2. Otherwise, obtain a list of causal-admissible sets subsumed in the complete set, and obtain maximal causal-admissible sets in the list. They are the corresponding 'most complete' sets that still satisfy causality.

To functionally express this, we make use of:

Definition 6 (Maximal lower causal-admissible sets). *Let* $\mathsf{mla} : \mathcal{F} \times 2^{\mathcal{A}} \rightarrow 2^{2^{\mathcal{A}}}$ *be such that, for any* $F \equiv (A, R, R^{dep})$ $(\in \mathcal{F})$, *any* $A_1 \subseteq A$, *and any* $A_x \in \mathsf{mla}(F, A_1)$, *all the following conditions hold.*

1. $A_x \in \Phi(\mathsf{ad}, F)$.
2. $A_x \subseteq A_1$.
3. Every $A_y \in \Phi(\mathsf{ad}, F)$ *that satisfies both 1. and 2. is not greater than* A_x.

We say that $A_x \in \mathsf{mla}(F, A_1)$ *is* A_1*'s maximal lower causal-admissible set.*

Example 4 (Illustration of maximal lower causal admissible sets (Continued)). Consider $a_1 \dashleftarrow a_2 \longleftrightarrow a_3$ (the argumentation F) again, then we have $\Phi(\mathsf{ad}, F) = \{\{\}, \{a_2\}, \{a_3\}, \{a_1, a_2\}, \{a_1, a_3\}\}$. Hence, as its maximal lower causal-admissible set, $\{\}$ and $\{a_1\}$ have $\{\}$; $\{a_2\}$ has $\{a_2\}$; $\{a_3\}$ has $\{a_3\}$; $\{a_1, a_i\}$ for each $i \in \{2, 3\}$ has $\{a_1, a_i\}$; $\{a_2, a_3\}$ has $\{a_2\}$ and $\{a_3\}$; and $\{a_1, a_2, a_3\}$ has $\{a_1, a_2\}$ and $\{a_1, a_3\}$.

For the other example $a_1 \dashleftarrow a_2 \leftrightarrows a_3$ we have $\Phi(\mathsf{ad}, F) = \{\{\}\}$ since all the three arguments are in a causality loop (Cf. Proposition 1). ♣

Since the complete semantics of any $F^{\mathsf{D}} \in \mathcal{F}^{\mathsf{D}}$ is guaranteed to exist, it is of interest to learn if mla can also be relied upon for that property of existence of a causal-complete set. We fortunately have:

Lemma 1 (Existence of mla). *For any* $F \equiv (A, R, R^{dep})$ $(\in \mathcal{F})$ *and any* $A_1 \subseteq A$, *there is some* $A_x \subseteq A$ *with* $A_x \in \mathsf{mla}(F, A_1)$.

Proof. $\{\}$, the least element of the meet semi-lattice $(2^{2^A}, \subseteq)$, is causal-admissible in F. □

Furthermore, mla is order-preserving:

Lemma 2 (Order preservation of mla). *For any $F \equiv (A, R, R^{dep})$ $(\in \mathcal{F})$, any $A_1, A_2 \subseteq A$ and any $A_x \in \mathsf{mla}(F, A_1)$, if $A_1 \subseteq A_2$, then all the following hold.*

1. *There is some $A_y \in \mathsf{mla}(F, A_2)$ with $A_x \subseteq A_y$.*
2. *For every $A_y \in \mathsf{mla}(F, A_2)$, if $A_x \not\subseteq A_y$, then A_x and A_y are not comparable in $\subseteq$.*

Proof. For *1.*, since $A_1 \subseteq A_2$ holds, it holds that $A_x \subseteq A_2$. We have two situations. If $A_x \in \mathsf{mla}(F, A_2)$, then there is nothing to show. If, on the other hand, there is some $A_x \subset A_3 \subseteq A_2$ with $A_3 \in \Phi(\mathsf{ad}, F)$ such that every $A_4 \in \Phi(\mathsf{ad}, F)$ with $A_4 \subseteq A_2$ is not greater than A_3, then obviously $A_x \subseteq A_4$, and $(A_y \equiv)A_4 \in \mathsf{mla}(F, A_2)$ by the definition of mla, as required.

For *2.*, from the definition of mla and from *1.*, there exists no $A_z \in \mathsf{mla}(F, A_2)$ with $A_z \subset A_x$. □

Hence, it is indeed sensible that we use mla for characterisation of causal- sets (Definition 7), with intuitive results of the relation among them (Theorem 1).

Definition 7 (Causal-sets). *For any $F \equiv (A, R, R^{dep})(\in \mathcal{F})$, we say that $A_1 \subseteq A$ is causal-complete (, causal-preferred, causal-stable, causal-grounded) iff there is some $A_x \in \mathsf{D}(\mathsf{co}, (A, R))$ (, $\in \mathsf{D}(\mathsf{pr}, (A, R))$, $\in \mathsf{D}(\mathsf{st}, (A, R))$, $\in \mathsf{D}(\mathsf{gr}, (A, R))$) such that $A_1 \in \mathsf{mla}(F, A_x)$.*

We denote the set of all causal-complete (, causal-preferred, causal-stable, causal-grounded) sets of F by $\Phi(\mathsf{co}, F)$ (, $\Phi(\mathsf{pr}, F)$, $\Phi(\mathsf{st}, F)$, $\Phi(\mathsf{gr}, F)$).

Theorem 1 (Relation among causal-sets). *All the following hold for any $F \equiv (A, R, R^{dep})(\in \mathcal{F})$ and any $A_x \in A$.*

1. *If $A_x \in \Phi(\mathsf{pr}, F)$, then $A_x \in \Phi(\mathsf{co}, F)$. Moreover, A_x is maximal in $\Phi(\mathsf{co}, F)$.*
2. *If $A_x \in \Phi(\mathsf{gr}, F)$, then $A_x = \bigcap_{A_y \in \Phi(\mathsf{co}, F)} A_y$.*
3. *$\Phi(\mathsf{sem}, F)$ exists for any $\mathsf{sem} \in \{\mathsf{co}, \mathsf{pr}, \mathsf{gr}\}$.*
4. *It is not necessary that $\Phi(\mathsf{st}, F)$ exists.*
5. *$\Phi(\mathsf{sem}, F) \subseteq \Phi(\mathsf{ad}, F)$ for $\mathsf{sem} \in \{\mathsf{ad}, \mathsf{co}, \mathsf{pr}, \mathsf{gr}\}$ and for also $\mathsf{sem} = \mathsf{st}$ if $\Phi(\mathsf{st}, F)$ exists.*

Proof. For *1.*, note that $\mathsf{D}(\mathsf{pr}, (A, R)) \subseteq \mathsf{D}(\mathsf{co}, (A, R))$, which readily proves the first part. The second part is by Lemma 2. For *2.*, note that $A_z \in \mathsf{D}(\mathsf{gr}, (A, R))$ is such that $A_z = \bigcap_{A_w \in \mathsf{D}(\mathsf{co}, (A, R))} A_w$. Thus, it is also by Lemma 2. For *3.*, if nothing else, $\{\}$, the least element in $(2^{2^A}, \subseteq)$ becomes a member of $\Phi(\mathsf{sem}, F)$ (Cf. Lemma 1). For *4.*, it suffices to refer to the well-known fact that $\mathsf{D}(\mathsf{st}, (A, R))$ may not exist. For *5.*, it suffices to refer to Definition 7. □

Definition 8 (Acceptability semantics). *For any $F \equiv (A, R, R^{dep})(\in \mathcal{F})$, we say that $\Phi(\mathsf{co}, F)$ (, $\Phi(\mathsf{pr}, F)$, $\Phi(\mathsf{st}, F)$, $\Phi(\mathsf{gr}, F)$) is the causal-complete (, causal-preferred, causal-stable, causal-grounded) semantics of F.*

Theorem 2 (Relation between D and Φ). *All the following hold for any $F \equiv (A, R, R^{dep})(\in \mathcal{F})$.*

1. $\Phi(\mathsf{ad}, F) \subseteq D(\mathsf{ad}, F)$.
2. *It is not necessary that* $\Phi(\mathsf{sem}, F) \subseteq D(\mathsf{sem}, F)$ *for* $\mathsf{sem} \in \{\mathsf{co}, \mathsf{pr}, \mathsf{gr}, \mathsf{st}\}$.
3. *For any* $\mathsf{sem} \in \{\mathsf{co}, \mathsf{pr}, \mathsf{gr}, \mathsf{st}\}$ *and any* $A_x \in \Phi(\mathsf{sem}, F)$, *there exists some* $A_y \in D(\mathsf{sem}, F)$ *with* $A_x \subseteq A_y$.
4. *Existence of* $\mathsf{D}(\mathsf{st}, F)$ *does not materially imply existence of* $\Phi(\mathsf{st}, F)$.

Proof. For *1.*, it is by Definition 5. For *2.*, it is by the fact that not necessarily $A_1 \in \mathsf{mla}(F, A_1)$ for $A_1 \in \mathsf{D}(\mathsf{co}, (A, R))$. For *3.*, if $\mathsf{sem} \in \{\mathsf{co}, \mathsf{pr}, \mathsf{gr}\}$, by *4.* of Theorem 1, we have the existence of $\Phi(\mathsf{sem}, F)$ if $\mathsf{sem} \in \{\mathsf{co}, \mathsf{pr}, \mathsf{gr}\}$. By Definition 7. If $\mathsf{sem} = \mathsf{st}$, if $\Phi(\mathsf{st}, F)$ does not exist, then *3.* is vacuous; otherwise, again by Definition 7. For *4.*, it suffices to consider 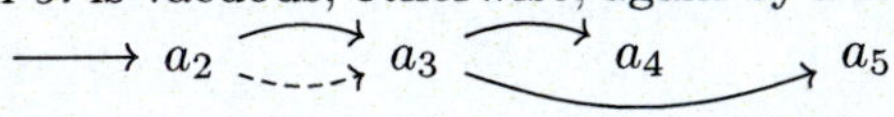 as $F_x \equiv (A_x, R_x, R_x^{dep})$ $(\in \mathcal{F})$ for which $\mathsf{D}(\mathsf{st}, (A_x, R_x)) = \{\{a_1, a_3\}\}$, for which, however, $\Phi(\mathsf{st}, F_x)$ clearly does not exist. □

4 Conclusion

We presented causal abstract argumentation which can be alternatively termed bipolar argumentation with causal support interpretation. Extension-based acceptability semantics were formulated, and the relation among them as well as the relation to acceptability semantics of abstract argumentation were identified. With this argumentation, we can respect the means-purpose relation among arguments, to preclude acceptance of a mission impossible. Compared to necessary and evidential support interpretations [13–15], our theory can deal with the desired causality relation generally, and can also cope with causality loops in computing acceptability semantics. Compared to (numerical) abstract persuasion argumentation [2,3], it handles causal dependency more concisely without needing a state transition system. Further, the causal-semantics have not been formulated in the dynamic argumentation theory, or in structured argumentation [11,12]. For future work, it can be interesting to also consider the labelling-based semantics [2,6].

Acknowledgements. We received comments from anonymous reviewers which helped improve the conclusion.

References

1. Arisaka, R., Ito, T.: Numerical abstract persuasion argumentation for expressing concurrent multi-agent negotiations. In: IJCAI Best of Workshops 2019 (2019, to appear)
2. Arisaka, R., Ito, T.: Broadening label-based argumentation semantics with may-must scales. In: CLAR, pp. 22–41 (2020)
3. Arisaka, R., Satoh, K.: Abstract Argumentation/Persuasion/Dynamics. In: PRIMA, pp. 331–343 (2018)
4. Bex, F.: An integrated theory of causal stories and evidential arguments. In: ICAIL, pp. 13–22 (2015)

5. Brewka, G., Strass, H., Ellmauthaler, S., Wallner, J., Woltran, S.: Abstract dialectical frameworks revisited. In: IJCAI (2013)
6. Caminada, M.: On the issue of reinstatement in argumentation. In: JELIA, pp. 111–123 (2006)
7. Cayrol, C., Lagasquie-Schiex, M.C.: On the acceptability of arguments in bipolar argumentation frameworks. In: ECSQARU, pp. 378–389 (2005)
8. Cayrol, C., Lagasquie-Schiex, M.-C.: Bipolarity in argumentation graphs: towards a better understanding. Int. J. Approx. Reason. **54**(7), 876–899 (2013)
9. Cohen, A., Parsons, S., Sklar, E.I., McBurney, P.: A characterization of types of support between structured arguments and their relationship with support in abstract argumentation. Int. J. Approx. Reason. **94**, 76–104 (2018)
10. Dung, P.M.: On the acceptability of arguments and its fundamental role in nonmonotonic reasoning, logic programming, and n-person games. Artif. Intell. **77**(2), 321–357 (1995)
11. Dung, P.M.: Assumption-based argumentation. In: Argumentation in AI, pp. 25–44. Springer (2009)
12. Modgil, S., Prakken, H.: A general account of argumentation with preferences. Artif. Intell. **195**, 361–397 (2013)
13. Nouioua, F., Risch, V.: Bipolar argumentation frameworks with specialized supports. In: ICTAI, pp. 215–218 (2010)
14. Nouioua, F., Risch, V.: Argumentation frameworks with necessities. In: SUM, pp. 163–176 (2011)
15. Oren, N., Norman, T.J.: Semantics for evidence-based argumentation. In: COMMA, pp. 276–284 (2008)

Analysis of Partial Semantic Segmentation for Images of Four-Scene Comics

Akira Terauchi[1(✉)], Naoki Mori[1], and Miki Ueno[2]

[1] Osaka Prefecture University, 1-1 Gakuen-cho, Naka-ku, Sakai, Osaka, Japan
terauchi@ss.cs.osakafu-u.ac.jp, mori@cs.osakafu-u.ac.jp
[2] Osaka Institute of Technology, 5-16-1 Omiya, Asahi-ku, Osaka, Osaka, Japan
miki.ueno@oit.ac.jp

Abstract. Ways of understanding human creations with the help of artificial intelligence (AI) have increased; however, those are still known as being one of the most difficult tasks. Our research challenge is to find ways to understand four-scene comics through AI. To achieve this aim, we used a novel dataset called "Four-scene Comics Story Dataset", which is the first dataset made by researchers and comic artists to develop AI creations. In this paper, we focused on the partial semantic segmentation of features such as eyes, mouth, or speech balloons. The semantic segmentation task of comics has been difficult because of the lack of annotated comic dataset. To solve this problem, we utilized the features of our dataset and easily created annotated dataset. For the semantic segmentation method, we used a model called DeepLabv3+. The effectiveness of our experiment is confirmed by computer simulations showing the segmentation result of test images from four-scene comics.

Keywords: Comic engineering · Deep learning · Semantic segmentation · Four-scene comics story dataset

1 Introduction

Against the background of advancement of machine learning technologies such as deep learning, the study for understanding creations by computers has recently attracted attention. Some examples of such creations are automatic generation of novels and illustrations by artificial intelligence. However, as creation is a high-level intellectual activity, no results beyond superficial imitation have currently been obtained.

In the field of image computation, semantic segmentation is one of the popular challenges. However, there have been few attempts until now to segment comic data due to a lack of annotated comic dataset. In this study, we solved this problem with our novel dataset, and created annotated dataset for easy semantic segmentation.

Y. Dong et al. (Eds.): DCAI 2020, AISC 1237, pp. 51–59, 2021.
https://doi.org/10.1007/978-3-030-53036-5_6

The purpose of this study is to obtain a good segmentation results for comics. Fully Convolutional Network (FCN), U-net, PSPNet, DeepLab, etc. [1–4] have been proposed as eligible methods for the semantic segmentation of an image. In this paper, in a hitherto-untried experiment, we investigated the partial semantic segmentation of comics into features such as eyes, mouth, or speech balloons.

2 Related Work

2.1 DeepLabv3+

In this study, we used the DeepLabv3+ [5] as the semantic segmentation model. DeepLabv3+ is based on a encoder-decoder structure and uses spatial pyramid pooling [6,7]. In addition, it extends the spatial pyramid pooling with atrous convolution, and can capture contextual information at multiple scales. Comics are different from natural images in that the size of targets such as eyes and mouth is not uniform. Therefore, in this study, we used the DeepLabv3+ for the robust segmentation about the scale change.

Atrous Convolution. Atrous convolution, also called delated convolution [8], is a convolution scheme that can explicitly control the resolution of features computed by deep convolutional neural networks. It adjusts the field-of-view of the filters to capture multiscale information. Atrous convolution has a convolution rate r, and when $r = 1$, it matches the standard convolution operation.

Atrous Spatial Pyramid Pooling. Atrous Spatial Pyramid Pooling (ASPP) is a pooling method that extends spatial pyramid pooling with atrous convolution. It sets the atrous convolution rate r variably in order to take into account the nonuniformity of scale. Layers convolved at the various rates are superimposed to form a single feature map by convolution in the 1×1 depth direction.

Depthwise Separable Convolution. In the depthwise separable convolution in DeepLabv3+, atrous convolution is first performed for each channel and 1×1 convolution is performed to combine the output from depth convolution. This operation greatly reduces the computational cost and the number of parameters while maintaining performance.

2.2 Comic Engineering

Among the fields related to the understanding of human creations, comic engineering is a field that targets comics composed of pictures and characters. Therefore, comic engineering is an area that targets multimodal data with aspects of both natural language processing and image processing.

Studies that have used comics are already available. Narita et al. [9] used Manga109 dataset [10,11], and proposed a comic-specific image retrieval system.

This system retrieved comics that had pictures that resembled sketches drawn by people. In another study used Manga109, Toru et al. [11] proposed a method for object detection, and confirmed its effectiveness by applying the detection method on eBDtheque [12]. Fujino et al. [13] focused on four-scene comics. They solved an order recognition task for understanding structures in comics. All these studies analyzed either only the pictures in the comics, or the annotations closely related to the pictures.

However, we require more annotations related to the comic contents to analyze them. In our paper, we used the Four-scene comics story dataset, and focused on the semantic segmentation with image processing. Details regarding the dataset are mentioned in the next section.

3 Dataset

The Manga109 dataset comprises 104 types of story comics and five types of four-scene comics. In our study, we used Four-scene comics story dataset [14]. This dataset, which has various characteristics, is the world's first research dataset for R&D on artificial intelligence (AI) and creative objects, with the researchers actively involved in its development. Unlike a dataset such as Manga109, which was created using commercial comics, the Four-scene comics story dataset consists of four-scene comics drawn by several comic artists for the same story. When commercial comics are used, in addition to problems such as copyright, there is little information on the author's sensitivity to their work being handled by a computer. Furthermore, there is a problem of commercial comics not being suitable for research aimed at understanding the meaning of comics. For example, it is necessary to label the emotions of comic characters with reader annotations because the emotions of comic characters are not specified, but it is possible that the annotated label may differ from the artist's intention. Moreover, it is difficult to study the diversity of expressions in comics, because it is rare to see multiple comic artists drawing for the same story. The Four-scene comics story dataset was created to solve the above-mentioned problems and provides a great advantage in AI research.

The Four-scene comics story dataset is created by overlapping layers in which each frame constitutes a picture. Examples of the layers include an eye layer, nose layer, mouth layer, outline layer, a body layer. These layers are so designed that they can be extracted separately, which means that necessary information can be easily extracted and used for the image analysis of the comics. Moreover, Ueno classified the comics drawn by the different artists into five touch types: Gag Touch, Shojo Touch, Shonen Touch, Seinen Touch, and Moe Touch. In this study, we perform the task of segmenting these comics parts with a computer: Shonen Touch, Seinen Touch, and Moe Touch. Figure 1 shows examples of comic images of the three touch types.

4 Experimental Setup

In the experiment, we perform semantic segmentation of comics, so it is necessary to convert part-image data such as eyes or mouth to label images before applying DeepLabv3+ to our dataset. Therefore, using the advantages of the dataset, we first created extracted images for each part. Then, the label images were created by extracting the non-transparent area of the part image as a mask, and giving each pixel a value equal to the index given for each class. Figure 2 shows an example of the label images created following this process. Notably, the index color image is shown for the purpose of visualization. Figure 2c shows the color map. Here, a significant point is that the labels for the semantic segmentation are created semi-automatically.

(a) Shonen Touch

(b) Seinen Touch

(c) Moe Touch

Fig. 1. Touch examples (The meaning of Japanese in the above scene below.) The woman says "You always look cool.", and the man replied "You think so?". (a) ©Drawn by Shiki Suzuki, Scenario written by Saki Harimura and at Spoma Inc. and Miki Ueno at Toyohashi University of Technology. (b) ©Drawn by Toshihito Yuzawa, Scenario written by Saki Harimura and at Spoma Inc. and Miki Ueno at Toyohashi University of Technology. (c) ©Drawn by Umeko Muneta, Scenario written by Saki Harimura and at Spoma Inc. and Miki Ueno at Toyohashi University of Technology.

5 Numerical Experiment

In this experiment, we performed a semantic segmentation task for each part in a four-scene comic. In addition, we compared the segmentation results between each touch by examining the *mIoU* values obtained by experiments. *mIoU* is a representative metric in semantic segmentation and takes values from 0 to 1, indicating the degree of overlap between the prediction region and the correct region. In this experiment, the problem of non-uniformity between the classes was solved by setting the weight for each class in the loss function. The weights to be set were determined based on the number of pixels belonging to each class. Table 1 shows the class weights set in this experiment. Table 2 shows the experimental settings.

The experiment was performed by assigning a numerical value to each class, i.e., the eye class (class No. 1), the mouth class (No. 2), the speech balloon class (No. 3), and the background class (No. 0; for other regions). We conducted experiments using two of the four classes in addition to experiments using all four classes to examine the difference in difficulty of segmentation among the parts. Of the 240 images, 216 were used for training and 24 for evaluation.

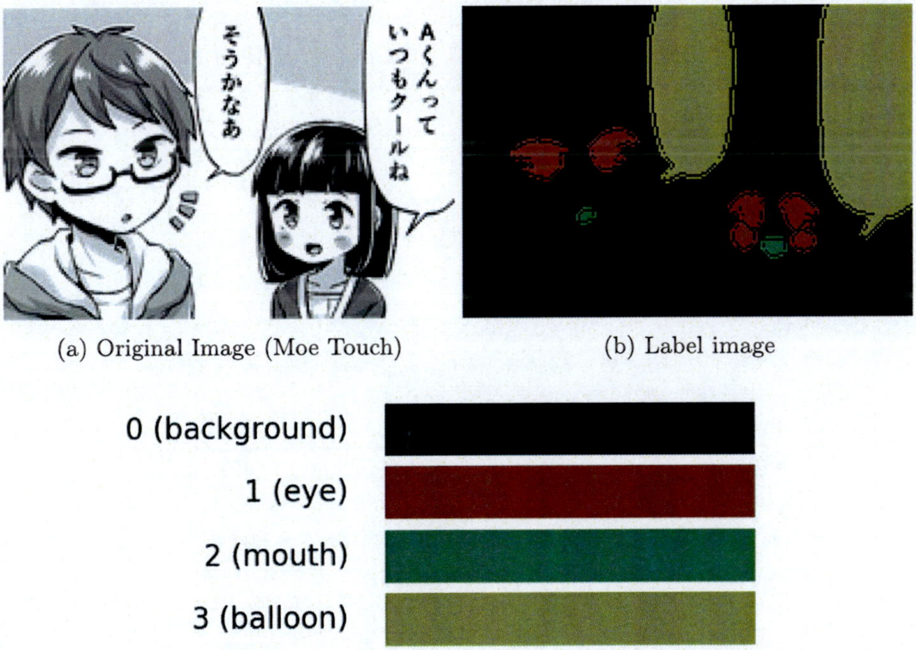

(a) Original Image (Moe Touch)

(b) Label image

(c) Color map

Fig. 2. An example of label images

In addition, the training and evaluation images were set so that the images of each touch were equally included.

Result

Figure 3 shows the example of prediction results for the evaluation images obtained in the experiment that included all four classes. Table 3 shows the $mIoU$ values of training, evaluation, Moe Touch, Seinen Touch, and Shonen Touch for each experiment. The average $mIoU$ value of Moe Touch, Seinen Touch and Shonen Touch corresponds to the $mIoU$ value for evaluation. In the experiment using four classes, the $mIoU$ of the evaluation images is 0.6921, and it can be seen from Fig. 3 that the approximate position and shape of the eyes, mouth, and speech balloon have been captured in the untrained images. In addition, it was found that the predicted area tended to be slightly larger than the correct area. This tendency was common to all three touches. Moreover, $mIoU$ for the Shonen Touch was lower than that for the other two touches in the all the experiments. This can be considered that the parts of the Shonen Touch were drawn much smaller than in the other two touches, and were greatly affected by the large segmentation area of prediction.

Table 1. Class weights

Background	1
Eye	10
Mouth	20
Balloon	5

Table 2. Experimental settings

Optimization method	Momentum
Loss function	Cross entropy
Number of epochs	5×10^6
Batch size	10
Atrous rates	6, 12, 18
Initial weight	Pascal voc
Crop size	(136, 196)

Table 3. $mIoU$ value for each experiment

Genre	Four classes	Back-eye	Back-mouth	Back-balloon
Training	0.7832	0.7671	0.7444	0.9520
Evaluation	0.6921	0.6983	0.6983	0.9175
Moe	0.7087	0.7103	0.7900	0.9335
Seinen	0.6936	0.6836	0.7619	0.9197
Shonen	0.6603	0.6997	0.5976	0.8961
Baseline	0.2500	0.5000	0.5000	0.5000

Next, we compared the difficulty of segmentation for the different parts. From Table 3, it can be seen that the $mIoU$ value of the speech balloon area is pretty high (0.9175) for evaluation. This is because the shape of the speech balloon is

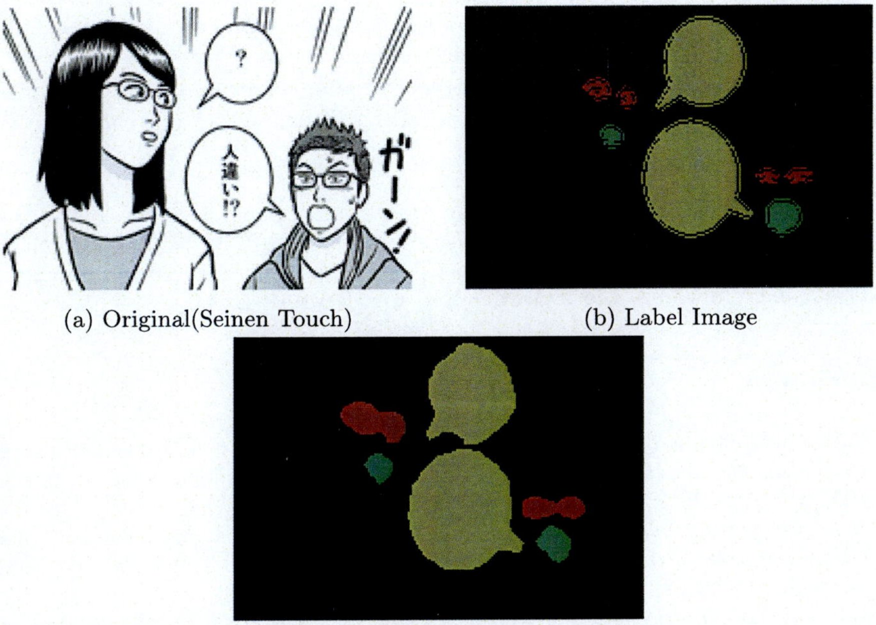

(a) Original(Seinen Touch)

(b) Label Image

(c) Prediction

Fig. 3. An example of test prediction (The meaning of the above scene). The man says "Wrong Person!? Oh my God!"

almost the same regardless of the touch of the comic, and the area of the speech balloon is larger than the other two parts.

In this experiment, we confirmed that the eyes and mouth were correctly detected, even in special cases where they overlapped a speech balloon. Conversely, examples of images for which the segmentation was not successful include those where the pupil is not drawn in the glasses, and images with small or jagged balloons. It is considered that these problems can be solved by including various-shaped parts in the training images or data augmentation.

6 Conclusion

In this study, using DeepLabv3+ and the Four-scene comics story dataset, we performed the semantic segmentation of eyes, mouth and speech balloons, which had not been attempted so far. Moreover, by observing the differences in the $mIoU$ values, the difference in the difficulty of segmentation for each part and each touch could be examined.

Future research topics include setting the class weights more properly or using the loss function considering class imbalances such as Focal Loss or Class

Balanced loss, adding more labels such as face or body, and data augmentation for stable segmentation. Furthermore, as an application of this technology, it is expected that comics parts will be extracted using segmented data as a mask and used for tasks such as sentiment analysis of characters or reading order estimation.

Acknowledgment. We thank the comic artists and Spoma Inc. for cooperating with this research. A part of this work was supported by ACT-I, JST. Grant Number: JPMJPR17U4, JSPS KAKENHI Grant, Grant-in-Aid for Scientific Research(C), 26330282, and Grant-in-Aid for Scientific Research(B), 19H04184.b.

References

1. Long, J., Shelhamer, E., Darrell, T.: Fully convolutional networks for semantic segmentation. In: The IEEE Conference on Computer Vision and Pattern Recognition (CVPR), June 2015
2. Badrinarayanan, V., Kendall, A., Cipolla, R.: SegNet: a deep convolutional encoder-decoder architecture for image segmentation. IEEE Trans. Pattern Anal. Mach. Intell. **39**(12), 2481–2495 (2017)
3. Ronneberger, O., Fischer, P., Brox, T.: U-Net: convolutional networks for biomedical image segmentation. In: Navab, N., Hornegger, J., Wells, W.M., Frangi, A.F. (eds.) Medical Image Computing and Computer-Assisted Intervention – MICCAI 2015, pp. 234–241. Springer, Cham (2015)
4. Zhao, H., Shi, J., Qi, X., Wang, X., Jia, J.: Pyramid scene parsing network. In: CVPR (2017)
5. Chen, L.-C., Zhu, Y., Papandreou, G., Schroff, F., Adam, H.: Encoder-decoder with atrous separable convolution for semantic image segmentation. In: ECCV (2018)
6. Grauman, K., Darrell, T.: The pyramid match kernel: discriminative classification with sets of image features. In: Tenth IEEE International Conference on Computer Vision (ICCV 2005 Volume 1, vol. 2, pp. 1458–1465, October 2005
7. Lazebnik, S., Schmid, C., Ponce, J.: Beyond bags of features: spatial pyramid matching for recognizing natural scene categories. In: 2006 IEEE Computer Society Conference on Computer Vision and Pattern Recognition (CVPR 2006), vol. 2, pp. 2169–2178, June 2006
8. Yu, F., Koltun, V.: Multi-scale context aggregation by dilated convolutions, November 2015
9. Narita, R., Ogawa, T., Matsui, Y., Yamasaki, T., Aizawa, K.: Sketch-based manga retrieval using deep features. In: The 31st Annual Conference of the Japanese Society for Artificial Intelligence, JSAI2017, p. 3H1OS04a2 (2017)
10. Matsui, Y., Ito, K., Aramaki, Y., Fujimoto, A., Ogawa, T., Yamasaki, T., Aizawa, K.: Sketch-based manga retrieval using Manga109 dataset. Multimed. Tools Appl. **76**(20), 21811–21838 (2017)
11. Ogawa, T., Otsubo, A., Narita, R., Matsui, Y., Yamasaki, T., Aizawa, K.: Object detection for comics using Manga109 annotations. CoRR, abs/1803.08670 (2018)
12. Guérin, C., Rigaud, C., Mercier, A., Ammar-Boudjelal, F., Bertet, K., Bouju, A., Burie, J.-C., Louis, G., Ogier, J.-M., Revel, A.: eBDtheque: a representative database of comics. In: Proceedings of the 12th International Conference on Document Analysis and Recognition (ICDAR), pp. 1145–1149 (2013)

13. Fujino, S., Mori, N., Matsumoto, K.: Recognizing the order of four-scene comics by evolutionary deep learning. In: 15th International Conference on Distributed Computing and Artificial Intelligence, DCAI 2018, Toledo, Spain, June 20–22 2018, pp. 136–144 (2018)
14. Ueno, M.: Creators and artificial intelligence: enabling collaboration with creative processes and meta-data for four-scene comic story dataset. In: The 32nd Annual Conference of the Japanese Society for Artificial Intelligence, JSAI2018, p. 4Pin116 (2018)

Two Agent-Oriented Programming Approaches Checked Against a Coordination Problem

Eleonora Iotti, Giuseppe Petrosino, Stefania Monica, and Federico Bergenti(✉)

Dipartimento di Scienze Matematiche, Fisiche e Informatiche,
Università degli Studi di Parma, 43124 Parma, Italy
{eleonora.iotti,stefania.monica,federico.bergenti}@unipr.it,
giuseppe.petrosino@studenti.unipr.it

Abstract. This paper discusses two approaches to agent-oriented programming and compares them from a practical point of view. The first approach is exemplified by Jadescript, which is an agent-oriented programming language that has been recently proposed to simplify the adoption of JADE. The second approach is exemplified by Jason, which is currently one of the most popular agent-oriented programming languages. Jason can be configured to use JADE to support the distribution of agents, which ensures that the discussed comparison between the two approaches can also take into account the performance of implemented multi-agent systems. In order to devise a quantitative comparison, the two considered languages are used to solve the same coordination problem, and obtained implementations are compared to discuss advantages and drawbacks of the two approaches.

Keywords: Jadescript · Jason · Agent-oriented programming

1 Introduction

The development of *AOP* (*Agent-Oriented Programming*) languages (e.g., [23]) and the study of *AOSE* (*Agent-Oriented Software Engineering*) (e.g., [6]) are of primary importance for the community of researchers and practitioners interested in software agents and agent-based software development. In the last few years, a plethora of methodologies, languages, and tools were presented in the literature (e.g., [18,19]). AOP languages are generally recognized in such a body of literature as important tools for the development of agent technologies, in contrast to traditional (lower-level) languages, which are often considered (e.g., [1,10]) not suitable to effectively implement software agents and agent-based software systems with the desired characteristics (e.g., [3]).

One of the approaches to AOP that attracted much attention is the *BDI* (*Belief-Desire-Intention*) approach [22]. In brief, the BDI approach schematizes software agents in terms of human-like features, and it targets the design of agents as intelligent software entities that are able to plan how to effectively

Y. Dong et al. (Eds.): DCAI 2020, AISC 1237, pp. 60–70, 2021.
https://doi.org/10.1007/978-3-030-53036-5_7

bring about their intentions on the basis of their desires and their beliefs. The BDI approach is actually implemented in several frameworks and languages, which include Jason [12], *GOAL* (*Goal-Oriented Agent Language*) [16], *JAL* (*JACK Agent Language*) [25], and *3APL* (*An Abstract Agent Programming Language*) [15]. Note that the BDI approach is inherently declarative, and its various implementations are mostly derived from logic programming. The tight connection between the BDI approach and logic programming is often considered as a limitation for the popularization of the BDI approach because logic programming languages tend to have a steep learning curve for the average programmer. On the one hand, experts of agent technologies are comfortable with the BDI approach, but on the other hand, students and practitioners that try to adopt the BDI approach tend to be discouraged. This is the reason why the literature has recently witnessed the introduction of AOP languages that follow procedural approaches, or that mix declarative features with procedural features to obtain sophisticated hybrid languages (e.g., [13,14]).

In addition to AOP languages, several software frameworks are available to support the construction of agent-based software systems. As stated in a recent survey [18], among such frameworks, the primacy in terms of popularity is held by *JADE* (*Java Agent DEvelopment framework*) (e.g., [2]) in both the academia and the industry. JADE provides a rich programming interface for developers who wants to program software agents, and it also provides a runtime platform to execute *MASs* (*Multi-Agent Systems*) on distributed computing architectures. Unfortunately, the general perception is that a sophisticated tool like JADE is hard to learn for the newcomers to agent-based software development (e.g., [7]) because it requires advanced skills in both object-oriented and agent-oriented programming. *JADEL* (*JADE Language*) (e.g., [4,7]) is a recent *DSL* (*Domain-Specific Language*) that aims at simplifying the use of JADE by embedding in the language the view of agents and MASs that JADE proposes. Jadescript [8,9,20,21] leverages the experience on JADEL to propose a new AOP language designed to simplify the construction of agents and MASs. Jadescript is an entirely new language that provides linguistic constructs designed to concretize the abstractions that characterize agent-based software development in the attempt to promote their effective use.

The major contribution of this paper is to assess some of the characteristics of Jadescript by means of a comparison with another AOP language in terms of the effectiveness in the construction of a solution to an illustrative problem. Jason was chosen as the point of reference to assess the characteristics of Jadescript, and such a choice offers the opportunity to compare the procedural approach to AOP that Jadescript advocates with the BDI approach that Jason uses. In particular, Sect. 2 briefly describes the major features that Jadescript and Jason offer to the programmer to master the complexity of agent-based software development. Section 3 describes the illustrative problem chosen to compare the two languages. Section 4 reports on a quantitative assessment of the implemented solutions to the studied problem in terms of suitable metrics, and it also compares the two languages in terms of a subset of the criteria described in [11], which are accepted criteria for the comparison of AOP languages. Finally, Sect. 5 concludes the paper by summarizing major documented results.

2 Jadescript and Jason in Brief

This section summarizes the major features that Jadescript and Jason offer, and it highlights their commonalities and differences. Only the features that are relevant for the comparison discussed in Sect. 4 are presented.

Linguistic Features. Jason is a powerful implementation of AgentSpeak, and therefore it follows a declarative approach to AOP. The syntax that Jason uses is very similar to the syntax originally proposed for AgentSpeak. Beliefs, goals, intentions, and plans are specified for each agent in the MAS. Jason agents are immersed in an environment, which is defined explicitly by naming a MAS together with its infrastructure and agents.

Jadescript is grounded on JADE, and therefore it supports the construction of agents that execute in the scope of the containers of a regular JADE platform. A Jadescript agent has a list of behaviours to perform tasks, and behaviours are defined procedurally to handle messages and to react to events. Following a procedural approach to AOP, Jadescript allows the declaration of behaviours in terms of properties, procedures, functions, message handlers, and event handlers. In addition, Jadescript allows the declaration of ontologies to support the communication among agents. Ontologies are declared in Jadescript in terms of propositions, predicates, concepts, and actions.

In summary, Jason is a declarative language, while Jadescript has declarative features but it is essentially a procedural language. The declarative features that Jadescript provide are meant to support pattern matching and event-driven programming, but the actions that agents perform are defined procedurally.

Multi-agent Systems. Jason provides a clear way to define MASs. The name of the MAS must always be specified, and its infrastructure must be chosen. The most common infrastructures are immediately available, and they are called `Centralized` and `Jade`. When the `Centralized` infrastructure is used, all Jason agents execute within the same process. On the contrary, the `Jade` infrastructure is based on JADE, and agents can be easily distributed across JADE containers when such an infrastructure is used. In the Jason declaration of a MAS, a Java class that defines the actual environment can be referenced by means of the keyword `environment`. Finally, the most important part of the definition of a MAS in Jason is the list of the agents that populate the environment. Such agents have names and references to specific agent files for their definitions.

Currently, Jadescript does not provide a specific support for the construction of a MAS because it assumes that JADE is used to activate agents within the environment. Jadescript agents are launched using JADE, and the infrastructure is managed in terms of a regular JADE platform. In Jadescript, the knowledge about the environment is passed among agents through messages, and ontologies are essential in such an exchange of messages.

In summary, a relevant difference between Jadescript and Jason is in the support for the construction of MASs. Actually, Jadescript does not yet offer a means to construct a MAS, but it assumes that JADE is directly used.

Agents. The declaration of a Jason MAS references a set of agent files, which declaratively define for each agent the initial set of beliefs (and relative rules), the initial set of goals (and relative rules), and all available plans. The declaration of a Jadescript agent consists of a set of properties and a set of lifecycle event handlers. Lifecycle event handlers are used to programmatically define what an agent should do to react to the changes of its lifecycle state.

The approaches adopted by Jason and by Jadescript for the declaration of agents are strongly related. They declare agents in terms of internal beliefs or properties, supported languages or ontologies, and available behaviours or plans. On the contrary, the syntaxes adopted by the two languages are very different. Jason is based on self-contained agent files, and the rules for beliefs and goals are written using a Prolog-like notation. Jadescript keeps a small part of the code of an agent in the agent declaration because behaviours, ontologies, functions, and procedures are coded outside of the agent declaration to promote reusability. Similarly, Jadescript supports inheritance to promote reusability by providing the programmer with the ability to extend the definitions of agents, behaviours, and ontologies. Instead, Jason does not offer inheritance and related features.

Plans and Behaviours. Plans are basic courses of actions for Jason agents, and a plan in Jason consists of a triggering event, a context, which is a set of actual beliefs, and a body. The body of a plan includes the actions to be performed and the sub-goals to be achieved. A goal in Jason is strictly related to the environment because it describes a state of the environment that the agent wants to actualize.

Behaviours are the means that Jadescript offers to define the tasks that agents perform. Each agent can activate multiple behaviours whose order of execution during the agent lifecycle is not explicitly declared by the programmer. Activated behaviours are instead collected into an internal list, and their execution is autonomously scheduled by the agent with using a non-preemptive scheduler. Note that a behaviour can directly access the internal state of the agent that activated it by accessing its properties, functions, and procedures.

Jadescript behaviours can be matched with Jason plans, but Jason plans are declarative while Jadescript behaviours are procedural. In particular, Jadescript behaviours are not explicitly related to the goals that they help to achieve.

Events. Events in Jason are related to the beliefs and the goals of each agent. Actually, an event can add or delete beliefs and goals, and it can also trigger the activation of a plan to bring about a goal.

Jadescript behaviours normally include event handlers, and event handlers are executed when managed events occur. One of the most recent additions to the language is the support for pattern matching [21], which allows declaratively defining the structure of managed events for each event handler.

Both languages support event handlers in terms of reactions to events. In both cases, the activation of event handlers affects the states of agents. The main difference between Jadescript and Jason with respect to event handlers is the presence of goals, which are explicit in Jason and implicit in Jadescript. Finally, it is worth noting that in both languages events can be internal or external.

Ontologies. Ontologies are of primary importance in Jadescript because they are used to support communication, and all non-trivial Jadescript agents are expected to reference at least one ontology. Actually, ontologies are used in Jadescript to state a fact or a relation about elements of the domain of the problem. On the contrary, Jason does not provide a support for ontologies. Note that ontologies cannot be reduced to beliefs because ontologies are descriptions of the domain of the problem, and events can neither add nor remove ontologies.

3 The Santa Claus Coordination Problem

The Santa Claus coordination problem [24] is a well-known coordination problem that is expressed, in its simplest form, as follows. Santa Claus has nine reindeer and ten elves in its team. He sleeps waiting for a group formed by all of his reindeer or by three of his elves. When three elves are ready and awake Santa Claus, he must work with them on new toys. Similarly, when the group of reindeer is ready and awakes Santa Claus, they work together to deliver toys. It is important that Santa Claus is awakened only when a group with the correct size is formed. In addition, Santa Claus should give priority to the reindeer in case that there are both a group of elves and a group of reindeer ready.

The major reasons for choosing the Santa Claus coordination problem for the comparison between Jadescript and Jason can be listed as follows. First, the problem is simple, but not trivial, and a number of solutions are documented in the literature. In particular, one of such solutions has already been implemented in JADEL [17], and the Jadescript solution used for the experiments described in Sect. 4 is a rework of such an implementation. Second, a Jason implementation can be found in the official Jason distribution[1], and such an implementation was used for the experiments described in Sect. 4. Third, the problem depends on numeric parameters that can be used to vary the characteristic size of the problem and to support quantitative comparisons.

In order to support a fair comparison among the mentioned solutions in Jadescript and Jason, the architecture adopted in the Jason solution was used to design the Jadescript solution. In particular, the architecture of both solutions consists of Santa Claus, nineteen workers (ten elves and nine reindeer), and two secretaries to schedule appointments with workers. Secretaries are called Edna (assigned to the elves) and Robin (assigned to the reindeer), and their responsibility is to form groups and to awake Santa Claus when needed.

[1] Version 2.4 downloaded from the official site (http://jason.sourceforge.net).

The remaining of this section describes the Jadescript solution to the Santa Claus coordination problem. The Jadescript solution uses four types of message contents, as follows. The `workerReady` proposition is used by workers to inform their secretary that they are ready, and the secretary, in turn, uses the `groupFormed` proposition to inform Santa Claus that a group is formed. The `OK` proposition is used by Santa Claus to notify the beginning of the working phase to the chosen group of workers. Finally, the `done` proposition is used by workers to tell that they have completed their jobs.

In the Jadescript solution, the actions of elves and reindeer are defined by means of two behaviours, which are activated by workers at startup, as shown in Fig. 1 (lines 8 and 9). The behaviour `SendReady` is declared (line 11) as `one shot`, and, as such, it contains a single action defined by the procedural code that follows the keyword `do`. When a Jadescript agent activates a behaviour, the behaviour is added to the list of active behaviours of the agent, and then the agent tries to execute it using its non-preemptive scheduler. A `one shot` behaviour is removed from the list of active behaviours just after its execution, and therefore the `SendReady` behaviour is used just once to send a `workerReady` message to the appropriate secretary (line 12). On the contrary, the `WaitForOK` behaviour is declared as `cyclic` (line 14). Cyclic behaviours do not leave the list of the active behaviours of the agent until they are explicitly deactivated, and therefore they are suitable for cyclic tasks such as message handling. In particular, the Jadescript construct that begins with the keywords `on inform` (line 15) declares a message handler that is executed at the reception of a message with performative `inform` and content `OK`. When a message with such characteristics is received, the worker first starts its work (lines 17 and 19), then it replies with a `done` message to Santa Claus (line 20), and it finally sends a `workerReady` message to its secretary (line 21).

The remaining of this section discusses the behaviours used by the Jadescript solution to the Santa Claus coordination problem to implement the two secretaries and Santa Claus. The Jadescript source codes of agents and behaviours is not presented in the paper due to page restrictions.

In the adopted architecture, secretaries count how many workers are actually ready, and when a group of the needed size is available, they promptly inform Santa Claus using a specific message. When created, a secretary activates only one cyclic behaviour, namely `HandleWorkerMessages`, and it shares three properties, namely `groupSize`, `kind`, and `santa`, with the behaviour. When scheduled for execution, the behaviour first checks for `workerReady` messages from workers, then it collects the identifiers of the senders, and it finally sends a `groupFormed` message to Santa Claus. Note that the reindeer communicate exclusively with Robin while elves communicate exclusively with Edna, so the group formation processes are handled independently for the two groups.

Santa Claus uses two behaviours to implement its two possible states. The first behaviour, which corresponds to the first state, is the cyclic behaviour `WaitForGroups`. Such a behaviour first checks if there is a message in the queue that states that a group of reindeer is ready. If no groups of reindeer are ready, the behaviour performs a similar check for a group of elves. When a group is correctly formed, Santa Claus first sends an OK message to all the workers in the group, and then it changes its state by deactivating the WaitForGroups behaviour to activate the WaitJobCompletion behaviour. The WaitJobCompletion behaviour is used to define the actions to perform when Santa Claus is in the second state. In such a state, Santa Claus waits for all the workers to send done messages, then the WaitJobCompletion behaviour is deactivated in favor of the WaitForGroups behaviour to perform a transition back to the first state.

```
agent Worker uses ontology SantaClausProblem
  property secretary as aid
  property myKind as text

  on create with kind as text, secretaryName as text do
    secretary = aid(secretaryName)
    myKind = kind
    activate behaviour WaitForOK
    activate behaviour SendReady

  one shot behaviour SendReady do
    send inform workerReady to secretary

  cyclic behaviour WaitForOK
    on inform when content is OK do
      if myKind = "reindeer" do
        do sleep with millis = 3000+rand(1000)
      else do
        do sleep with millis = 1000+rand(1000)
      send inform done to sender of message
      activate behaviour SendReady
```

Fig. 1. The Worker agent and its SendReady and WaitForOK behaviours in Jadescript.

4 Quantitative Comparison

The quantitative comparison between Jadescript and Jason presented in this section is based on three solutions to the Santa Claus coordination problem. The considered implementations are: the Jadescript implementation presented in Sect. 3, the Jason implementation on the Centralized infrastructure, and the Jason implementation on the Jade infrastructure. Note that the use of JADE allows comparing execution times. Also note that the execution time of the Jason implementation on the Centralized infrastructure is reported as baseline.

In order to unambiguously measure execution times, only Santa Claus, Robin, and the reindeer were activated. Hence, the settings for a single execution are described in terms of the number of reindeer R and the number of works to be done W. Under such an assumption, the execution times measure how well the implementations can handle the increase of the number of agents, and how well agents are capable of handling message exchanges. Two types of experiments were performed: in the first type, execution times were measured when the number of works is kept low and the number of reindeer increases, while in the second type, execution times were measured when the number of reindeer

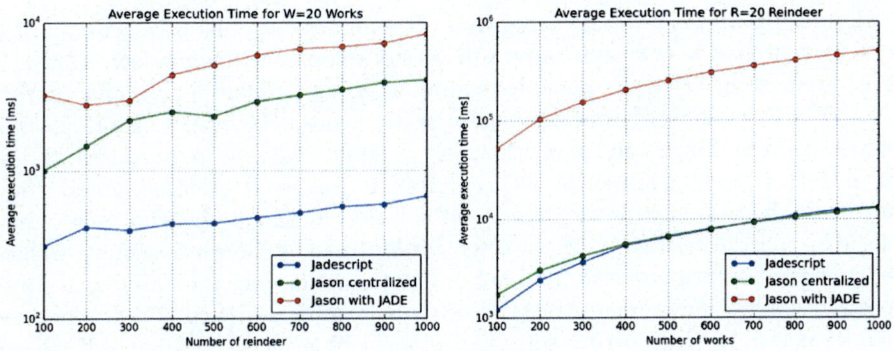

(a) (b)

Fig. 2. Plots of execution times for the three experiments as (a) $W = 20$ and the number of reindeer increases, and (b) $R = 20$ and the number of works increases.

is kept low and the number of works increases. For the first type of experiment, a fixed number of $W = 20$ works is chosen, and reindeer range from $R = 100$ to $R = 1000$. For the second type of experiment, only $R = 20$ reindeer are considered, but the works to be done range from $W = 100$ to $W = 1000$. For all considered executions, the termination condition was expressed in terms of the number of works done.

All experiments were repeated for 100 iterations for each considered configuration, and the average execution times over the 100 iterations were recorded. For all experimented configurations, the measurement of execution times started at the activation of Santa Claus, and it stopped when Santa Claus worked W times with reindeer and a group of reindeer was formed for the $(W+1)$–th time. Such a choice allowed to ignore the time needed to start up and shut down the platform, so that only the actual execution times of agents were recored. The experiments were executed on an Apple MacBook Air mid-2013 with a 1.3 GHz Intel Core i5, 3 MB L3 shared cache, 4 GB LPDDR3 1600 MHz RAM, Java version 12, Jason version 2.4, and JADE version 4.3.0.

The execution times for all experiments are shown in Fig. 2. When the number of reindeer increases and $W = 20$, the Jadescript solution is much faster than the Jason solutions, as shown in Fig. 2(a). When the number of works increases and $R = 20$, the performance of the Jason solution with the `Centralized` infrastructure is very similar to the performance of the Jadescript solution, as shown in Fig. 2(b). On the contrary, Fig. 2(b) shows that the Jason solution with the `Jade` infrastructure is more than ten time slower that the Jadescript solution.

Even if the measured execution times shown in Fig. 2 offer an interesting point of view to compare the two considered languages, other criteria must be taken into account to propose a fair comparison. For example, a comparison between the Jadescript solution and the Jason solutions in terms of *LOCs* (*Lines Of Code*) shows that Jason solutions require only 45 LOCs instead of 71 LOCs.

Besides LOCs, it is worth noting that [11] enumerates a list of accepted criteria to evaluate tools for agent-based software development. Only a few of the criteria proposed in [11] are applicable to support a comparison between Jadescript and Jason. Criterion 1(d) in [11] focuses on the simplicity of AOP tools. In this perspective, the Jason solutions are surely elegant and they require less LOCs. On the other hand, Jadescript is very close to an agent-oriented pseudocode and this is the reason why the Jadescript solution looks simpler and more readable. Also, Jason heavily relies on operators instead of keywords, which makes the Jadescript solution definitely more understandable than the Jason solutions, especially for the newcomers to AOP. Finally, criterion 1(i) in [11] emphasizes the relevance of the support for software engineering principles. Some well-known AOSE methodologies for the BDI approach are applicable to Jason, but they are not directly supported by the language. The Jadescript solution, instead, could be enhanced by means of available constructs in the language for inheritance (e.g., by specializing `Worker` agents as `Reindeer` agents and `Elf` agents) and modularization (e.g., by moving behaviour declarations outside of agent declarations) to better support maintainability, composability, and reusability.

5 Conclusion

Jadescript has been recently introduced as an AOP language that targets the complexity of building agent-based software systems with JADE. A few examples of the use of Jadescript have already been proposed [8,9,20,21] to present the language to the community of researchers and practitioners interested in agent-based software development, but a quantitative comparison with another AOP language was lacking. This paper compares Jadescript and Jason on a specific problem, but comparisons with other languages and on other problems have already been planned for the near future.

As far as performance is concerned, Fig. 2(a) shows that Jadescript helps the programmer in fine-tuning the performance of the MAS, and Fig. 2(b) shows that Jadescript ensures an effective use of JADE. Therefore, from the reported performance comparison, it is evident that Jadescript is preferable to Jason for the construction of agent-based software systems intended to scale to real-world applications (e.g. [5]) with a large number of agents.

References

1. Bădică, C., Budimac, Z., Burkhard, H.D., Ivanovic, M.: Software agents: Languages, tools, platforms. Comput. Sci. Inf. Syst. **8**(2), 255–298 (2011)
2. Bellifemine, F., Bergenti, F., Caire, G., Poggi, A.: JADE – A Java Agent DEvelopment framework. In: Multi-Agent Programming, pp. 125–147. Springer (2005)
3. Bergenti, F.: A discussion of two major benefits of using agents in software development. In: Petta, P., Tolksdorf, R., Zambonelli, F. (eds.) Engineering Societies in the Agents World III. Lecture Notes in Artificial Intelligence, vol. 2577, pp. 1–12. Springer, Heidelberg (2003)

4. Bergenti, F.: An introduction to the JADEL programming language. In: Proceedings of the 26th IEEE International Conference on Tools with Artificial Intelligence (ICTAI 2014), pp. 974–978. IEEE (2014)
5. Bergenti, F., Caire, G., Gotta, D.: Large-scale network and service management with WANTS. In: Industrial Agents: Emerging Applications of Software Agents in Industry, pp. 231–246. Elsevier (2015)
6. Bergenti, F., Gleizes, M.P., Zambonelli, F. (eds.): Methodologies and Software Engineering for Agent Systems: The Agent-Oriented Software Engineering Handbook. Springer, Boston (2004)
7. Bergenti, F., Iotti, E., Monica, S., Poggi, A.: Agent-oriented model-driven development for JADE with the JADEL programming language. Comput. Lang. Syst. Struct. **50**, 142–158 (2017)
8. Bergenti, F., Monica, S., Petrosino, G.: A scripting language for practical agent-oriented programming. In: Proceedings of the 8th International Workshop on Programming Based on Actors, Agents, and Decentralized Control (AGERE 2018), pp. 62–71. ACM (2018)
9. Bergenti, F., Petrosino, G.: Overview of a scripting language for JADE-based multi-agent systems. In: Proceedings of the 19th Workshop "From Objects to Agents" (WOA 2018). CEUR Workshop Proceedings, vol. 2215, pp. 57–62 (2018)
10. Bordini, R.H., Braubach, L., Dastani, M., El Fallah Seghrouchni, A., Gomez-Sanz, J.J., Leite, J., O'Hare, G., Pokahr, A., Ricci, A.: A survey of programming languages and platforms for multi-agent systems. Informatica **30**(1), 33–44 (2006)
11. Bordini, R.H., Dastani, M., Dix, J., El Fallah Seghrouchni, A.: Multi-Agent Programming. Springer, Boston (2005)
12. Bordini, R.H., Hübner, J.F., Wooldridge, M.: Programming Multi-agent Systems in AgentSpeak Using Jason. Wiley, Hoboken (2007)
13. Challenger, M., Mernik, M., Kardas, G., Kosar, T.: Declarative specifications for the development of multi-agent systems. Comput. Stand. Inter. **43**, 91–115 (2016)
14. Fichera, L., Messina, F., Pappalardo, G., Santoro, C.: A Python framework for programming autonomous robots using a declarative approach. Sci. Comput. Program. **139**, 36–55 (2017)
15. Hindriks, K.V., De Boer, F.S., Van der Hoek, W., Meyer, J.J.: Agent programming in 3APL. Auton. Agent. Multi Agent Syst. **2**(4), 357–401 (1999)
16. Hindriks, K.V., Dix, J.: GOAL: A multi-agent programming language applied to an exploration game. In: Shehory, O., Sturm, A. (eds.) Agent-Oriented Software Engineering, pp. 235–258. Springer, Heidelberg (2014)
17. Iotti, E., Bergenti, F., Poggi, A.: An illustrative example of the JADEL programming language. In: Proceedings of the 10th International Conference on Agents and Artificial Intelligence (ICAART 2018), vol. 1, pp. 282–289. ScitePress (2018)
18. Kravari, K., Bassiliades, N.: A survey of agent platforms. J. Artif. Soc. Soc. Simul. **18**(1), 11 (2015)
19. Müller, J.P., Fischer, K.: Application impact of multi-agent systems and technologies: A survey. In: Shehory, O., Sturm, A. (eds.) Agent-Oriented Software Engineering, pp. 27–53. Springer, Heidelberg (2014)
20. Petrosino, G., Bergenti, F.: An introduction to the major features of a scripting language for JADE agents. In: Ghidini, C., Magnini, B., Passerini, A., Traverso, P. (eds.) Advances in Artificial Intelligence (AI*IA 2018), pp. 3–14. Springer, Cham (2018)

21. Petrosino, G., Bergenti, F.: Extending message handlers with pattern matching in the Jadescript programming language. In: Proceedings of the 20th Workshop "From Objects to Agents" (WOA 2019). CEUR Workshop Proceedings, vol. 2404, pp. 113–118 (2019)
22. Rao, A.S., Georgeff, M.P.: BDI agents: From theory to practice. In: Proceedings of the 1st International Conference on Multiagent Systems (ICMAS 1995), vol. 95, pp. 312–319. AAAI (1995)
23. Shoham, Y.: An overview of agent-oriented programming. In: Software Agents, vol. 4, pp. 271–290. MIT Press (1997)
24. Trono, J.A.: A new exercise in concurrency. ACM SIGCSE Bull. **26**(3), 8–10 (1994)
25. Winikoff, M.: JACK intelligent agents: An industrial strength platform. In: Bordini, R.H., Dastani, M., Dix, J., El Fallah, Seghrouchni A. (eds.) Multi-Agent Programming, pp. 175–193. Springer, Boston (2005)

Context-Aware Information for Smart Retailers

Ichiro Satoh(✉)

National Institute of Informatics, 2-1-2 Hitotsubashi, Chiyoda-ku, Tokyo 101-8430, Japan
ichiro@nii.ac.jp

Abstract. Smart stores use ubiquitous digital signages to show information, images and videos, on their items, e.g., selling prices and advertisements. However, most signages tend to always or periodically show the same information to anyone. This paper presents a framework for easily developing novel digital signage whose content are aware of contextual changes in the real world. To spatially bind items and information, the framework automatically deploys programs for displaying contents at digital signage close to the items. It was constructed as a general-purpose middleware system for selecting and deploying programs with information according to context.

1 Introduction

RFID tags are widely used for item management, e.g., inventory management and accounting. However, existing item managements with RFID tags explicitly or implicitly assume information on items to be maintained in the central database or to register the same information on items in individual companies. Therefore, the flow of each item in the supply chain does not match the flow of information on the item. For this reason, for example, when dealing with a new item, each retailer must add or update information on the item its item database. RFID tags enable us to trace the flow of items in the supply chain. The goal of this paper is to propose an approach to distributing information on items along with the flow of items, i.e., supply chains. Suppose a supply chain for consumer products, e.g., electric lights. They are attached with RFID tags, whose identifier is bound to context-aware content.

- **In warehouse:** While the light was in a warehouse, information on the light, e.g., its product number, serial number, date of manufacture, size, and weight, should be displayed on a stationary or mobile terminal nearby the light.
- **In a store's showcase:** While the light is being showcased in a retailer, content for advertising it should be selected according to contextual information, e.g., location and time, in addition to customers' historical activities, e.g., purchase histories, and then be displayed at a digital signage nearby it.
- **In a store's checkout counter:** When a customer carried the light to the cashier of the retailer, the cashier machine detects the presence of the item and then show information on the light e.g., its price, and order another light for the factory as an additional order.
- **In house:** When a light was bought and transferred to the house of its buyer, instruction for the light should be automatically displayed at a display in the house.

Y. Dong et al. (Eds.): DCAI 2020, AISC 1237, pp. 71–80, 2021.
https://doi.org/10.1007/978-3-030-53036-5_8

Although our initial target is in a small-scale supply chain from warehouse to house, our proposed framework is available as a large-scale one consisting of factories, warehouses, retailers, and buyers in large-scale supply chains.

This work is part of a research collaboration with one of the largest retail companies in Japan.[1] The overall research collaboration is to explore a novel approach to displaying context-aware information on items attached with RFID tags on portable terminals used in supply chains, where the information is maintained in a central database system. Nevertheless, this paper focuses a framework for binding items and information on items even when the items are distributed through a supply chain. The framework is constructed as a context-aware middleware system for executing and migrating mobile agents, where a mobile agent is an autonomous program that can travel from computer to computer under its own control. In this framework, mobile agents can be spatially bound to physical entities, e.g., items and terminals, so that they are treated as virtual counterpart objects for the entities. When each item is distributed from a point to another point in a supply chain, the framework migrates mobile agents for displaying context-aware information on the item at stationary or portable terminals in a store, nearby the current location of the item. Using RFID tag systems, it detects the presence of items at places, e.g., warehouse and stores, and then instructs agents bound to the items to deploy at terminals, including digital signage.

The remainder of this paper is organized as follows. In Sect. 2, we outline our basic idea behind the approach. Section 3 presents the design and implementation of the proposed framework. In Sect. 4, we describe its current status. Section 5 surveys related work and Sect. 6 provides a summary.

2 Basic Approach

This section outlines our framework for binding between items and information by using RFID tag and mobile agent technologies. Each item is attached with at least one RFID tag. Each item is bound to at least one content. The framework detects the presence of items by using RFID technology and spatially bind between items and autonomous programs for displaying information on the items by using mobile agent technology. To realize the example scenario illustrated in the previous section, the framework needs to satisfy the following requirements.

- To detect the presence of portable and stationary terminals, including RFID tag readers, near from items, the framework introduces two approaches. When an RFID tag reader detects RFID tags, the framework can know that the reader and tags are near and then display information on the screen of the reader. We also assume that terminals to be attached with RFID tags. When an RFID tag reader recognize both the tags of items and the tag of a terminal, the framework treats the terminal to be near the items.
- Information on an item should change dependently on the context of the item, such as in the warehouse, at showcase in a store, during accounting in the store, at the customer's home, and then be displayed on terminals. Although terminals are equipped

[1] The company's name cannot be disclosed due to the company's request.

with embedded computers with storage, its computational resources, including storages, are limited. Therefore, they cannot store all content inside them.
- Since some content may be highly interactive for customers, it should response to user actions quickly. However, such terminals are often connected to its server through a poor network in the sense that the network is narrow band and is often disconnected due to various reasons, e.g., radio interference with microwave ovens used in the retailers. Therefore, terminal itself plays its content without communicating with other systems to avoid the effect of communication latency.
- To support large-scale context-aware systems, they need to be managed in a non-centralized manner. Mobile agents can be managed without any centralized servers.

To satisfy these requirements, our framework detects the locations of items and customers by using RFID tags or two-dimensional barcodes, where we assume that the tags or barcodes have their own unique identifiers. To dynamically deploy content for items at terminals, the framework uses mobile agent technology to easily deploy and provide dynamic content terminals. This is because each mobile agent-based content can define programs to play its visual/audio content and interact with users inside it. After arriving at its destination, a mobile agent can continue working without losing the results of working, e.g., the content of instance variables in the agent's program, at the source computer, even when networks between the source and destination computers are disconnected. In this framework, mobile agents for content can be classified into two types:

- **Immutable content** is a mobile agent whose state cannot be modified after it is created. Therefore, it does not need to return its home server to update its changes while running on a terminal.
- **Mutable content** is a mobile whose state can be modified after it is created. It is required to return to the home server or informs its changes to the server.

Mobile agent-based content can directly interact with the user, because they are executed on computers nearby them, whereas conventional approaches, e.g., RPC and HTTP-based ones, must have network latency between computers and remote servers, because their content are executed on remote servers. Terminals used in stores and warehouses tend to have the limited amount of their storages. Mobile agents are useful to save their storages, because they need to be present at such terminals only while their contents are played.

3 Design and Implementation

The framework itself is constructed as a general-purpose middleware system. It consists of three parts: context information managers, runtime systems for mobile agents, and mobile agent-based content as shown in Fig. 1. The first provides a layer of indirection between passive RFID tag readers and agents. It manages one or more RFID tag readers to monitor contexts in the real world and provides neighboring runtime systems with up-to-date contextual information of its target entities. It is implemented based on our previous location management systems with RFID tags [7]. The second is constructed as a distributed system consisting of multiple computers in addition to servers. The third corresponds to immutable or mutable mobile agents and is defined as conventional Java-based software components.

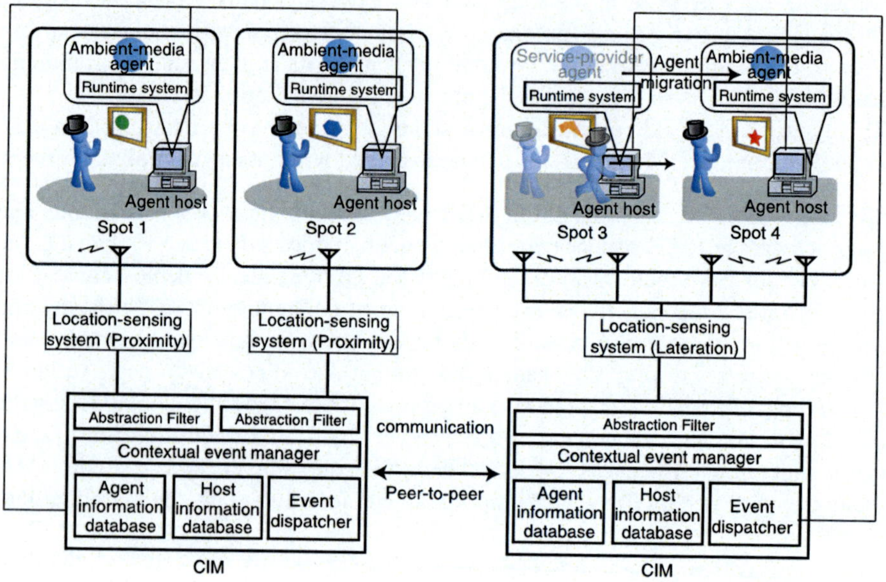

Fig. 1. System structure.

3.1 Context Information Manager

Each context information manager (CIM) manages one or more sensing systems to monitor context in the real world, e.g., the presences of RFID tags attached with items. The current implementation of each CIM supports passive or active RFID-tag systems. It has a database for mapping the identifiers of RFID tags and agents for content and context, e.g., the locations of items, e.g., in warehouse, retailer, and home, and can exchange information with other CIMs through a peer-to-peer communication protocol. To abstract away the differences between the underlying locating systems, each CIM maps low-level positional information from each of RFID tag systems into information in a symbolic model of the location. When each CIM detects the presence of tags attached to items, it searches the identifiers of the agents that are bound to the tag's identifier of its detecting tag from its database. If it successfully finds the corresponding agents, it sends a *fetch* message to the runtime system that the agents are present at, where each *fetch* message is to instruct the agents to migrate to runtime systems near from the tag. Otherwise, it multicasts a *query* message to other CIMs though a UDP multicast mechanism, where the *query* message is to ask other CIMs to find the identifiers and locations of the agents that are bound to the tag's identifier. When it receives the results from the CIMs that knows the agents bound to the tag's identifier, it sends *fetch* messages to the CIMs to instruct the agents to migrate to runtime systems near from the tag. Since agents may require special resources, e.g., large-size displays, the framework enables each agent to specify its necessary resources that runtime systems need to provide. It uses a language based on CC/PP (composite capability/preference profiles) [13]. The language is used to specify the capabilities of terminals and the requirements of mobile agents in an XML notation. For example, a description may contain information on the following properties of a terminal, e.g., screen size, number of colors, CPU, memory, input devices, secondary storage, presence/absence of loud-speakers, etc.

3.2 Runtime System

Each runtime system has two-forms of functionality: one for advertising its capabilities and another for executing and migrating mobile agents from/to other runtime systems. When a runtime system receives a *discovery* message with the identifier of a newly arriving tag from a CIM, it replies with one of the following three responses: (i) if the identifier in the message is identical to the identifier of the tag to which it is attached, it returns profile information on its capabilities to the CIM; (ii) if one of agents running on it is tied to the tag, it returns its network address and the requirements of the agent; and (iii) if neither of the above cases applies, it ignores the message. Each runtime systems is responsible for executing and migrating agents to other runtime systems running on different computers through a TCP channel using mobile-agent technology (Fig. 2). It is built on the Java virtual machine (Java VM) version 1.8 or later versions, which conceals differences between the platform architectures of the source and destination computers. It governs all the agents inside it and maintains the life-cycle state of each agent. When the life-cycle state of an agent changes, e.g., when it is created, terminates, or migrates to another runtime system, its current runtime system issues specific events to the agent.

When an agent is transferred over the network, not only the code of the agent but also its state is transformed into a bitstream by using Java's object serialization package and then the bit stream is transferred to the destination. Since the package does not support the capturing of stack frames of threads, when an agent is deployed at another computer, its runtime system propagates certain events to instruct it to stop its active threads. Arriving agents may explicitly have to acquire various resources, e.g., video and sound, or release previously acquired resources.

The framework allows agents to communicate with other servers via TCP, UDP, and remote method invocation (RMI), when the agents invoke our original TCP, UDP, and RMI libraries on behalf of Java's original libraries. It provides agents with references to other servers, where each reference is a mobility-transparent address to another server and forwards the data that an agent sends to the server or receives from the server from or to the agent. Agents communicate with other servers specified as references, even when they migrate to other computers.

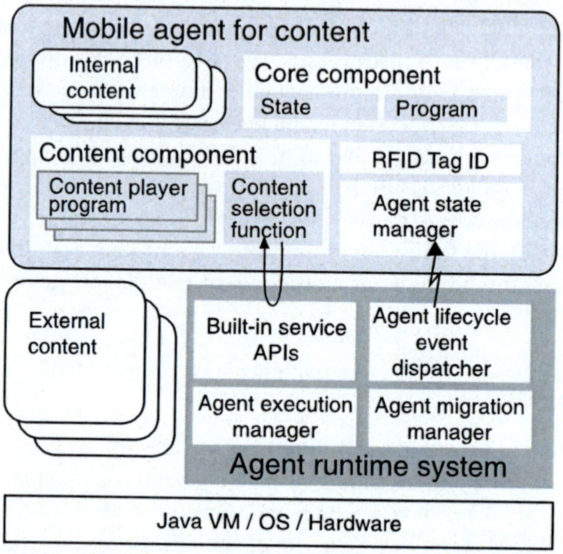

Fig. 2. Architecture of runtime system for service-provider agent.

3.3 Mobile Agent-Based Content

Each agent can be defined as conventional Java-based software components, e.g., JavaBeans. It is bound to at most one RFID tag, whose identifier is unique. When the presence of RFID tags at a space is detected by RFID readers, a CIM discovers mobile agents bound to the tag and then instructs the agents to migrate to a computer inside the space via the current runtime system of the agent. The program of each agent is responsible for selecting and playing the content maintained in the agent. Each agent maintains information, including visual content, on the item that the tag is attached to and program for playing the information at its current computer. Since each agent itself can define how to play its information, the information does not need to be represented in common formats.

The information to be played needs to be selected by context. In addition, relocating all information to the terminal increases the communication traffic and exhausts the storage of the terminal. Therefore, each agent consists of more than one component, where each component is a set of Java objects. Components in an agent can be classified into two kinds, i.e., core and content components. The first is responsible for storing and processing the common data among other components in the agent. The second is responsible for storing contents, e.g., text, image, video, and sound, and playing the content. When deploying at terminal, each agent makes a copy of it, where each copy consists of the core component and some content components, which should be displayed at the terminal

Thus, each agent can dynamically generate its own copy in response to specific information. To do this, the framework is to introduce a mechanism for selectively duplicating the state and the program code in the duplication of the agent and reconfiguring the agent. It enables the state of each agent, e.g., program variables in heap area, to be duplicated to its clone agent by using Java's object serialization package for marshaling components. The package does not support the capturing of stack frames of threads. Instead, when an agent is duplicated, the runtime system issues events to it to invoke their specified methods, which should be executed before the agent is duplicated, and it then suspends their active threads. The program can play its content by using its own player program. The current implementation supports (rich) text data, html, image data, e.g., JPEG and GIF, video data, e.g., animation GIF and MPEG, and sound data, e.g., WAV and MP3. Since the annotation part is defined as Java-based general-purpose programs, we can easily define interactions between visitors and agents.

4 Current Status

Although the current implementation was not built for performance, we measured the cost of migrating a null agent (a 10-KB agent, zip-compressed) and an annotation agent (1.8-MB agent, zip-compressed) from a source computer to a destination computer that was recommended by the CIM. The latency of discovering and instructing an agent attached to a tag after the CIM had detected the presence of the tag was 310 ms and the respective cost of migrating the null and annotation agent between two computers over a TCP connection was 32 ms and 175 ms. This evaluation was operated with embedded computers, where each of them was a Raspberry Pi (ARMv8-based SoC processor (Broadcom BCM2837B0, Cortex-A53, 1.4 GHz)), 1 GB DRAM memory, and SD card storage). Migrating containers included the cost of opening a TCP-transmission, marshaling the agents, migrating them from their source computers to their destination computers, unmarshaling them, and verifying security.

Next, we outline how to implement the example scenario discussed in the first section. In the scenario, we suppose that appropriate information is presented to terminals, including digital signage, through a supply chain of an electric light according to context in the light. Our framework permits a context-aware system to have more than one CIM. Each agent is bound to at most one RFID tag through at least one CIM. We also assume each item to be attached with an RFID tag, which is widely used in existing supply chains and has its own unique identifier.

When the lights are in the warehouse, information for stock management should be displayed on a terminal carried by staffs or near from the lights. We assume that staffs in the warehouse have portable RFID readers. When a staff reads an RFID tag attached to the light with the RFID reader, the identifier of the tag is sent to a CIM and then the CIM tries to detect agents bound to the tag. The CIM informs the agents to migrate to a RFID reader or terminal carried by him/her. The agents locally run its own program and content inside his/her reader or portable terminals and can connect to other servers via their current runtime systems.

We assume that soon after an electric light is displayed at a showcase and is near digital signage in a retailer, clerks of the retailer read the identifier of the RFID tag attached to the light to detect the locations of the light. The CIM of the retailer tries to discover the agent bound to the tag according to the context of the light, where the context is in a showcase in the retailer. Therefore, it returns an agent that is bound to the light and is available in a show case. For example, an agent for advertising the light is deployed at a digital signage close its target light, which could display its advertising content, including selling price, to attract purchases by customers who visit the retailer. When a customer brings the light to the cash register to buy it, the cashier reads the RFID tag attached to the light and the CIM finds the agent corresponding to the identifier of the tag. The CIM then instructs the agent to migrate to the cashier terminal and execute itself to show information on the light, e.g., selling price. The agent requests the retailer's stock management system to decrease the number of stocks of the light via its runtime system. When a light was bought and transferred to the house of its buyer, an agent that advises how it was to be used is deployed at a terminal in the house (Fig. 3).

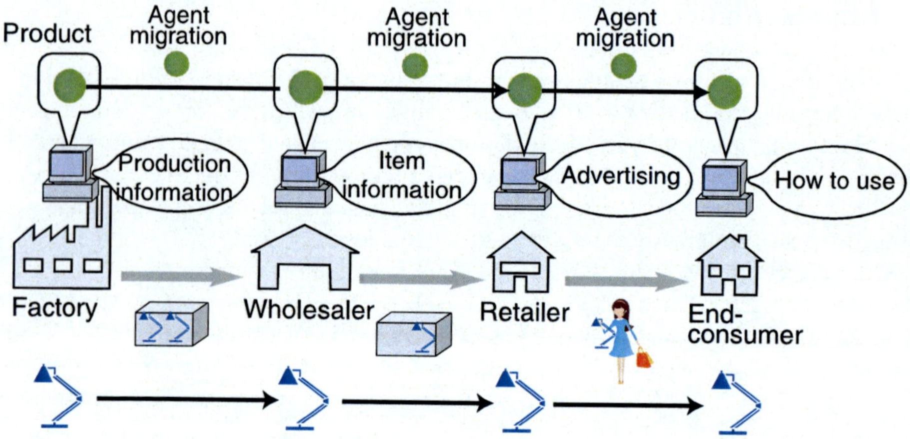

Fig. 3. Forwarding agents to digital signage when user moves.

5 Related Work

Terminals in retailers, e.g., digital signage, have been explored in commercial development for not only for advertising and sales promotions but also in information sharing, entertainment and so on [6]. For example, Walmart in USA has 27000 displays installed in most stores with 140 million impressions per day. Display placement and contents are designed to promote customers to buy items.

There have been several attempts to provide users with context-aware information in the right situation and at the right location. Rather than taking instructions step by step from the human user, context-aware systems know how to anticipate a changing situation or an emerging need and trigger the right response at the precise time. Several retailers had experiments on location-aware services in their stores, because retailers believed location-based technology can improve their customers' experiences, drive revenue and increase operational efficiency. Among them, beacon technology, including iBeacon, which uses Bluetooth connections to send messages to people on their mobile device, triggered by the person's proximity to the beacon, become popular [2]. Beacons are small enough to attach to a wall or countertop. Many retailers have used beacons to send in-store shoppers' coupons and other offers.

Several researchers have explored active media with the aim of enabling users to watch/listen to context-aware information, e.g., annotation about exhibits at the right time and in the right place. Watson et al. [12] have proposed the term *u-commerce* (or *ubiquitous commerce*), which is defined as the use of ubiquitous networks to support personalized and uninterrupted communications and transactions between a firm and various stakeholders to provide a level of value over, above and beyond traditional commerce. A number of architectures and prototypes of u-commerce systems have been described in the literature [5]. The *Shopper's Eye* [4] proposed a location-aware service with wirelessly enabled portable terminals, e.g., PDAs and smart phones. As a shopper travels about, his or her personal terminal transmits messages, which include information about his or her location, shopping goals, preferences, and related purchase history. When this information is received, stores create a customized offer of goods and services. The Impulse project [11] is a PDA-based system whereby customers may add products to a list, indicating preferences such as warranty terms, merchant reputations, availability, time limits for the purchases and preferred price. When a potential customer enters shopping zones, individual stores, or shopping malls, his/her agent engages nearby merchants in a silent exchange seeking items on the list and opens negotiations on the terms of the sale, alerting the shopper if a deal has been agreed on. It is envisaged that the merchant gains valuable information about customers' purchasing behavior during the negotiation process.

Several research projects proposed context-aware recommendation approaches based on content-based and collaborative filtering approaches [1] to predict the rating or rank that each user would give to items, where collaborative filtering is a recommendation method for each active user is received by comparing with the preferences of other users who have rated the product in similar way to the active user. Context-aware recommendation approaches incorporate contextual information, e.g., weather, location, mood, and season, in addition to traditional factors, the number of buying items, to rate or rank them based on users' historical activities with historical rating pattern of the user or other similar users [3].

6 Conclusion

We designed and implemented a framework for providing context-aware content on terminals in warehouses, stores, and houses. It was characterized at dynamically deploying mobile agent-based components consisting of content data and program code for displaying the data, so that it can display content in various formats. Using RFID tag systems, it detects the presence of items at places, e.g., warehouse and retailer's stores, and then instructs agents bound to the items to deploy at terminals, including digital signage. It has contributions to bind between items and information on items through a supply chain. The framework was constructed as a general-purpose middleware system for mobile agents. Although we had many experiences on context-aware services in real spaces [9], it was evaluated on its basic performance.

Finally, we would like to identify further issues that need to be resolved in the future. We plan to carry out experiments with real users in a supermarket in future.[2]

References

1. Adomavicius, G., Tuzhilin, A.: Toward the next generation of recommender systems: a survey of the state-of-the-art and possible extensions. IEEE Trans. Knowl. Data Eng. **17**(6), 34–749 (2005)
2. Apple: Getting started with iBeacon (2014). https://developer.apple.com/ibeacon/Getting-Started-with-iBeacon.pdf
3. Chen, A.: Context-aware collaborative filtering system: predicting the user's preference in the ubiquitous computing environment. In: Proceedings of International Symposium on Location- and Context-Awareness (LoCA 2005). LNCS, vol. 3479, pp. 244–253. Springer (2005)
4. Fano, A.: Shopper's eye: using location-based filtering for a shopping agent in the physical world. In: Proceedings of International Conference on Autonomous Agents, pp. 416–421, ACM Press (1998)
5. Galanxhe-Janaqi, H, Nah, F.F.-H.: U-commerce: emerging trends and research issues. Ind. Manag. Data Syst. **104**(9), 744–755 (2004)
6. Harrison, J., Andrusiewicz, A.: An emerging marketplace for digital advertising based on amalgamated digital signage networks. In: Proceedings of IEEE International Conference on E-Commerce, pp. 149–156 (2003)
7. Satoh, I.: Context-aware agents to guide visitors in museums. In: Proceedings of 8th International Conference Intelligent Virtual Agents (IVA 2008). Lecture Notes in Artificial Intelligence (LNAI), vol. 5208, pp. 441–455, September 2008
8. Satoh, I.: Mobile agents. In: Handbook of Ambient Intelligence and Smart Environments, pp. 771–791. Springer (2010)
9. Satoh, I.: Experiences in context aware-services. In: Proceedings of 10th International Symposium on Ambient Intelligence (ISAmI 2019) Advances in Intelligent Systems and Computing, vol. 1006, pp. 45–53. Springer (2020)
10. Schaeffler, J.: Digital Signage: Software, Networks, Advertising, and Displays: A Primer for Understanding the Business. Taylor and Francis, New York (2008)
11. Tewari, G., Youll, J., Maes, P.: Personalized location-based brokering using an agent-based intermediary architecture. Decis. Support Syst. **34**(2), 127–137 (2003)
12. Watson, R.E., Pitt, L.F., Berthon, P., Zinkhan, G.M.: U-commerce: expanding the universe of marketing. J. Acad. Mark. Sci. **30**(4), 333–347 (2002)
13. World Wide Web Consortium (W3C), Composite Capability/Preference Profiles (CC/PP) (1999). http://www.w3.org/TR/NOTE-CCPP

[2] We initially planned to have an experiment at a supermarket in May, 2020, but we must postpone it due to COVID-19.

Morphometric Characteristics in Discrete Domain for Brain Tumor Recognition

Jesús Silva[1](✉), Jack Zilberman[1], Narledis Núñez Bravo[2], Noel Varela[3], and Omar Bonerge Pineda Lezama[4]

[1] Universidad Peruana de Ciencias Aplicadas, Lima, Peru
jesussilvaUPC@gmail.com, jack.zilberman@upc.pe
[2] Universidad Simón Bolívar, Barranquilla, Colombia
nnunez3@unisimonbolivar.edu.co
[3] Universidad de la Costa, Barranquilla, Colombia
nvarela2@cuc.edu.co
[4] Universidad Tecnológica Centroamericana (UNITEC), San Pedro Sula, Honduras
omarpineda@unitec.edu

Abstract. World Health Organization (WHO) classifies brain tumors by their level of aggressiveness into four grades depending on their aggressiveness or malignancy as I to IV respectively [1]. From this classification of primary brain tumors, the four categories can be considered in two groups: Low Grade (LG) and High Grade (HG), in which the LG group is composed of grade I and II brain tumors, while the HG group is composed of grades III and IV brain tumors [2]. This paper focuses on the morphometric analysis of brain tumors and the study of the correlation of tumor shape with its degree of malignancy.

Keywords: Morphometric characteristics · Brain tumor · Recognition · Degree of malignancy

1 Introduction

This paper focuses on the morphometric analysis of brain tumors applying morphometric descriptors in a discrete domain. The development of new mathematical tools and algorithms allows the estimation and quantification of some morphological aspects of brain tumors for a better understanding of them with the information obtained from MRIs and their relationship with biological characteristics [3].

This work focuses on the morphometric analysis of brain tumors applying morphometric descriptors in a discrete domain. The development of new mathematical tools and algorithms allows the estimation and quantification of some morphological aspects of brain tumors for a better understanding of them with the information obtained from the MRI and its relationship with the biological characteristics [4].

The segmentation of brain tumors is crucial for the diagnosis and control of tumor growth, in all cases it is necessary to quantify the tumor volume in order to measure and

Y. Dong et al. (Eds.): DCAI 2020, AISC 1237, pp. 81–88, 2021.
https://doi.org/10.1007/978-3-030-53036-5_9

implement an objective analysis [5]. In [6] they applied pattern classification methods to separate two different types of brain tumors, primary metastatic gliomas (MET). They also proposed pattern recognition techniques for the classification of gliomas. The proposed classification method combines conventional MRI sequences, this method consisted of extraction of characteristics such as tumor shape, intensity characteristics, as well as invariant rotation of texture characteristics. They reported a precision, sensitivity and specificity for the classification of low and high grade neoplasms of 88%, 85% and 96%, respectively.

[7] presents the hypothesis of GBM and METs have different three-dimensional (3D) morphological attributes based on their physical characteristics. They identified a distinct boundary surface between healthy and pathological tissue on the tumor surface. The morphometric characteristics of the shape index and curvature were calculated for each tumor surface and used to construct a morphometric model of GBM and MET.

Another work to discriminate between MET and GBM based on the morphometric analysis was proposed by [8], they propose a form analysis as an indicator to discriminate these 2 types of brain pathologies that allowed discrimination. Cross validation resulted in a classification with an accuracy of 95.8%.

Tumor morphometric analysis also includes classification of images based on tissue histology, in terms of different components, which provides a series of indexes of tumor composition.

In [9] they propose two methods of tissue classification based on morphometric statistics of various locations and scales based on the coincidence of the space pyramid and the coincidence of the linear space pyramid.

Morphometric studies are not only applied to brain tumors, [10] proposed a quantitative morphometric analysis of hepatocellular carcinoma, where the main objective is to analyze the quantitative relationship between tumor morphology and malignant potential in liver tumors.

While [11] showed an inverse correlation between the compactness and the degree of tumor infiltration through its analysis and quantification of the invasion of carcinoma of the cervix based on a 3D tumor reconstruction of the tissue.

Preliminary research presents preliminary results of morphometry studies in brain tumors, where one of the main proposed descriptors is discrete compactness. Preliminary results show an inverse relationship between discrete compactness and grade of malignancy in primary tumors [12].

2 Materials and Methods

This section presents the acquisition protocol and the definition of regions of interest (ROI). Subsequently, the proposed morphometric descriptors in the discrete domain are presented.

2.1 Data Base

The database consists of 40 multi-contrast MRI (Magnetic Resonance Imaging) in patients with glioma, of which 20 belong to the BG group (histological diagnosis:

astrocytoma or oligoastrocytoma) and 20 to the AG group (anaplastic astrocytoma and glioblastoma multiforme tumors). This image data set was obtained from the Multimodal Brain Tumor Segmentation Challenge Organization (BRATS) [7]. They were acquired at four different centers over the course of several years, using MRI with different field strengths (1.5T and 3T) and the implementation of different image sequences.

The image datasets in this repository share the following four MRIs: T1-weighted image, T1-contrast weighted image (T1C), T2-weighted image, and T2-FLAIR weighted image (FLAIR) [10].

2.2 Pre-processing and ROI Definition

The images are pre-processed to homogenize the data. All subject image volumes were co-registered at the T1C MRI and resampled at an isotropic resolution of 1 mm in a standardized axis orientation [13]. The BRATS image repository dataset also contains expert manual annotation of four region types [14]: edema, unenhanced nucleus, necrotic nucleus, and active nucleus. Figure 1 shows the BG and AG glioma regions.

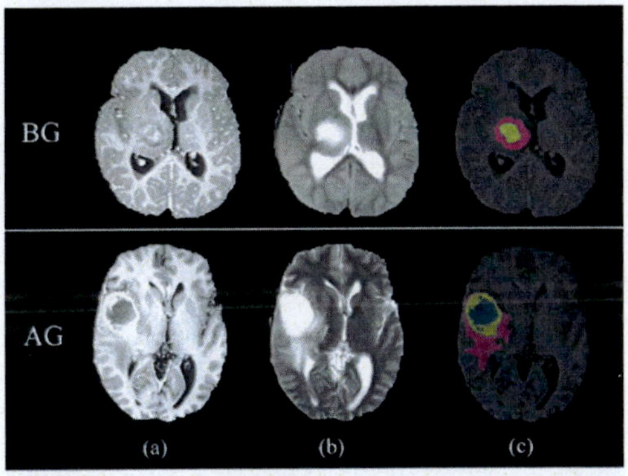

Fig. 1. Example of axial view of MRI of BG (first row) and AG (second row) patients with the presence of glioma. (a) T1C-weighted image with contrast enhancement, (b) T2-weighted image, and (c) regions of the brain tumor labeled as edema (in purple), active nucleus (in yellow), and necrotic nucleus (in blue).

The segmented regions were processed by Moore's Neighborhood algorithm for edge detection in order to obtain the discretized 2D contours for each tumor region. Figure 2 shows an example of annotated regions (edema, tumor core, active core and necrotic core) of a GA glioma patient in the first row and their corresponding discrete contours in the second row.

2.3 Form Descriptors

This section describes the main characteristics of each of the discrete morphological descriptors used to analyze the tumor regions, which include volume, surrounding surface area, contact surface area, discrete compactness, discrete tortuosity and volume ratio of the tumor core to the edema [15].

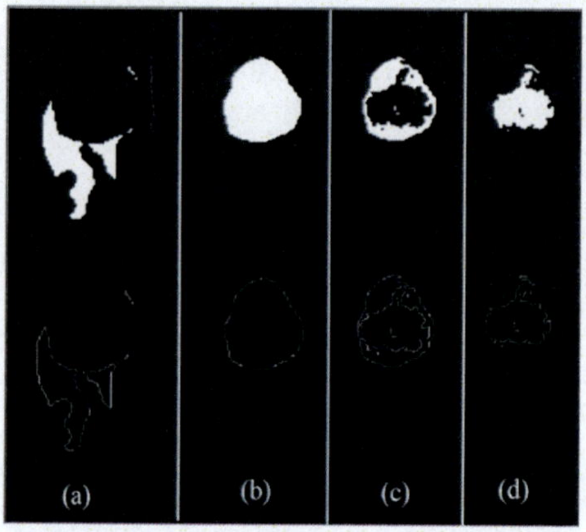

Fig. 2. Examples of tumor regions segmented in the first row and their corresponding contours detected in the second row are shown; the corresponding tumor regions are (a) edema, (b) tumor, (c) active nucleus and (d) necrotic tumor.

2.4 Contact Surface Area

In the discrete domain and according to Bribiesca, the area of the surrounding surface (A) is expressed in Eq. (1), as follows [16]:

$$A = 6an - 2Ac \tag{1}$$

Where Ac is the contact surface area, a is the face area of a voxel, and n is equal to 6 in Eq. 1.

2.5 Discrete Curvature

The discrete curvature of a discrete shape at a Q-vertex is the tangent line that forms the contingency angle ω, which is the change of slope between the continuous segments of straight line at that point.

2.6 Discreet Tortuosity

The tortuosity τ of a curve represented by a chain defined by Bribiesca [5] as the sum of all absolute values of the elements of the chain, is expressed in Eq. (2):

$$\tau = \sum_{i=1}^{n} |a_i|. \tag{2}$$

In this case, the tortuosity is measured from the discrete contours of the tumor regions, where the straight-line segments of these contours have a length according to the size and the number of voxels. In a two-dimensional image there may be one or more contours, while in a 3D image there may also be more contours in each cut. The discrete τ for 3D objects is defined as the sum of all absolute values of the curvatures of all concatenated contours present in the image.

2.7 Volumetric Ratio

The objective was to analyze the volume relationship (rv) between tumor and edema in gliomas as a descriptor that can be correlated with the degree of malignancy.

3 Results

The values of volume (V), contact area surface (Ac), surrounding surface area (A), discrete compactness (Cd), discrete tortuosity (τ) and volumetric ratio (rv), between tumor region and edema, were obtained for all the segmented ROIs of the 20 patients diagnosed with BG gliomas (edema and tumor core) and the 20 patients with AG gliomas (edema, tumor core, active region and necrotic tumor). All the computational algorithms to quantify these discrete morphometric descriptors were implemented in an interactive environment for visualization and programming.

Table 1 shows the table of mean values and standard deviation of these descriptors, obtained from all segmented regions, for BG and AG gliomas. An increase in tumor volume is not necessarily related to the degree of malignancy of the gliomas, but rather to the degree of tumor expansion and the homogeneity of the peripheral surface of the tumors [17].

Table 1. Mean values and standard deviation of edema, tumor core, active core and necrosis for BG and AG glioma regions.

Descriptor	Edema	Edema	Tumor	Tumor	Active core		Necrotic	
Grade	BG	AG	BG	AG	BG	AG	BG	AG
V (cm^3)	7.58 ± 9.46	15.25 ± 9.16	6.52 ± 5.64	10.05 ± 6.71	–	7.46 ± 5.86	–	3.70 ± 3.46
Width (cm^2)	1083.1 ± 1373.9	2208.6 ± 1342.1	951.8 ± 821.0	1478.3 ± 991.8	–	1070.1 ± 850.5	–	524.4 ± 499.2
A (cm^2)	107.74 ± 98.82	157.07 ± 70.90	52.72 ± 54.69	58.32 ± 29.83	–	98.72 ± 58.60	–	59.83 ± 45.97
Cd	0.13 ± 0.05	0.08 ± 0.03	0.06 ± 0.02	0.05 ± 0.02	–	0.11 ± 0.05	–	0.18 ± 0.11
T	34.89 ± 3.72	38.11 ± 3.01	32.79 ± 4.62	36.98 ± 4.10	–	48.23 ± 8.10	–	44.54 ± 7.65
rv	–	–	1.30 ± .90	0.98 ± 0.05	–	–	–	–

Although the results in Table 1 show that the mean volume value for AG gliomas is greater than the mean volume value for BG gliomas, this does not necessarily have a correlation with the degree of glioma malignancy. The results show a high variation in volume values (V) for both BG and AG gliomas [17].

It is the same case for the area of the surrounding surface (A) and the surface of the contact area (Ac), where the variation between the values of the areas presents a high deviation. The relationship between the volume and the surface area of the contact area is directly proportional, as shown in Fig. 3, where the graphs of this relationship are presented for the cases of the edema and tumor core regions in all patients with AG and BG gliomas of malignancy.

Figure 4 shows the comparison of the mean values of compactness and τ for regions with edema and tumor nucleus for BG and AG gliomas respectively.

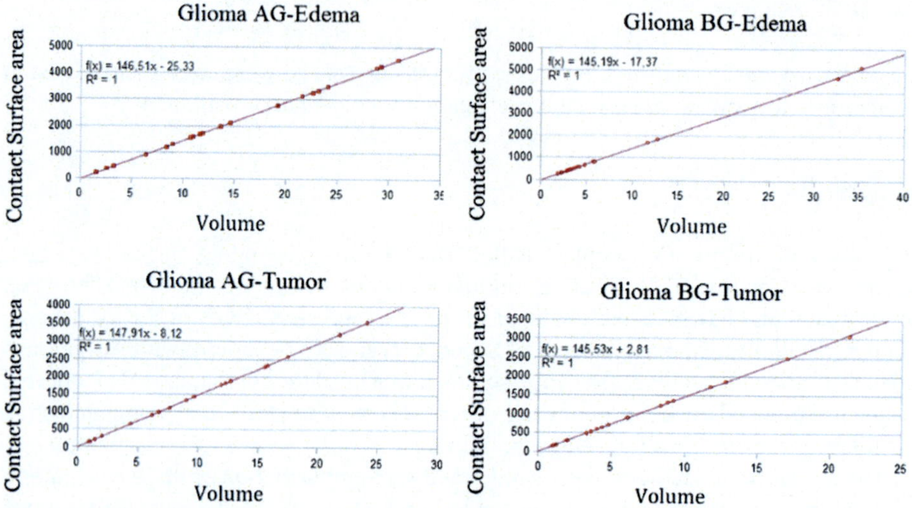

Fig. 3. Relationship between the volume (V) and the surface of the contact area (Ac) in the case of edema and regions of the tumor nucleus, for both BG and AG gliomas.

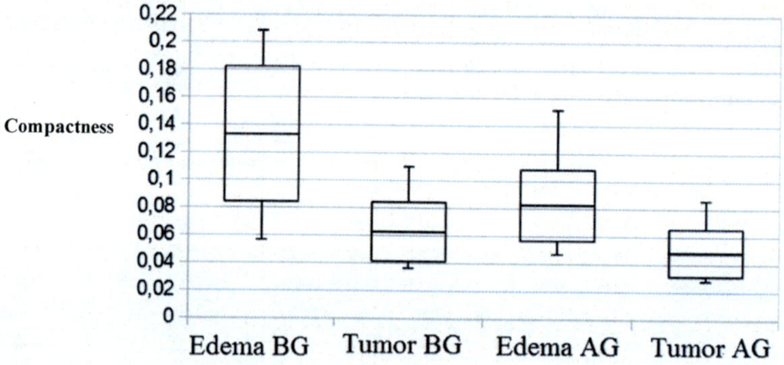

Fig. 4. Representation of the space of morphometric characteristics: Discrete compactness.

The results show an inverse correlation between the discrete compactness and the degree of malignancy of gliomas ($r = -0.3458$, $\rho = 0.0352$); likewise, the results show low direct correlation between tortuosity versus the degree of malignancy in tumors ($r = 0.4758$, $\rho = 0.0055$). The other morphometric descriptors such as volume, enveloping surface area and contact area showed a significant correlation with the degree of malignancy in brain tumors.

4 Discussion and Conclusion

This paper presents some morphometrics characteristics such as volume, surrounding surface area, contact area, compactness, tortuosity and volume relationship (region of the tumor versus the region of the edema) to analyze the forms of the brain tumor and its correlation with the degree of malignancy. All the descriptors were extracted of the ROI of the edema, tumor core, active core and necrotic core regions and were implemented completely in the discrete domain to adapt them to voxel units instead of classic measures. In this approach, it was found that discrete compactness and tortuosity can be morphometric descriptors capable of distinguishing between BG and AG gliomas.

Figures 3 and 4 show that BG gliomas have a slightly higher compactness value than AG gliomas, but a lower tortuosity value compared to an AG glioma. However, the difference in the discrete values of compactness and tortuosity between the gliomas of BG and AG are low. It should be noted that tumor segmentation has some limitations associated with pre-processing and segmentation methods for MRI sequences, because the images were co-recorded and resampled at 1 mm isotropic resolution in a standardized axial orientation, which could introduce some smoothing on the tumor surface.

Similarly, the MRI sequences were annotated by delineating the tumor regions in every three axial sections, interpolating the segmentation as well as by using morphological operators and region growth techniques, so that in this case the tumor regions they do not correspond entirely to the structure of the actual tumor. Since the discrete values of compactness and tortuosity are sensitive to the tumor surface, it is necessary to obtain these discrete morphometric descriptors in a reliable and robust way from the segmented region and, consequently, the correlation coefficients of the degree of tumor malignancy with compactness. discreet and tortuous could be improved.

References

1. Saba, T., Mohamed, A.S., El-Affendi, M., Amin, J., Sharif, M.: Brain tumor detection using fusion of hand crafted and deep learning features. Cogn. Syst. Res. **59**, 221–230 (2020)
2. Blanchet, L., Krooshof, P., Postma, G., Idema, A., Goraj, B., Heerschap, A., Buydens, L.: Discrimination between metastasis and glioblastoma multiforme based on morphometric analysis of MR images. Am. J. Neuroradiol. **32**(1), 67–73 (2011). http://www.ajnr.org/content/early/2010/11/04/ajnr.A2269
3. Gamero, W.M., Agudelo-Castañeda, D., Ramirez, M.C., Hernandez, M.M., Mendoza, H.P., Parody, A., Viloria, A.: Hospital admission and risk assessment associated to exposure of fungal bioaerosols at a municipal landfill using statistical models. In: International Conference on Intelligent Data Engineering and Automated Learning, pp. 210–218. Springer, Cham, November 2018
4. Özyurt, F., Sert, E., Avcı, D.: An expert system for brain tumor detection: fuzzy C-means with super resolution and convolutional neural network with extreme learning machine. Med. Hypotheses **134**, 109433 (2020)
5. Wu, Q., Wu, L., Wang, Y., Zhu, Z., Song, Y., Tan, Y., Wang, X.F., Li, J., Kang, D., Yang, C.J.: Evolution of DNA aptamers for malignant brain tumor gliosarcoma cell recognition and clinical tissue imaging. Biosens. Bioelectron. **80**, 1–8 (2016)
6. Kharrat, A., Mahmoud, N.E.J.I.: Feature selection based on hybrid optimization for magnetic resonance imaging brain tumor classification and segmentation. Appl. Med. Inf. **41**(1), 9–23 (2019)
7. Sharif, M., Amin, J., Raza, M., Yasmin, M., Satapathy, S.C.: An integrated design of particle swarm optimization (PSO) with fusion of features for detection of brain tumor. Pattern Recogn. Lett. **129**, 150–157 (2020)

8. Chang, H., Borowsky, A., Spellman, P., Parvin, B.: Classification of tumor histology via morphometric context. In: 2013 IEEE Conference on Computer Vision and Pattern Recognition, pp. 2203–2210, June 2013
9. Moitra, D., Mandal, R.: Review of brain tumor detection using pattern recognition techniques. Int. J. Comput. Sci. Eng. **5**(2), 121–123 (2017)
10. Einenkel, J., Braumann, U.D., Horn, L.C., Pannicke, N., Kuska, J.P., Schhütz, A., Hentschel, B., Hockel, M.: Evaluation of the invasion front pattern of squamous cell cervical carcinoma by measuring classical and discrete compactness. Comput. Med. Imaging Graph **31**, 428–435 (2007)
11. Gomathi, P., Baskar, S., Shakeel, M.P., Dhulipala, S.V.: Numerical function optimization in brain tumor regions using reconfigured multi-objective bat optimization algorithm. J. Med. Imaging Health Inf. **9**(3), 482–489 (2019)
12. Chen, S., Ding, C., Liu, M.: Dual-force convolutional neural networks for accurate brain tumor segmentation. Pattern Recogn. **88**, 90–100 (2019)
13. Kistler, M., Bonaretti, S., Pfahrer, M., Niklaus, R., Büchler, P.: The virtual skeleton database: an open access repository for biomedical research and collaboration. J. Med. Internet Res. **15**(11), e245 (2013). http://www.jmir.org/2013/11/e245/
14. Amin, J., Sharif, M., Gul, N., Yasmin, M., Shad, S.A.: Brain tumor classification based on DWT fusion of MRI sequences using convolutional neural network. Pattern Recogn. Lett. **129**, 115–122 (2020)
15. Kim, B., Tabori, U., Hawkins, C.: An update on the CNS manifestations of brain tumor polyposis syndromes. Acta Neuropathol. **139**, 703–715 (2020). https://doi.org/10.1007/s00401-020-02124-y
16. Viloria, A., Bucci, N., Luna, M., Lis-Gutiérrez, J.P., Parody, A., Bent, D.E.S., López, L.A.B.: Determination of dimensionality of the psychosocial risk assessment of internal, individual, double presence and external factors in work environments. In: International Conference on Data Mining and Big Data, pp. 304–313. Springer, Cham, June 2018
17. Thivya Roopini, I., Vasanthi, M., Rajinikanth, V., Rekha, M., Sangeetha, M.: Segmentation of tumor from brain MRI using fuzzy entropy and distance regularised level set. In: Nandi, A.K., Sujatha, N., Menaka, R., Alex, J.S.R. (eds.) Computational Signal Processing and Analysis, pp. 297–304. Springer, Singapore (2018)

Filtering Distributed Information to Build a Plausible Scene for Autonomous and Connected Vehicles

Guillaume Hutzler(✉), Hanna Klaudel, and Abderrahmane Sali

Université Paris-Saclay, Univ Evry, IBISC, 91020 Evry, France
{guillaume.hutzler,hanna.klaudel}@ibisc.univ-evry.fr

Abstract. To make their decisions, autonomous vehicles need to build a reliable representation of their environment. In the presence of sensors that are redundant, but not necessarily equivalent, that may get unreliable, unavailable or faulty, or that may get attacked, it is of fundamental importance to assess the plausibility of each information at hand. To this end, we propose a model that combines four criteria (relevance, trust, freshness and consistency) in order to assess the confidence in the value of a feature, and to select the values that are most plausible. We show that it enables to handle various difficult situations (attacks, failures, etc.), by maintaining a coherent scene at any time despite possibly major defects.

Keywords: Autonomous vehicles · Plausibility · Confidence · Sensor fusion

1 Introduction

Driving autonomy promises many benefits such as facilitating travel by reducing traffic jams and the number of accidents for which man is mainly responsible, or improving the comfort of the users during their travels. The control of autonomous vehicles is carried out in three stages: perception, decision, action. The perception stage corresponds to the acquisition of data by sensors, coupled to the analysis of these data to build the scene (integrated view of the environment). The decision stage corresponds to the selection of actions to be performed by the vehicle based on the scene, the state of the vehicle and its "intentions" (current maneuver, mission to perform). The action stage corresponds to the realization of the actions chosen based on the various actuators of the vehicle.

In a man-controlled vehicle, none of the sensors and features that they measure (speed, engine rpm, position, etc.) are crucial to the driving activity, they are only an assistance to help the driver in smoothly conducting his vehicle. In that case, the perception is mainly achieved by the senses of the driver. In an autonomous vehicle however, the relevance and reliability of environmental information are crucial to the decision-making process: a noisy perception or

Y. Dong et al. (Eds.): DCAI 2020, AISC 1237, pp. 89–101, 2021.
https://doi.org/10.1007/978-3-030-53036-5_10

a rough reconstruction of the scene can lead to a bad decision and therefore a higher risk of accident. To tackle this issue, manufacturers use various and multiple sensors to obtain some redundancy in order to have more certainty about the environment. A problem arises when one has to choose between various values, measured by different sensors or combinations of sensors, for the same physical measure (position, speed, etc.). Sensors may have different margins of errors, or may be weather sensitive, and they may be subject to malfunctions, faults, or even attacks, that may lead to the production of erroneous values. Filtering the values in order to identify the most plausible one in the present context is thus a major stake in building (iteratively and in real-time) a reliable scene and therefore to increase the robustness of the system. This task is very similar to what happens in avionics where integrating data from different sources is necessary in order to build a view of the environment and create so-called *situational awareness* [9].

In this paper, we first specify the context and related works in Sect. 2. We then define some terminology, explain the criteria used to evaluate the plausibility, and present our algorithm in Sect. 3. We finally explain in Sect. 4 how it was implemented and comment about a selection of results that we obtained.

2 Context and Related Work

An autonomous vehicle is a vehicle that has the capability to perceive the surrounding environment, and to make decisions that are transmitted to actuators, so as to drive without human intervention [12]. To this end, it is equipped, among other things, with a set of sensors allowing to analyze the current situation with sufficient accuracy to make a good decision. Sensors may be divided into categories relative to their ranges [21]: proximity sensors (e.g. ultrasounds), short-range sensors (e.g. cameras, short-range radars), medium range sensors (e.g. LiDARs, medium range radars), long Range Sensors (e.g. long-range radars) and location sensors (e.g. GPS). Autonomous vehicle manufacturers offer a wide range of sensors in order to respect a cost/accuracy/range compromise.

Communicating vehicles are vehicles that use communication technologies to exchange data (e.g. position, speed, intent) with each other [2]. Transmissions travel either directly between vehicles (Vehicle to Vehicle or V2V communication), or through an infrastructure (Vehicle to Infrastructure or V2I communication). Since vehicles may exchange information about their position, speed, steering wheel angle, etc., we may consider the reception of messages from other vehicles as an additional sensor, whose information can be shared with other sensors.

2.1 Data Fusion

Since the same information can be inferred from various sensors, we do have redundancy of information, hence the need to merge the data to increase the robustness of the system, which is called data fusion [5]. There are a lot of different approaches in data fusion, and a lot of different classifications altogether, depending in particular on the type of data that are processed (raw data, features, decisions), and the type of data that are produced.

In our case, we are interested in both redundant and complementary data, that we fuse in a *feature in-feature out* (FEI-FEO) approach (also known as feature fusion, symbolic fusion, information fusion or intermediate level fusion), according to Dasarathy's classification [6]. In this approach, a set of features is processed so as to "improve, refine or obtain new features" [5]. We thus assume that we work with features, i.e., values that have already passed through the data analysis process at the lower level. The vehicle has to make a single decision, and the process is thus centralized.

The main constraints that we have to face are the following: the fusion has to occur in *real-time* since the vehicle does not have the possibility to stop so as to make its next decision; the result of the fusion has to be as precise and *reliable* as possible since the safety issues, for the passengers and the other users of the road are high; the fusion process has to be *fault-tolerant* since sensors may be subject to failures and/or attacks. In addition, we wish the fusion process to be *verifiable* [1] and *explainable* [8].

2.2 Related Work

We briefly mention the most popular approaches to data fusion, and how they relate to our case.

Probabilistic models need a priori knowledge (Bayesian inference [15,19]), and are vulnerable to attacks (Dempster-Shafer [7,11]). Artificial intelligence approaches based on neural networks [13] need a large amount of learning data and also lack explainability. This is also the case for other approaches based on genetic programming that tackle the issue of fault tolerance [3]. Similarly, approaches based on fuzzy logic [14,17] are not suitable because the system depends on an inference engine designed by human experts, and thus stability is not guaranteed.

Since part of the information available to the vehicle comes from other vehicles, we may also have to take into account the trust that we have in this data. Trust-based approaches [4,10,22] provide tools primarily applied to multi-agent systems, based on reputation.

3 Proposed Approach

In this section, we propose a model allowing to build a coherent scene for the *Ego* vehicle in real-time, enabling it to make well-informed decisions. This vehicle receives features from a set of sensors, and also from other vehicles surrounding it. The model has to take into account the uncertainty of the information related to the environment and the agents, and to withstand possible attacks.

3.1 Terminology and Outline of the Study

At this point, it is useful to define the terminology that we will use in the remaining of the paper.

- A *feature* is a measure that describes some aspect of an entity in the environment. In the case of autonomous vehicles, features may be the position or speed of the vehicle.
- The *scene* of a vehicle is a set of features that represents both its state and the environment.
- The *Ego* vehicle is the reference vehicle, which tries to build its own scene.
- The features are produced by *information sources*, which may include a variety of elements: it may be a sensor whose raw data are processed to produce the corresponding feature, a collection of complementary features that are combined to produce the target feature, or a prediction about the probable value of a given feature computed from a past scene.

In order to understand the method and algorithm more easily, we will first specify the outline of the study in terms of *sensors*, *features* and *sources of information.*

The reference vehicle, *Ego*, has the following sensors (gathered in Table 1 with their characteristics, as provided by the manufacturers): three ultrasonic sensors: two on the sides and one behind; four cameras: one behind, three forward (one in the middle and two on the mirrors); a LiDAR; a GPS; and a V2V communication sensor. The features that are taken into account are the *Position* (longitudinal and lateral with respect to the road) and *Speed* (longitudinal and lateral with respect to the road). The sources of information can be:

- an individual sensor: each sensor of the *Ego* vehicle (GPS, LiDAR, Camera, Ultrasonic sensors) is considered as a source of information that computes one or several features;
- the V2V communication module: the surrounding vehicles may communicate their features (position, speed) to the *Ego* vehicle;
- complementary features: a feature can be computed from the values of two or more complementary features. For example, in order to have the absolute position of a vehicle V_i, we can combine the relative position of the vehicle V_i, given by a camera of *Ego*, with the absolute position of *Ego* given by the GPS.
- the prediction module: the value of the features may be estimated thanks to the information in the previous scene(s).

Table 2 presents the sources of information to be taken into consideration when calculating the *Position* features. The checked boxes indicate that the corresponding source of information is to be considered for the calculation of this feature, while δ represents the margin of error of the source.

Table 1. Ego's sensors with their respective tolerances δ.

Sensor	Feature	Sampling	Angle	Range	δ
GPS	Pos_{abs}, $Speed_{abs}$	25 Hz	–	–	1 m
LiDAR	Pos_{rel}, $Speed_{rel}$	20 Hz	360°	150 m	1%
Camera	Pos_{rel}, $Speed_{rel}$	20 Hz	120°	200 m	0.5%
Ultrasound	Pos_{rel}	20 Hz	120°	4 m	0.05%
V2V	Pos_{abs}, $Speed_{abs}$	1 Hz	360°	1 km	1 m

Table 2. Sources of information to calculate the *Position* features of *Ego* and vehicles V_i.

Source of information	Pos_{abs} Ego	Pos_{abs} V_i	Pos_{rel} V_i	δ
GPS	✓			1 m
V2V communications		✓		1 m
LiDAR			✓	1%
Camera			✓	0.5%
Ultrasound			✓	0.05%
(GPS, V2V communications)			✓	1 m + 1 m
(GPS, LiDAR)		✓		1 m + 1%
(GPS, camera)		✓		1 m + 0.5%
(GPS, ultrasound)		✓		1 m + 0.05%
(V2V communications, LiDAR)	✓			1 m + 1%
(V2V communications, camera)	✓			1 m + 0.5%
(V2V communications, ultrasound)	✓			1 m + 0.05%
Prediction	✓	✓	✓	0

3.2 Criteria for Selecting Sources

Since we have, potentially, lots of different ways to compute the same feature, the problem consists in selecting, at any time, the most reliable and plausible source of information. To this end, we define criteria that characterize both the quality of the source in general (relevance and trust), and the quality of a given information in particular (freshness and consistency). These criteria are used to compute a global *Confidence* for each information:

- **Relevance**: Is a given source of information well suited to measure a given feature?
- **Trust**: Did the source provide information that was considered as correct in a recent past?
- **Freshness**: Is the information considered as recent or out of date?
- **Consistency**: Is the information consistent with the previous values of the same source?

Relevance of the Source. Each sensor has been designed for a given purpose. But one sensor, which is very relevant to measure a given feature and used mainly in this way, may also be used, although in a less relevant manner, to measure another feature. This relevance is represented by a static percentage defined beforehand for each information source and each feature.

For example, to define the position of Ego, we can use either the GPS or the combination of values coming from a LiDAR (which provides the relative position of a vehicle A with respect to vehicle Ego) with the absolute position communicated by A, or make a prediction based on the last known position and speed of Ego. We can consider that the GPS is perfectly suited to the measurement of this feature, since it has been designed to this end. It can also be assumed that the combination of the relative and absolute positions of A is less suitable because of the potential errors during communication, and also because information received from A may be considered as less trustworthy. Finally, the prediction is a default choice when other sources of information are considered as faulty and should be rated as the least relevant source of information.

Relevance thus defines a partial order between the sources of information for each feature (100% being the most relevant) to favor specific sources of information with respect to others.

Trust in the Source. Trust in the source is a percentage that reflects the quality of the measurements provided by the source during a limited time window corresponding to a near past. When a source provides a measure that is considered as correct, the trust increases, and conversely it diminishes if the measure is considered incorrect, according to Eq. (1).

Trust in the source at iteration t, expressed as a percentage, depends on its value at iteration $t-1$ and on the distance between the tolerance interval of the value coming from the source (IT_s) and the tolerance interval of the value that has been selected (IT_r) from all possible sources for this feature (a tolerance is fixed for each feature). It is assumed that a penalty must be greater than a reward, in order to quickly detect a malfunction of a source or misleading information sent by a malicious vehicle. It is thus not necessary to rely on any reputation system: a new vehicle will be given at first a medium trust; if the information sent by the vehicle is coherent, the trust will quickly increase, but if it is not, the source will be strongly penalised and quickly discarded.

$$
\begin{aligned}
Trust_0 &= 100 \\
Trust_t &= \begin{cases} \min(100, Trust_{t-1} + \sigma^{++}) & \text{if } R \in IT_s \\ \min(100, Trust_{t-1} + \sigma^{+}) & \text{if } R \notin IT_s \text{ and } IT_r \cap IT_s \neq \emptyset \\ \max(0, Trust_{t-1} + \sigma^{-}) & \text{otherwise} \end{cases}
\end{aligned}
\tag{1}
$$

with $\sigma^- < 0 < \sigma^+ < \sigma^{++}$ and $|\sigma^-| > |\sigma^{++}|$.

Freshness of Information. The sources of information are not synchronous. Each source produces features at its own pace, and each value is given a measure of freshness: a value that has just been received at the time of calculation has a freshness of 100%, a value that has an age that exceeds a threshold is considered out of date (see Eq. 2). We consider a fixed freshness for a time d, before a linear decay, according to a gradient a that may be variable depending on the feature.

$$Freshness(age) = \begin{cases} 100 & \text{if } age \leq d \\ 100 - a(age - d) & \text{otherwise} \end{cases} \tag{2}$$

To prevent the system from relying solely on its predictions, especially in the case where no reliable source is available, the prediction value is given a freshness, which is the age of the scene from which the prediction has been computed.

Consistency of Information. To ensure consistency between the various scenes selected at each moment t, we add a percentage that takes into account the predicted value val_p, the measured value val_m, and the margin of error $toleranceError_m$ associated with val_m. The predicted value is calculated from a bounded time window, using a linear regression over the values measured by a sensor. The distance between val_p and val_m is then calculated. If this distance is less than a certain threshold, a percentage associated with this distance is assigned, otherwise it is considered that this new measured value val_m is not consistent with the predicted value val_p:

$$Consistency = 200 - \frac{|val_p - val_m|}{toleranceError_m + 1} \cdot 100 \text{ with } 0 \leq Consistency \leq 100 \tag{3}$$

Confidence. Based on the four aforementioned criteria, we calculate a global measure of *Confidence* for each source of information according to the following formula:

$$Confidence = \frac{Relevance \cdot Trust \cdot Freshness \cdot Consistency}{100^4} \cdot 100 \tag{4}$$

3.3 The Algorithm

At each iteration and for each feature, given the set of values provided for this feature by all the sources of information, the algorithm selects the unique value R in the following steps:

1. calculate, for each source, the value of the feature and a measure of *Confidence* in the value.
2. select, among all the sources, the one for which the *Confidence* in the value is the best and store the (plausible) value R of the selected source. In case of equality, we choose the source whose value is higher in the following order: *Relevance* > *Trust* > *Freshness*. In case no source of information has a *Confidence* above a predefined threshold ζ, the emergency stop is launched.

3. update the trust in all the sources, by attributing rewards or penalties, according to Eq. (1). This enables to quickly disregard faulty or malicious sources, but also to allow a source to have temporary failures (e.g. the GPS in a tunnel), without definitely blacklisting it. *Trust* in Prediction is a special case: it is given the trust value of the source selected at this iteration.

One may argue that some sources of information rely, for the calculation of a given feature, on the values of other features, which may create mutual dependencies between the different sources of information. To overcome this problem, if some required feature is not available yet, we can use instead the predicted value for this feature.

4 Implementation and Experimentation

We did a large number of simulations using GAMA [18], a free and open-source agent-based simulation environment, with its "vehicleBehaviorExtension" [16], which we have enriched with a set of sensors. Using this framework, we implemented the plausibility algorithm described in this paper, as well as the decision algorithm proposed in [1]. We have also implemented a simplified version of the action module, which makes it possible to obtain flexible longitudinal and lateral trajectories.

The aim of the experimentation was to validate that the Ego vehicle could always maintain a coherent scene at any time, whatever the perturbations or attacks arising on the sensors. As a first step, we assumed that internal sensors could not be hacked, and that the only attacks could come from communications. We consider in our scenario the following anomalies:

- Noise: a random value in the range of the error margin $[-\delta, +\delta]$ of the information source is added to the actual value;
- Breakdown: a source of information does not provide information at time t;
- Fault: the actual value is replaced by a random one.

Our test scene contains two vehicles (Ego and A). Ego has an initial speed of 35 m/s and is controlled by our decision algorithm. Vehicle A has only a V2V sensor, has an initial speed of 28 m/s, and is controlled by IDM (intelligent driving model) [20]. The road consists of two lanes with a length of 1 km. Both vehicles are on the same road lane, Ego behind A at some distance, so that Ego has to overtake A before the end of the lane.

Our algorithm uses a set of parameters, which have a direct impact on the performance of the system. For now, we did not achieved any parametric study but selected these values empirically, in order to obtain desired properties (e.g. the time after which a source is disregarded, or considered again after recovering). In the following, we quote the most important ones, with the corresponding values:

- the rewards and penalties σ^{++}, σ^{+}, σ^{-}, are set respectively to 25, 10 and -30. They have an impact on the time the system takes to eliminate a faulty source of information and to reconsider a previously eliminated source;

- the source filter threshold $\zeta = 30\%$;
- the delay d and the gradient a a in the calculation of freshness: d is equal to the period of the sensor; we assume that this delay remains at 100% as long as the information is available in the period, while beyond this period, it decreases rapidly with gradient $a = 8$.

4.1 Results of Experiments and Discussion

We first studied the nominal case, i.e., the case with no faults except the noise associated with the sensors. Figure 1 represents the evolution of the selected values of the relative position of vehicle A with respect to Ego. It is very similar to that of actual values, confirming the choices that were made by the algorithm. The change of sources of information around 12k cycles corresponds to the loss of perception of the vehicle A during an overtaking. At first, A is seen by the front camera, then by the right camera, and finally by the rear camera.

We then inject targeted and random faults. Targeted injection aims at analyzing the ability of the solution to withstand potentially critical cases, while random injection tests the solution empirically.

Targeted Fault Injection. Figure 2 shows the impact of fault injection to camera (1) (primary source) and camera (2) (secondary sensor). We injected two faults in each sensor, which consist in random values being provided by the sensors, in the intervals [500 ms, 900 ms] and [2100 ms, 2800 ms] for camera (1) and [750 ms, 1050 ms] and [1900 ms, 2400 ms] for camera (2). One may observe that the system rejects the faulty source as soon as the error is injected (camera (1) is rejected at time 500 ms and camera (2) is rejected at time 750 ms). The system selects a third source while waiting for the faulty sources to resume, which is the case for camera (1) at time 1900 ms, 1 s after the end of the first fault. At time 2100 ms, camera (1) is rejected again as a second fault occurs, and since camera (2) is also faulty at that time, camera (3) resumes. Camera (2) resumes at time 3400 ms and camera (1) resumes at time 3800 ms, 1s after the end of their respective fault interval.

Random Fault Injection. In this configuration, we inject faults and failures at random times.

- For failures, which may occur at any time, we observe that if occurrence probability is under 50%, the system always manages to find a reliable source of information and that the number of switches increases proportionally to the probability of failure.
- For faults, which consist in random values appearing at random moments, if probability of occurrence is under 5% for each information source, we observe that the system always finds a reliable source and a value to select. Indeed, the frequency of injection is very low (on average two injections per second), which allows the system to penalize and then reward the source on average in

two cycles. Similarly, since injections do not last long, trust is restored very quickly, and the system thus remains operating.

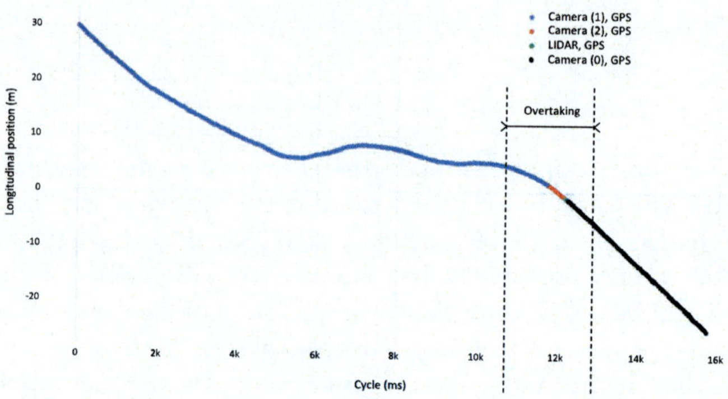

Fig. 1. Selected value R of relative position of vehicle A with respect to Ego: nominal case

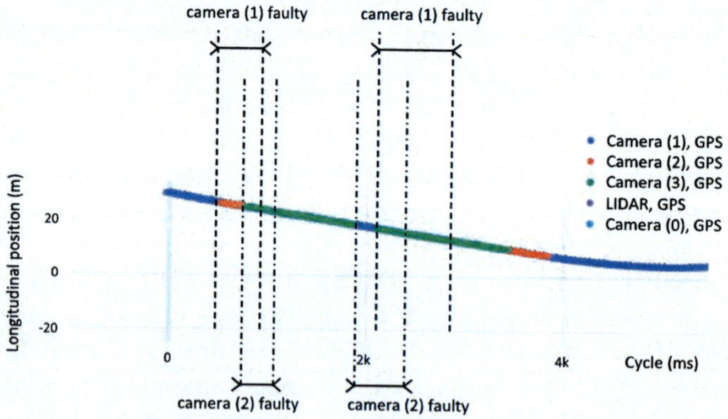

Fig. 2. Relative position of vehicle A wrt Ego: fault injection into camera (1) and camera (2)

Limitations. Although our proposal gives promising results in most situations, its operation is quite sensitive to the choice of parameters. It may happen for example that the system becomes unstable: when an incorrect value has once been selected, the system penalizes all sources of information that are giving correct values; therefore, once the fault injection phase is over, it cannot find a reliable source to select, either because of a low trust, or because of a low consistency.

This issue may be overcome by implementing several improvements to the method: the first one would be to compute the consistency of a feature with more elaborated methods than a simple linear regression. Kalman filters are good candidates since they would allow to take noise and errors into account; a second possible improvement is to also assess the consistency of the value of a feature with respect to the values provided by the other sources of information for the same feature; A third improvement would be, instead of selecting a single value as the most plausible one, to average the values of all the sources of information that appear to be consistent in the measure of the feature, which is common in the field and would limit the impact of incorrect values.

Finally, a parametric study is necessary in order to fine-tune the algorithm. By understanding precisely the role of each parameter, this will enable on the one hand to improve and stabilize the results, and on the other hand to give the user the choice between different driving options, all of which guaranteeing a safe behavior.

5 Conclusion and Perspectives

The objective of this project was to design an algorithm to increase the robustness of the decision of an autonomous vehicle, taking into account the redundancy of the information collected by the sensors. To this end, we proposed a mechanism to select in real time, among the values coming from the various sensors, the most plausible ones to be used in the construction of the scene, in order to have the best possible decision making.

The approach is partially inspired by trust-based approaches, and is adapted to the particular context of our project and its constraints. This resulted in a hybrid model that analyzes the operation of sensors (or more generally sources of information), as well as the variation of the measured or computed values. This model takes into account the relevance of a source for a feature, the trust in the source, the freshness of the information and the consistency of the information in time.

In order to assess the validity of the proposal, we conducted an intensive campaign of experiments. Only a few results are presented in this article, but the model has demonstrated a highly satisfying behaviour and has met the expectations in a large range of conditions, either in the nominal case or in the presence of various forms of failures.

Some limitations have clearly been identified, but one of the main strength of the algorithm lies in its modular design, which enables a lot of adaptations and tuning for the computation of each of the four criteria. In the process of developing the algorithm, some parts of the model have deliberately been kept simple (and even naive) in a first step and the parameterization has been done very quickly and empirically. This was to validate the general principle of the computation of a global measure of confidence in the values provided by the various sources of information, and the selection of the most plausible one, without focusing on details of implementation or specific optimizations. Now that the approach has proved to be valid, we will concentrate on these details, notably to improve the computation of the consistency of the values.

Acknowledgment. This work has received the financial support of SystemX, under research project CTI "Cybersecurity for Intelligent Transport".

References

1. Arcile, J., Devillers, R., Klaudel, H.: VerifCar: a framework for modeling and model checking communicating autonomous vehicles. Auton. Agent. Multi Agent Syst. **33**(3), 353–381 (2019)
2. Arena, F., Pau, G.: An overview of vehicular communications. Future Internet **11**(2), 27 (2019)
3. Bentley, P., Lim, S.L.: Fault tolerant fusion of office sensor data using cartesian genetic programming. In: IEEE Symposium Series on Computational Intelligence (SSCI), pp. 1–8, November 2017
4. Bhuiyan, T., Josang, A., Xu, Y.: Trust and reputation management in web-based social network. In: Web Intelligence and Intelligent Agents, pp. 207–232 (2010)
5. Castanedo, F.: A review of data fusion techniques. Sci. World J. **704504**(10), 2013 (2013)
6. Dasarathy, B.V.: Sensor fusion potential exploitation-innovative architectures and illustrative applications. Proc. IEEE **85**(1), 24–38 (1997)
7. Dempster, A.P.: A generalization of Bayesian inference. J. Roy. Stat. Soc.: Ser. B (Methodol.) **30**(2), 205–232 (1968)
8. Došilović, F., Brcic, M., Hlupic, N.: Explainable artificial intelligence: a survey. In: 41st International Convention on Information and Communication Technology, Electronics and Microelectronics (MIPRO), May 2018
9. Frey, T., Aguilar, C, Engebretson, K., Faulk, D., Lenning, L.: F-35 information fusion. In: Aviation Technology, Integration, and Operations Conference, June 2018
10. Granatyr, J., Botelho, V., Lessing, O.R., Scalabrin, E.E., Barthàs, J.-P., Enembreck, F.: Trust and reputation models for multiagent systems. ACM Comput. Surv. (CSUR) **48**(2), 27 (2015)
11. Jiang, W., Wei, B., Qin, X., Zhan, J., Tang, Y.: Sensor data fusion based on a new conflict measure. Math. Prob. Eng. **5769061**(01), 2016 (2016)
12. Jo, K., Kim, J., Kim, D., Jang, C., Sunwoo, M.: Development of autonomous car–part I: d istributed system architecture and development process. IEEE Trans. Industr. Electron. **61**(12), 7131–7140 (2014)
13. Kolanowski, K., Świetlicka, A., Kapela, R., Pochmara, J., Rybarczyk, A.: Multisensor data fusion using elman neural networks. Appl. Math. Comput. **319**, 236–244 (2018)
14. Majumder, S., Pratihar, D.K.: Multi-sensors data fusion through fuzzy clustering and predictive tools. Expert Syst. Appl. **107**, 165–172 (2018)
15. Rubin, D.B.: Bayesian inference for causal effects: the role of randomization. Ann.Stat. **6**(1), 34–58 (1978)
16. Sobieraj, J.: Méthodes et outils pour la conception de Systèmes de Transport Intelligents Coopératifs. Ph.D. thesis, Université Paris-Saclay (2018)
17. Soleymani, S.A., Abdullah, A.H., Zareei, M., Anisi, M.H., Vargas-Rosales, C., Khan, M.K., Goudarzi, S.: A secure trust model based on fuzzy logic in vehicular ad hoc networks with fog computing. IEEE Access **5**, 15619–15629 (2017)
18. Taillandier, P., Gaudou, B., Grignard, A., Huynh, Q., Marilleau, N., Caillou, P., Philippon, D., Drogoul, A.: Building, composing and experimenting complex spatial models with the gama platform. GeoInformatica **23**(2), 299–322 (2019)
19. Taylor, C.N., Bishop, A.N.: Homogeneous functionals and Bayesian data fusion with unknown correlation. Inf. Fus. **45**, 179–189 (2019)
20. Treiber, M., Hennecke, A., Helbing, D.: Congested traffic states in empirical observations and microscopic simulations. Phys. Rev. E **62**(2), 1805 (2000)

21. Yan, C., Xu, W., Liu, J.: Can you trust autonomous vehicles: contactless attacks against sensors of self-driving vehicle. In: DEF CON, vol. 24 (2016)
22. Zhang, J.: A survey on trust management for VANETs. In: 2011 IEEE International Conference on Advanced Information Networking and Applications, pp. 105–112. IEEE (2011)

A Lightweight Pedestrian Detection Model for Edge Computing Systems

Wen-Hui Chen[1](✉), Han-Yang Kuo[1], Yu-Chen Lin[2], and Cheng-Han Tsai[3]

[1] Graduate Institute of Automation Technology, National Taipei University of Technology, Taipei, Taiwan
whchen@ntut.edu.tw

[2] Department of Automatic Control Engineering, Feng Chia University, Taichung, Taiwan

[3] Mechanical and Mechatronics Systems Laboratories, Industrial Technology Research Institute, Hsinchu, Taiwan

Abstract. Most vision-based pedestrian detection systems adopted deep learning approaches with the framework of convolutional neural networks (CNNs) to reach the state-of-the-art detection accuracy. As CNN-based approaches are computationally intensive, the deployment of those systems to a resource-limited edge device is a challenging task, especially for a real-time application such as intelligent vehicles. In this study, we proposed a lightweight high-performance edge computing solution to achieve rapid and accurate performance in pedestrian detection. Experimental results showed that the proposed framework can effectively reduce the miss rate of the YOLO-tiny detection model from 48.8% to 26.2% while achieving an inference speed of 31 frames per second (FPS) tested on the Caltech pedestrian dataset.

Keywords: Edge computing · Convolutional neural networks · Pedestrian detection

1 Introduction

According to the World Health Organization (WHO) report, road traffic accidents were the leading cause of death and injury. It indicated that several hundred thousand people lost their lives on roads each year [1]. The growth of automobiles over the past decade has contributed to the rise of the accident rate. With the pervasion of digital technology, advanced driver assistance systems (ADAS) have gained popularity in the automotive industry in recent years. Nowadays, many modern vehicles have equipped with some sort of ADAS functions to provide drivers with safer, better, and more comfortable driving experience.

Pedestrian detection is one of the key ADAS functions to safety control and collision avoidance. The main challenges of pedestrian detection include human body articulation, occlusion, the changes of illumination and angle of view, and varying in appearance as well as scales. For example, people can look different in different clothes or postures.

Y. Dong et al. (Eds.): DCAI 2020, AISC 1237, pp. 102–112, 2021.
https://doi.org/10.1007/978-3-030-53036-5_11

In addition, lighting variations can also influence the image pixel values of an object, leading to the challenge of the object detection tasks in computer vision.

The design of robust features that can precisely describe a pedestrian became a key step to build a detection system. Most detection algorithms in the early period of detection research heavily relied on feature design accompanied by a learning classifier. Haar-like features [2] and the histogram of oriented gradients (HoG) [3, 4] were used in well-known milestone detectors around 2001 and 2005, respectively. Based on HoG features, deformable part models [5, 6] were the state-of-the-art pedestrian detection framework before convolutional neural networks (CNN) came into play [7]. Traditional detection approaches were considered as the handcrafted features-based models. Although they have demonstrated some promising results, they still lacked the generalized discriminative ability.

The rise of CNN ignited the progress of object detection and has become the mainstream approach for pedestrian detection since 2012 [8–10]. The strength of CNN-based deep learning models is its ability to feature learning that can automatically extract higher-level features from input data, which sets deep learning-based approaches apart from traditional machine learning models. However, CNN-based learning models were computationally extensive, so there was an issue when deploying developed models to a resource-limited device.

In this study, we proposed a lightweight high-performance pedestrian detection architecture for practical considerations. The remainder of this paper was organized as follows. Section 2 provided a review of the related work on pedestrian detection, followed by a detailed description of the proposed approach in Sect. 3. The experimental setup and results were presented in Sect. 4. Finally, conclusions were drawn in Sect. 5.

2 Related Works

2.1 Deep Learning-Based Detection Approaches

The success of deep learning that outperforms conventional approaches in various applications has made it become an active research topic in recent years. For example, deep learning approaches won computer vision competitions on ImageNet [10] and speech recognition on the popular TIMIT dataset [11].

In object detection, deep learning-based approaches outperform traditional detection models and have become the mainstream since the resurgence of neural networks in 2012. Pedestrian detection in this period also achieved state-of-the-art results by using CNN-based deep learning approaches. R-CNN is the first CNN-based approach applied to object detection tasks [12]. It used selective search to find object candidate boxes, called object proposals and re-scaled them to a fixed size in order to feed into a CNN model to extract object features. Then, a linear SVM was used to predict the presence of the object.

In 2015, an improvement of R-CNN called Fast R-CNN was proposed to avoid the redundant feature computation problem in the R-CNN model by calculating the feature map from the entire image only once [13]. It trains a detector as well as a bounding box regressor over the same network. Shortly after the Fast R-CNN was proposed, an end-to-end detector, called Faster R-CNN debuted to overcome the speed bottleneck of

Fast R-CNN by introducing Region Proposal Networks to generate object proposals with a unified network framework [14]. There were other variants based on Faster R-CNN such as the introduction of Feature Pyramid Networks to detect objects with various scales [15]. Deep learning-based approaches can be divided into two categories: one-stage and two-stage models. The R-CNN family is the most representative of two-stage CNN based detection models.

You Only Look Once (YOLO) [16, 17] and Single Shot MultiBox Detector (SSD) [18] were two typical representatives for one-stage detection models. Well-known for its simple and fast detection speed, YOLO was first proposed in 2015. Unlike two-stage models, YOLO performs the detection task using features from the whole image without generating region proposals. Although YOLO can achieve high detection speed, it suffers from a drop of localization accuracy compared to two-stage detection models. In addition, YOLO could fail in detecting small objects because each grid cell lies in the detected image only contains a limited predetermined set of anchor boxes.

The study of deep learning and the expansion of its applications are still growing rapidly as many researchers attempt to apply it to traditional domains as well as new fields. However, the performance brought by deep learning comes with the cost of computation and memory requirements, leading to a gap between the development and deployment of deep learning algorithms, especially for real-time applications on embedded systems.

2.2 Design of Lightweight Detection Models

High performance CNN-based detection models have deeper convolutional layers, so the model itself has huge parameters and the learning algorithm requires many floating-point operations. Those algorithms were originally designed for standard desktop computers with graphics processing units to accelerate and perform model training and inference. Therefore, there was an issue when deploying the detection model to resource-limited embedded systems, especially for real-time applications.

Cloud computing is a common way to meet both the computation and memory requirements in the development of CNN-based approaches. However, cloud computing has latency, scalability, and privacy issues. Hence, it is not suitable for applications that require real-time inference, such as analyzing data from intelligent vehicles. Edge computing is a distributed paradigm and a feasible way to deal with these issues. However, how to move deep learning algorithms to edge nodes that have limited computing power, memory, and storage space posts another challenge to be dealt with.

Many efforts have been made in providing a solution to deep learning algorithms running on mobile or embedded devices in recent years. Some of which rely on specific hardware chips dedicated to accelerate the inference speed; others provide the design of new architecture of CNN models with less parameters, reduced layers, and fewer operations to alleviate the memory and storage requirements, while preserve performance accuracy, such as SqueezeNet and MobileNets [19–21]. In addition to lightweight architecture design, model compression [22, 23] is another way to develop lightweight models for resource-limited platforms.

Although the above approaches provide a feasible way to obtain lightweight models through model compression, it is not a trivial task for average people to code manually into an edge device. Hence, some companies provided users with software development kits to alleviate the difficulties of implementing compressed models to edge devices. For example, Google released Tensorflow Lite for users to easily deploy deep learning models on mobile and IoT devices [24]. Nvidia Jetson TX2 boards that come with software tools for users to deploy deep learning models. Xilinx released the Deep Neural Network Development Kit (DNNDK) for its edge boards. The OpenVINO toolkit for Intel processors and some heterogeneous chips is also an option to deploy deep learning models to edge devices.

3 The Proposed Approach

There are some well-known deep learning-based detection models, such as YOLOv3, SSD, Faster R-CNN, and RetinaNet. To fairly evaluate the performance of those models, we implemented and tested them on the same platform. The specification of the test environment was listed in Table 1. The detection performance was tested on the Caltech pedestrian dataset and followed the scenarios provided by the dataset. There are four different scenarios according to the pedestrian scale in an image, including *near* (80 pixels or more in height), *medium* (between 30 and 80 pixels), *far* (30 pixels or less), and *reasonable* (50 pixels or more in height or no more than 35% occluded). Figure 1 shows the miss rate versus false positive per image (FPPI) curves for various detection models at the *reasonable* scale. Table 2 listed the performance of various detection models at the *reasonable* scale.

As we focus on real-time applications, the algorithm that has fast inference speed is preferred. From the test results, we can observe that the YOLO family takes the lead in inference speed but needs more improvement in accuracy compared to other detection algorithms. In addition, it can be seen that an image with high resolution can increase the accuracy but slow down the inference speed.

Table 1. Specification of the test environment.

CPU	Intel® Core™ i5-7500 3.40 GHz
Memory	2133 DDR4 16G
GPU	RTX 2070
Software packages	Python 3.6, cuda 10.0, cudnn 7.5, Tensorflow-gpu 1.13.1, Keras 2.2.4, opencv-python 4.1.0, numpy 1.16.2, Torch 1.0.1

YOLO is a fast object detection algorithm. Since its debut in 2014, some improvements have been made to YOLOv2 and YOLOv3 for better detection accuracy. The inference time of the latest version, YOLOv3 can reach 22 ms at 320 by 320 image size. However, the test speed released from their inventors was based on the standard computer with a GPU acceleration hardware platform. The YOLOv3-tiny model is a downsized version of YOLOv3, which is more suitable for running on resource-limited systems but pays the cost of accuracy degradation. The original YOLOv3-tiny network

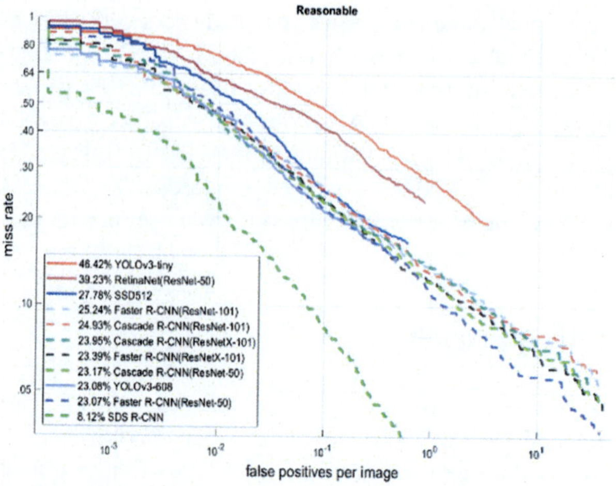

Fig. 1. The miss rates vs. false positive per image (FPPI) for various detection models at the *reasonable* scale.

Table 2. Comparison of different detection models at the reasonable scale.

Models	Miss rate	FPS
YOLOv3-tiny	46.4%	250
RetinaNet (ResNet-50)	39.2%	37.3
RetinaNet (ResNet-101)	39.1%	25.2
SSD512	27.8%	23.3
YOLOv3-608	23.1%	38.5
Faster R-CNN (ResNet-50)	23.0%	15.2
Faster R-CNN (ResNet-101)	25.2%	13.3
Faster R-CNN (ResNetX-101)	23.4%	2
Cascade R-CNN (ResNet-50)	23.2%	10.8
Cascade R-CNN (ResNet-101)	24.9%	9.8
Cascade R-CNN (ResNetX-101)	24.0%	2
SDS R-CNN	8.1%	4.98

architecture was listed in Table 3. YOLOv3-tiny constructs a feature pyramid at two different scales, which can raise a problem when applied to pedestrian detection [25].

We implemented different feature layers of YOLOv3-tiny and tested them at different image sizes to empirically determine the best model parameters to design a lightweight detection model. We observed that more feature layers do not guarantee to have the best miss rate but need more computations. We modified the feature pyramid of original YOLOv3-tiny to extract features with five different scales to improve the detection accuracy. The proposed modified detection architecture was listed in Table 4.

Table 3. The architecture of the YOLOv3-tiny model.

Layer	Type	Filters	Size/Stride	Input	Output
0	Convolution	16	3 × 3/1	416 × 416 × 3	416 × 416 × 16
1	Maxpool		2 × 2/2	416 × 416 × 16	208 × 208 × 16
2	Convolution	32	3 × 3/1	208 × 208 × 16	208 × 208 × 32
3	Maxpool		2 × 2/2	208 × 208 × 32	104 × 104 × 32
4	Convolution	64	3 × 3/1	104 × 104 × 32	104 × 104 × 64
5	Maxpool		2 × 2/2	104 × 104 × 64	52 × 52 × 64
6	Convolution	128	3 × 3/1	52 × 52 × 64	52 × 52 × 128
7	Maxpool		2 × 2/2	52 × 52 × 128	26 × 26 × 128
8	Convolution	256	3 × 3/1	26 × 26 × 128	26 × 26 × 256
9	Maxpool		2 × 2/2	26 × 26 × 256	13 × 13 × 256
10	Convolution	512	3 × 3/1	13 × 13 × 256	13 × 13 × 512
11	Maxpool		2 × 2/1	13 × 13 × 512	13 × 13 × 512
12	Convolution	1024	3 × 3/1	13 × 13 × 512	13 × 13 × 1024
13	Convolution	256	1 × 1/1	13 × 13 × 1024	13 × 13 × 256
14	Convolution	512	3 × 3/1	13 × 13 × 256	13 × 13 × 512
15	Convolution	21	1 × 1/1	13 × 13 × 512	13 × 13 × 21
16	YOLO				
17	Route 13				13 × 13 × 256
18	Convolution	128	1 × 1/1	13 × 13 × 256	13 × 13 × 128
19	Up-sampling		2 × 2/1	13 × 13 × 128	26 × 26 × 128
20	Route 19 8				26 × 26 × 384
21	Convolution	256	3 × 3/1	26 × 26 × 384	26 × 26 × 256
22	Convolution	21	1 × 1/1	26 × 26 × 256	26 × 26 × 21
23	YOLO				

4 Experimental Results

The proposed model was deployed on the Nvidia TX2 platform. Figure 2 shows a snapshot of the test environment. The comparison of performance for the proposed model and the YOLOv3-tiny detection model at different feature layers was shown in Fig. 3. The proposed approach, captioned as YOLOv3-tiny-modified in Fig. 3, obtained 26.16% in miss rate outperforming YOLOv3 and YOLOv3-tiny models at various feature layers structure.

Accuracy and inference speed are a trade-off for most deep learning models. In order to verify the inference speed of the proposed approach on embedded devices, we conducted experiments by deploying the detection algorithms on the TX2 platform and

Table 4. The architecture of the proposed model.

Layer	Type	Filters	Size/Stride	Input	Output
0	Convolution	16	3 × 3/1	416 × 416 × 3	416 × 416 × 16
1	Maxpool		2 × 2/2	416 × 416 × 16	208 × 208 × 16
2	Convolution	32	3 × 3/1	208 × 208 × 16	208 × 208 × 32
3	Maxpool		2 × 2/2	208 × 208 × 32	104 × 104 × 32
4	Convolution	64	3 × 3/1	104 × 104 × 32	104 × 104 × 64
5	Maxpool		2 × 2/2	104 × 104 × 64	52 × 52 × 64
6	Convolution	128	3 × 3/1	52 × 52 × 64	52 × 52 × 128
7	Maxpool		2 × 2/2	52 × 52 × 128	26 × 26 × 128
8	Convolution	256	3 × 3/1	26 × 26 × 128	26 × 26 × 256
9	Maxpool		2 × 2/2	26 × 26 × 256	13 × 13 × 256
10	Convolution	512	3 × 3/1	13 × 13 × 256	13 × 13 × 512
11	Maxpool		2 × 2/1	13 × 13 × 512	13 × 13 × 512
12	Convolution	1024	3 × 3/1	13 × 13 × 512	13 × 13 × 1024
13	Convolution	256	1 × 1/1	13 × 13 × 1024	13 × 13 × 256
14	Convolution	256	3 × 3/1	13 × 13 × 256	13 × 13 × 256
15	Route 14 10				13 × 13 × 768
16	Convolution	35	1 × 1/1	13 × 13 × 768	13 × 13 × 35
17	YOLO				
18	Route 13				13 × 13 × 256
19	Up-sampling		2 × 2/1	13 × 13 × 256	26 × 26 × 256
20	Ele-wise Sum 8				26 × 26 × 256
21	Convolution	256	3 × 3/1	26 × 26 × 256	26 × 26 × 256
22	Convolution	21	1 × 1/1	26 × 26 × 256	26 × 26 × 21
23	YOLO				
24	Route 21				26 × 26 × 256
25	Convolution	128	3 × 3/1	26 × 26 × 256	26 × 26 × 128
26	Up-sampling		2 × 2/1	26 × 26 × 128	52 × 52 × 128
27	Ele-wise Sum 6				52 × 52 × 128
28	Convolution	128	3 × 3/1	52 × 52 × 128	52 × 52 × 128
29	Convolution	21	1 × 1/1	52 × 52 × 256	52 × 52 × 21
30	YOLO				
31	Route 28				52 × 52 × 128

(*continued*)

Table 4. (*continued*)

Layer	Type	Filters	Size/Stride	Input	Output
32	Convolution	64	3 × 3/1	52 × 52 × 128	52 × 52 × 64
33	Up-sampling		2 × 2/1	52 × 52 × 64	104 × 104 × 64
34	Ele-wise Sum 4				104 × 104 × 64
35	Convolution	64	3 × 3/1	104 × 104 × 64	104 × 104 × 64
36	Convolution	21	1 × 1/1	104 × 104 × 64	104 × 104 × 21
37	YOLO				

made a comparison of the inference speed in terms of FPS with existing approaches. The experimental results were listed in Table 3. From Table 3, we can observe that the proposed approach not only can obtain a relatively low miss rate but also can reach 31.3 FPS in inference speed (Table 5).

Fig. 2. A snapshot of the experimental setup.

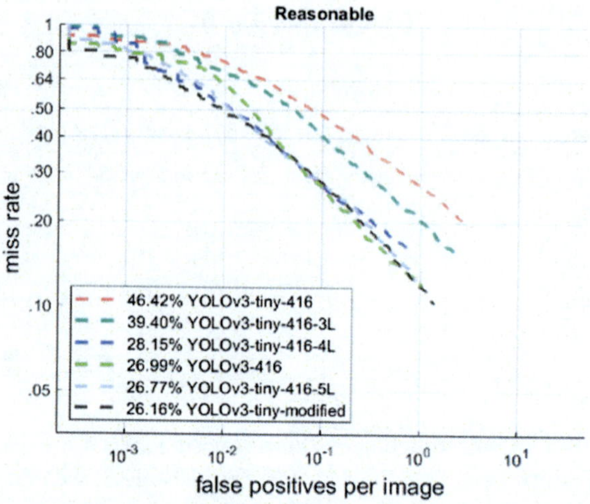

Fig. 3. Results on the Caltech dataset for various YOLOv3 detection models.

Table 5. Performance comparison of detection models on different platforms.

Models	Image size	Platform	Miss rate	FPS
HOG + SVM [26]	1280 × 720	Xilinx ZC702	52.0%	60
HOGLBP + LinSVM [27]	1242 × 375	Nvidia Tegra X1	38.7%	20
YOLOv3	608 × 608	Jetson TX2	23.1%	4.9
YOLOv3	416 × 416	Jetson TX2	27.0%	4.9
YOLOv3-tiny	416 × 416	Jetson TX2	48.8%	33.3
CFPN-1 [28]	416 × 416	Jetson TX2	40.5%	43.5
The Proposed Approach	416 × 416	Jetson TX2	**26.2%**	**31.3**

5 Conclusion

We have developed a lightweight object detection model based on the YOLOv3-tiny architecture for taking advantage of its fast running speed, but improve its detection accuracy to make it more usable. Tested on the Caltech pedestrian dataset, experimental results showed that the proposed approach offers potential contributions to edge computing. First, our model can successfully reach an inference speed at 31 FPS running on the TX2 platform, which is very promising and provides the possibility to be used for real-time applications. Second, the proposed detection algorithm has a relatively low miss rate compared to similar approaches, making a lightweight detection device feasible for real-world applications.

Acknowledgment. This work is supported in part by the Taiwan Ministry of Science and Technology (MOST) under Grant No. MOST 108-2218-E-035-013 and Industrial Technology Research Institute (ITRI) under Grant No. 3000615964.

References

1. World Health Organization. More than 270 000 pedestrians killed on roads each year. https://www.who.int/mediacentre/news/notes/2013/make_walking_safe_20130502/en/. Accessed 11 Oct 2019
2. Viola, P., Jones, M.: Rapid object detection using a boosted cascade of simple features. In: IEEE Conference on Computer Vision and Pattern Recognition (CVPR), Kauai, USA, p. I (2001)
3. Dalal, N., Triggs, B.: Histograms of oriented gradients for human detection. In: IEEE Conference on Computer Vision and Pattern Recognition (CVPR), San Diego, USA, pp. 886–893 (2005)
4. Viola, P., Jones, M. J., Snow, D.: Detecting pedestrians using patterns of motion and appearance. In: 9th IEEE International Conference on Computer Vision, Nice, France, pp. 734–741 (2003)
5. Felzenszwalb, P., McAllester, D., Ramanan, D.: A discriminatively trained, multiscale, deformable part model. In: IEEE Conference on Computer Vision and Pattern Recognition (CVPR), Anchorage, USA, pp. 1–8 (2008)
6. Felzenszwalb, P., Girshick, R., McAllester, D., Ramanan, D.: Object detection with discriminatively trained part based models. IEEE Trans. Pattern Anal. Mach. Intell. **32**(9), 1627–1645 (2010)
7. Krizhevsky, A., Sutskever, I., Hinton, G.: ImageNet classification with deep convolutional neural networks. In: Advances in Neural Information Processing Systems (NIPS), Lake Tahoe, USA, pp. 1097–1105 (2012)
8. Ouyang, W., Wang, X.: Joint deep learning for pedestrian detection. In: IEEE International Conference on Computer Vision, Sydney, Australia, pp. 2056–2063 (2013)
9. Sermanet, P., Kavukcuoglu, K., Pedestrian, S., LeCun, Y.: Pedestrian detection with unsupervised multi-stage feature learning. In: IEEE Conference on Computer Vision and Pattern Recognition, Portland, USA, pp. 3626–3633 (2013)
10. Zhao, Z.Q., Zheng, P., Xu, S.T., Wu, X.: Object detection with deep learning: a review. IEEE Trans. Neural Netw. Learn. Syst. **30**(11), 3212–3232 (2019)
11. Graves, A., Mohamed, A., Hinton, G.: Speech recognition with deep recurrent neural networks. In: IEEE International Conference on Acoustics, Speech and Signal Processing (ICASSP), Vancouver, Canada, pp. 6645–6649 (2013)
12. Girshick, R., Donahue, J., Darrell, T., Malik, J.: Rich feature hierarchies for accurate object detection and semantic segmentation. In: IEEE Conference on Computer Vision and Pattern Recognition, Columbus, USA, pp. 580–587 (2014)
13. Girshick, R.: Fast R-CNN. In: IEEE International Conference on Computer Vision, Santiago, Chile, pp. 1440–1448 (2015)
14. Ren, S., He, K., Girshick, R., Sun, J.: Faster R-CNN: towards real-time object detection with region proposal networks. In: Advances in Neural Information Processing Systems (NIPS), Montreal, Canada (2015)
15. Lin, T., Dollar, P., Girshick, R., He, K., Hariharan, B., Belongie, S.: Feature pyramid networks for object detection, Honolulu, USA, pp. 936–944 (2017)
16. Redmon, J., Divvala, S., Girshick, R., Farhadi. A.: You only look once: unified, real-time object detection. In: IEEE Conference on Computer Vision and Pattern Recognition, Las Vegas, USA, pp. 779–788 (2016)
17. Redmon, J., Farhadi, A.: YOLOv3: an incremental improvement. arXiv:1804.02767v1 (2018)
18. Liu, W., Anguelov, D., Erhan, D., Szegedy, C., Reed, S., Fu, C., Berg, A.: SSD: single shot multibox detector. In: 14th European Conference on Computer Vision (ECCV), Amsterdam, Netherlands, pp 21–37 (2016)

19. Iandola, F.N., Han, S., Moskewicz, M.W., Ashraf, K., Dally, W.J., Keutzer, K.: SqueezeNet: AlexNet-level accuracy with 50x fewer parameters and <0.5MB model size. In: arXiv:1602.07360v4 (2016)
20. Howard, A.G., Zhu, M., Chen, B., Kalenichenko, D., Wang, W., Weyand, T., Andreetto, M., Adam, H.: MobileNets: efficient convolutional neural networks for mobile vision applications. arXiv:1704.04861 (2017)
21. Sandler, M., Howard, A., Zhu, M., Zhmoginov, A., Chen, L. C.: MobileNetV2: inverted residuals and linear bottlenecks. In: IEEE Conference on Computer Vision and Pattern Recognition, Salt Lake City, USA, pp. 4510–4520 (2018)
22. Cheng, Y., Wang, D., Zhou, P., Zhang, T.: A survey of model compression and acceleration for deep neural networks. arXiv:1710.09282 (2017)
23. Chen, G., Choi, W., Yu, X., Han, T., Chandraker, M.: Learning efficient object detection models with knowledge distillation. In: 31st Conference on Neural Information Processing Systems (NIPS), Long Beach, USA (2017)
24. TensorFlow Lite Homepage. https://www.tensorflow.org/lite. Accessed 17 Feb 2020
25. He, W., Huang, Z., Wei, Z., Li, C., Guo, B.: TF-YOLO: an improved incremental network for real-time object detection. Appl. Sci. **9**, 1–16 (2019)
26. Meus, B., Kryjak, T., Gorgon, M.: Embedded vision system for pedestrian detection based on HOG+SVM and use of motion information implemented in Zynq heterogeneous device. In: 23rd International Conference on Signal Processing: Algorithms, Architectures, Arrangements, and Applications, Poznan, Poland, pp. 406–411 (2017)
27. Campmany, V., Silva, S., Espinosa, A., Moure, J.C., Vázquez, D., López, A.M.: GPU-based pedestrian detection for autonomous driving. arXiv:1611.01642 (2016)
28. Chen, P., Hsieh, J., Gochoo, M., Wang, C., Liao, H.: Smaller object detection for real-time embedded traffic flow estimation using fish-eye cameras. In: IEEE International Conference on Image Processing (ICIP), Taipei, Taiwan, pp. 2956–2960 (2019)

Collaborative Recommendations in Online Judges Using Autoencoder Neural Networks

Paolo Fantozzi[1,2] and Luigi Laura[1,3](✉)

[1] Italian Association for Informatics and Automatic Calculus (AICA), Milan, Italy
[2] Sapienza University of Rome, Rome, Italy
fantozzi@diag.uniroma1.it
[3] International Telematic University Uninettuno, Rome, Italy
luigi.laura@uninettunouniversity.net

Abstract. Programming contests such as International Olympiads in Informatics (IOI) and ACM International Collegiate Programming Contest (ICPC) are becoming increasingly popular in recent years. To train for these contests, there are several Online Judges available, in which users can test their skills against a usually large set of programming tasks.

Thus, in order to help the learners, it is crucial to recommend them tasks that are challenging but not unsolvable. In this paper we present a Recommender System (RS) for Online Judges based on an Autoencoder (Artificial) Neural Network (ANN). We also discuss the results of a preliminary experimental evaluation of our approach, trained with the dataset of all the submissions in the Italian National Online Judge, used to train students for the Italian Olympiads in Informatics.

Keywords: Autoencoder neural networks · Recommender systems · Programming Contests

1 Introduction

Programming Contests (PCs) are competitions in which participants are faced a set of tasks that require writing computer programs. Recent literature have emphasized the importance and the effectiveness of PCs in the process of learning computer programming [2,3,5,9,13,20].

In order to train for PCs, learners use Online Judges (OJs), also known as Programming Online Judges, i.e., web based e-learning tools where a user can submit solutions to a programming task. The user choose a task from the many available; after reading its statement, that includes the required formatting of input and output or the use of a programming interface, the user he writes a code to solve the task. The code is submitted to the OJ, that verifies both the correctness, usually by testing it against a certain number of test cases, and the

Y. Dong et al. (Eds.): DCAI 2020, AISC 1237, pp. 113–123, 2021.
https://doi.org/10.1007/978-3-030-53036-5_12

efficiency, by checking that the running time and/or the memory usage is under some limit. In Fig. 1 is shown an example of a programming task.

However, choosing the right task is becoming a complex problem, and an example of an *information overloading scenario*, as observed in [22]: an unexperienced user has to choose from thousands programming tasks, many of which are probably beyond his current abilities. For example, University of Valladolid Online Judge has more than 200k users and 2k tasks, whilst SPOJ accounts approximately 600k users and 6k (public) tasks.

With so many available tasks, it is important to help users selecting their next task by using a Recommender System (RS). Traditionally, RS are broadly divided into two categories: Content Based ones, in which the recommendations derive from features of the items to be suggested, and Collaborative Filtering approaches, in which the suggestion is based on the items chosen by users *similar to the current one.*

As observed in [4], there are some peculiarities of Online Judges that prevent the use of a general Recommender System:

- users slowly improve their abilities, one task after the other, so the general concept of user *preferences* does not apply: recommending a movie or a novel differs significantly from recommending a task; a user will probably still like a novel after one year, whilst he might find a task too easy after the same amount of time.
- Users with *similar* skills, i.e. users to whom we might want to suggest the same set of tasks, might behave very differently in OJs, thus preventing us from considering them *similar.* For example, one might solve all the tasks involving a given skill, while the other might just solve one task, related to that skill, and then move on to tasks involving different skills.

In this paper we propose a task recommender system based on an Autoencoder Neural Network (ANN). We trained and tested the ANN using data from the Online Judge used in the Italian Olympiads in Informatics (Olimpiadi Italiane di Informatica - OII) [11], targeted at secondary school students training. The preliminary experimental results confirm the effectiveness of our approach.

This paper is organized as follows: the next section provides the necessary background related to programming contests, online judges, and recommender systems, whilst our approach is detailed in Sect. 3. In Sect. 4 we detail our experimental findings and concluding remarks are addressed in Sect. 5.

2 Related Works and Background

In this section we discuss related work and the necessary background concerning programming contests, online judges, and recommender systems.

2.1 Programming Contests and Online Judges

A programming contest is a competition in which contestants are faced with a set of programming tasks, also called problems, to be solved in a limited amount of time and/or with a limited amount of memory usage.

Olimpiadi Italiane di Informatica 2014

Fisciano, 18 – 20 settembre 2014 **taglialegna • IT**

Abbatti S.p.A. (which is the Italian brand of *tearDown INC*) is a big enterprise that works in the field of tree felling. In particular, it's been a few years since it started improving in tearing down *barky trees*, a peculiar kind of trees which is very tall and thick. This particular species grow in a very tidy way: the woods made of these trees are actually a long horizontal line of trunks, placed at one decameter (32, 8 ft) one another. Each one of the trees has a particular height, which is expressed by a positive number (decameters).

Tearing down one of these trees is a very hard thing to do and, although tearDown INC employes the most advanced technologies on the market, it is a very time consuming activity, since barky trees' bark is incredibly thick. The workers have the opportunity to chose in which direction (left or right) the tree should fall after the cut.

Each time a barky tree falls down it hits the trees that haven't been torn down which are placed on its falling trajectory; in other words, it tears down each tree which is closer than its height in the direction of the fall. Since the number of barky trees in this woods is huge this dynamic of the fall creates a domino effect.

In order to be the best enterprise in barky tree felling tearDown INC developed a system which is able to scan the whole wood, choosing which trees should be cut by workers and in which directions they should fall with the aim of cutting all the trees. It's important to recall that it's in the best interest of the enterprise to minimize the number of trees that need to be torn down by workers directly. Your role in this situation is to implement the system for tearDown INC.

In the following we can see an example of a solution of the instance shown in the picture above; it is enough to cut two trees:

Fig. 1. Above and the previous page: an example of a problem from a programming contest; this task is taken from the final contest of the 2014 edition of the Italian Olympiads in Informatics (OII).

A single task can be broken into different subtasks of increasing complexity: basic techniques might be enough to solve, within the given time and/or space limits, some of the subtasks whilst the most difficult ones might require very specific algorithmic techniques and data structures.

Popular programming contests are:

- The International Olympiads in Informatics (IOI), that are an annual programming competition for secondary school students patronized by UNESCO.
- The ACM International Collegiate Programming Contest (ICPC) is a multi-tier, team-based, programming competition operating under the auspices of ACM.
- The very recent International Olympiads in Informatics in Team (IOIT), that started in 2017, that are a team competition, like ACM ICPC, differently from IOI (individual competition). Currently there are only four nations involved: Italy, Romania, Russia, and Sweden.
- Google Code Jam, that is based on multiple online rounds that concludes in the World Finals.
- Facebook Hacker Cup, that is *an annual worldwide programming competition where hackers compete against each other for fame, fortune, glory and a shot at the coveted Hacker Cup.*

Online Judges are web based platforms that provide a large number of programming tasks to be solved. There are several popular OJ platforms, including the already mentioned University of Valladolid Online Judge, Sphere Online Judge (SPOJ), CodeChef, and Peking University Online Judge.

Yera and Toledo [22] present a brief survey on OJs, whilst more information on tools and techniques for automatic evaluation of solutions submitted to OJs can be found in [1,6].

2.2 Recommender Systems in OJs

As already observed in the introduction, despite the large amount of literature devoted to RS, the peculiarities of recommendation in OJs, where the relation user-item is way more complex than the typical RS cases, prevent from using standard techniques and forces the development of ad-hoc methods. This aspect is detailed in the paper of Audrito et al. [4], where the authors propose a first approach on building a RS by tackling the problem of ranking tasks in Online Judges.

Indeed, so far few research focused in the recommendation of tasks in OJs: we mention the traditional collaborative filtering method with a new similarity measure adapted to the case [17], and an approach based on fuzzy logic [22]. Caro and Jimenez considered user-based and similarity-based approaches In [7]. Di Mascio et al. proposed a framework that can allow recommendations and that can foster motivation in students by means of a lightweight, badge-based, gamified approach [12].

2.3 Recommender Systems and Artificial Neural Networks

The use of deep learning techniques for recommender systems is divided in two categories that we call *classical* and *hybrid.*

The classical approach uses the standard architectures of neural networks, applying them to this task. So, in this case, the most important part consists in the formulation of the problem, since that, if the input data are not suitable to be the input of that specific deep learning technique, then the result will be totally inaccurate.

The other approach is the hybrid one, that consists in using more than one type of architecture at the same time. This kind of approach is useful when the input data are not easily representable as a standard structure, like a user-item matrix. In [23] the authors use a Convolutional Neural Network to extract features from images and then an Autoencoder to build the recommender system on the features.

In [24], in order to build a recommender system for hashtag in tweets, the authors use at the same time some CNNs and some Recurrent Neural Networks (RNNs). In this work they use the CNNs to extract features from the image and then the RNNs to extract information from the text, combining them using different weights based on co-attention.

Since that a recommendation task is similar to a dimensionality reduction task, many of the state-of-the-art techniques use some kind of Autoencoder to map the input in a smaller space, that will be the representation of the correlations in the recommender system. In particular ,Sedhain et al. [15] introduce the using of a vanilla Autoencoder to build a recommender system. They use a partial masked input (the same techniques we use in this work) and try to reconstruct it in output, the elements added in the output will be the recommended elements. Strub and Mary [16] extends the work of Sedain et al. [15]: they use a denoising Autoencoder instead of the vanilla Autoencoder to build a more robust system.

Chen and de Rijke [8] follow a similar approach, but they use a Variational Autoencoder to perform top-N recommendation. They encode both the user ratings and some side information in the compact space in the Autoencoder.

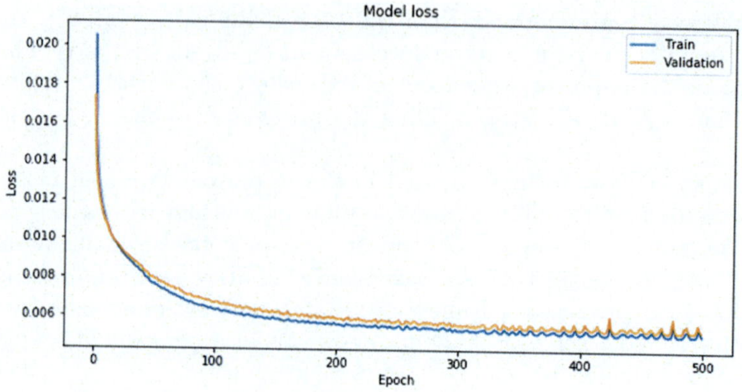

Fig. 2. Model loss.

Zhang et al. [25] generalize the Contractive Autoencoder paradigm into matrix factorization framework. Li et al. [14] combine a probabilistic matrix factorization with Marginalized Denoising Stacked Autoencoders to perform collaborative filtering. This work can be considered as a general framework to use these kinds of techniques; in this context, several works, including [18,19,21], can be viewed as special cases of this framework.

3 Recommending Tasks Using Autoencoder Neural Network

Our goal is to provide recommendations to the users regarding the next task to deal with among all the tasks in a system. To build this kind of system we assume that:

- if a user obtains a score for a task it means that it is the max score possible for that user in that task;
- a user should solve the problems sorted by their grade of difficulty for the user; thus, a user should never try to solve a problem that is much harder than the last one he solved;
- it is possible to deduce the score for a new problem based on the scores the user obtained in other problems;
- given a snapshot of the scores for the users it is not important the order followed by the user to solve the tasks to foresee the score for another task;

Based on the above assumptions, we decided to build a model that takes as input the current scores of a user and provides probable scores for the same user for other tasks. Then we can choose between the forecasted scores and suggest the task with the highest score between them.

If we consider the score as a judgment of the user for the item (i.e, the programming task), then we can just exploit the already known techniques for

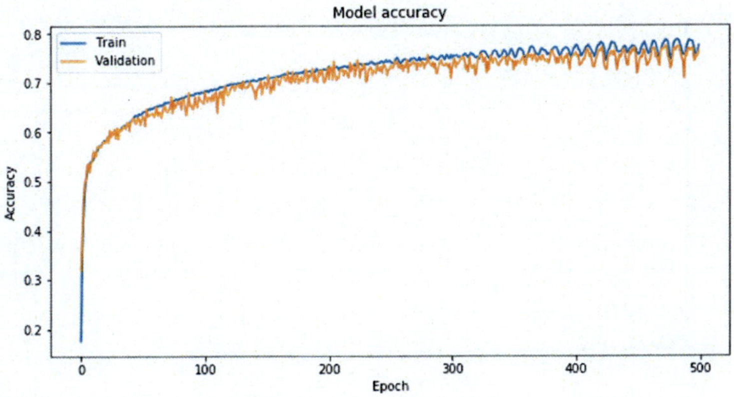

Fig. 3. Model accuracy.

recommending items. We chose to use an autoencoder to build the model. Since that we want to use just the scores of the users, without any information from other sources, we use a masking mechanism on the input: we mask a fraction of the scores in input as a non-solved task, and then we perform backpropagation from the complete scores. In this way, in the bottleneck layer, there should be a compact representation of the similarities of the tasks.

4 Experimental Evaluation

To test the method we have designed, we have taken the submission to the OII Training platform [11] in a defined time range. The submissions were in the form: `<user_id, task_id, datetime, score>`, where each submission corresponds to a possible solution to a task from a user that performs a certain score, based on many test cases. We filtered out all the scores equal to zero because we can't know if they were just users testing the behaviour of the platform. Then we considered only the best score for each task, for each user, to ignore all the attempts to solve the problem before the user found the solution.

We performed a preliminary set of experiments with the original data: we built a user $\times$ task matrix where each cell contains the best score of the user for the task. The result is a 3148×409 matrix with 43051 non-empty cells. The matrix has an average of 105 submissions for each task and 13 for each user. The max number of users which have submitted to the same task is 1070 and the max number of tasks with submissions from the same user is 336. We consider the zero valued cells as a problem with no submission from the user.

To use this matrix as a training set for this model, we duplicated the matrix and then we have randomly masked some positive scores with zero. The masked matrix will be the input to the model and the original matrix will be the output to reconstruct. The number of tasks masked for each user is a random number between 3 and 7, with the constraint that it should be anyway equal at most to the half of the submission for the user.

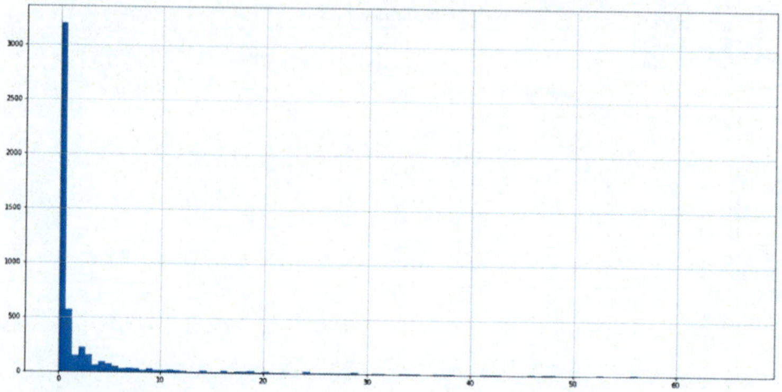

Fig. 4. Distribution of sum of squared errors (SSE).

After the preliminary experiments, it was clear that data was too small, thus we performed an operation of data augmentation. In particular, we have multiplied many times the same rows of the matrix with the result of a matrix with 8 times the rows of the original. Then each row has been randomly masked independently, so we unlikely had duplicated rows in the matrix. We have load all the data on a Google Colab instance with an available GPU. Then we have splitted the data on train and test set with a ratio of 0.8/0.2. We used Tensorflow to build a Sequential model with 5 layers: the input layer, a dense 64 neurons layer, a dense 16 neurons layer, a dense 64 neurons layer, and an output layer with dimension equal to the input layer. All the activations for the layers are ReLU with a constraint of a max value of 1.0 (the max value of the score). We used an Adam optimizer with a learning rate of 0.001 and a mean squared error loss function.

We trained the model for 500 epochs with a batch size of 128, and we have imposed a validation split of 0.2. The resulting learning curve is the shown in Fig. 2, whilst the accuracy curve measured is depicted in Fig. 3.

Since that the standard accuracy doesn't represent well the error of the model (since that we have a sparse matrix) we computed a sum of the squared errors on

Table 1. Statistics of the distribution of SSE (Fig. 4).

Mean	2.27
Std deviation	6.07
Min	0.00
25% (1st quartile)	0.00
50% (2nd quartile)	0.05
75% (3rd quartile)	1.38
Max	66.19

each samples in the test set. The resulting values follow the distribution shown in Fig. 4; in Table 1 we report some stats of the SSE distribution.

Overall, the results seem promising, and we plan to evaluate the effectiveness of this approach in the next Italian Training Camp for IOI, where we will ask learners to evaluate the tasks suggestions against a more classical recommender system built using the Simple Python RecommendatIon System Engine (SurPRISE - http://surpriselib.com); our long term goal is to implement this recommender system inside the OII Training platform [11].

5 Conclusions

In this paper we proposed the design of a recommender system for tasks suggestions in Online Judges, based on a Autoencoder Neural Network.

We trained the ANN with the data from the OJ used by the secondary school students training for the Italian Olympiads in Informatics (Olimpiadi Italiane di Informatica - OII) [10,11].

Our preliminary results seem promising, and we plan to carry on our investigations by asking students, in the next Italian Training Camp for IOI, to evaluate the suggestions made by this ANN based RS against a more classical one (SurPRISE). After this experimental evaluation, we plan to implement this RS inside the OII Training platform.

References

1. Ala-Mutka, K.M.: A survey of automated assessment approaches for programming assignments. Comput. Sci. Educ. **15**(2), 83–102 (2005)
2. Astrachan, O.: Non-competitive programming contest problems as the basis for just-in-time teaching. In: 2004 34th Annual Frontiers in Education, FIE 2004, vol. 1, pp. T3H/20–T3H/24, October 2004
3. Audrito, G., Demo, G.B., Giovannetti, E.: The role of contests in changing informatics education: a local view. Olympiads Inf. **6**, 3–20 (2012)
4. Audrito, G., Mascio, T.D., Fantozzi, P., Laura, L., Martini, G., Nanni, U., Temperini, M.: Recommending tasks in online judges. In: 9th International Conference on Methodologies and Intelligent Systems for Technology Enhanced Learning, MIS4TEL 2019, Avila, Spain, 26–28 June 2019. Advances in Intelligent Systems and Computing, vol. 1007, pp. 129–136. Springer (2019)
5. Blumenstein, M., Green, S., Fogelman, S., Nguyen, A., Muthukkumarasamy, V.: Performance analysis of game: a generic automated marking environment. Comput. Educ. **50**, 1203–1216 (2008)
6. Caiza, J., Del Alamo, J.: Programming assignments automatic grading: review of tools and implementations. In: INTED2013 Proceedings of 7th International Technology, Education and Development Conference, IATED, 4–5 March 2013, pp. 5691–5700 (2013)
7. Caro-Martinez, M., Jimenez-Diaz, G.: Similar users or similar items? Comparing similarity-based approaches for recommender systems in online judges. In: Aha, D.W., Lieber, J. (eds.) Case-Based Reasoning Research and Development, vol. 10339, pp. 92–107. Springer, Cham (2017)

8. Chen, Y., de Rijke, M.: A collective variational autoencoder for top-N recommendation with side information. In: Proceedings of the 3rd Workshop on Deep Learning for Recommender Systems, pp. 3–9 (2018)
9. Dagienė, V.: Sustaining informatics education by contests. In: International Conference on Informatics in Secondary Schools-Evolution and Perspectives, pp. 1–12. Springer (2010)
10. Di Luigi, W., Fantozzi, P., Laura, L., Martini, G., Morassutto, E., Ostuni, D., Piccardo, G., Versari, L.: Learning analytics in competitive programming training systems. In: 2018 22nd International Conference Information Visualisation (IV), pp. 321–325, July 2018
11. Di Luigi, W., Farina, G., Laura, L., Nanni, U., Temperini, M., Versari, L.: oii-web: an interactive online programming contest training system. Olympiads Inf. **10**, 195–205 (2016)
12. Di Mascio, T., Laura, L., Temperini, M.: A framework for personalized competitive programming training. In 2018 17th International Conference on Information Technology Based Higher Education and Training (ITHET), pp. 1–8, April 2018
13. Garcia-Mateos, G., Fernandez-Aleman, J.L.: Make learning fun with programming contests. In: Transactions on Edutainment II, pp. 246–257. Springer (2009)
14. Li, S., Kawale, J., Fu, Y.: Deep collaborative filtering via marginalized denoising auto-encoder. In: Proceedings of the 24th ACM International on Conference on Information and Knowledge Management, pp. 811–820 (2015)
15. Sedhain, S., Menon, A.K., Sanner, S., Xie, L.: AutoRec: autoencoders meet collaborative filtering. In: Proceedings of the 24th International Conference on World Wide Web, pp. 111–112 (2015)
16. Strub, F., Mary, J.: Collaborative filtering with stacked denoising autoencoders and sparse inputs (2015)
17. Toledo, R.Y., Mota, Y.C.: An e-learning collaborative filtering approach to suggest problems to solve in programming online judges. Int. J. Distance Educ. Technol. **12**(2), 51–65 (2014)
18. Van den Oord, A., Dieleman, S., Schrauwen, B.: Deep content-based music recommendation. In: Advances in Neural Information Processing Systems, pp. 2643–2651 (2013)
19. Wang, H., Wang, N., Yeung, D.-Y.: Collaborative deep learning for recommender systems. In: Proceedings of the 21th ACM SIGKDD International Conference on Knowledge Discovery and Data Mining, pp. 1235–1244 (2015)
20. Wang, T., Su, P., Ma, X., Wang, Y., Wang, K.: Ability-training-oriented automated assessment in introductory programming course. Comput. Educ. **56**, 220–226 (2011)
21. Wang, X., Wang, Y.: Improving content-based and hybrid music recommendation using deep learning. In: Proceedings of the 22nd ACM International Conference on Multimedia, pp. 627–636 (2014)
22. Yera Toledo, R., Caballero Mota, Y., Martínez, L.: A recommender system for programming online judges using fuzzy information modeling. Informatics, **5**(2), 17 (2018)
23. Zhang, F., Yuan, N.J., Lian, D., Xie, X., Ma, W.-Y.: Collaborative knowledge base embedding for recommender systems. In: Proceedings of the 22nd ACM SIGKDD International Conference on Knowledge Discovery and Data Mining, pp. 353–362 (2016)

24. Zhang, Q., Wang, J., Huang, H., Huang, X., Gong, Y.: Hashtag recommendation for multimodal microblog using co-attention network. In: IJCAI, pp. 3420–3426 (2017)
25. Zhang, S., Yao, L., Xu, X.: Autosvd++: an efficient hybrid collaborative filtering model via contractive auto-encoders. In: Proceedings of the 40th International ACM SIGIR Conference on Research and Development in Information Retrieval, SIGIR 2017, New York, NY, USA, pp. 957–960. Association for Computing Machinery (2017)

Natural Language Inference in Ordinary and Support Verb Constructions

Ignazio Mauro Mirto(✉)

Università di Palermo, 90128 Palermo, Italy
ignaziomauro.mirto@unipa.it

Abstract. The family of clause types known as 'support (or 'light') verb construction' (SVC) manifests a peculiar syntax-semantics interface if compared with ordinary verb constructions (OVC). If, in e.g. *She laughed*, the verb licenses an argument and assigns it a semantic role, syntacticians of every stripe nowadays agree that it is the noun *laugh*, in *She gave a laugh*, which fulfils the same function. The differences between the two types have been extensively discussed in the linguistics literature (systematic research started in the 1970s), less so in Computational Linguistics. This paper has two objectives. First, it will propose an innovative type of semantic role, which is termed *Cognate Semantic Role* (CSR) because the verb employed in the notation is etymologically related to the predicate licensing arguments. *She laughed* and *She gave a laugh* therefore express *the same* role >the-one-who-laughs<, assigned by *laughed* and *a laugh* respectively. Second, it will introduce a tool capable of extracting CSRs automatically from both OVCs and SVCs; thus a device will be used for detecting the construction type. CSRs offer a number of advantages for the formalization of entailments and paraphrases and for Machine Translation.

Keywords: Non-verbal predicates · Semantic roles · Entailments · Paraphrasing

1 Introduction

The following pages propose a tool capable of extracting meaning automatically. Consider the following sentences:

(1) I cani spaventano dei gatti 'Dogs frighten some cats'
(2) I cani si spaventano dei gatti 'Cats frighten dogs'

On the surface, (1) and (2) are identical in all aspects, excluding the presence in (2) of a third person clitic *si*, a reflexive pronoun. However, (1) and (2) express opposite meanings: in the former, dogs frighten cats whilst in the latter cats frighten dogs.

The tool introduced below is capable of returning the semantic core of a sentence. Such a core is expressed by the following pairings between (a) the arguments (NPs and PPs) and (b) the semantic role which each argument carries:

Y. Dong et al. (Eds.): DCAI 2020, AISC 1237, pp. 124–133, 2021.
https://doi.org/10.1007/978-3-030-53036-5_13

Table 1. A semantic representation of sentence (1)

Head of NP		Semantic role
CANI Dogs	→	>chi/cosa **spaventa** qualcuno< >(s)he who/what **frightens** someone<
GATTI Cats	→	>chi viene **spaventato**< >(s)he who is **frightened**<

Table 2. A semantic representation of sentence (2)

Head of NP/PP		Semantic role
CANI Dogs	→	>chi viene **spaventato**< >(s)he who is **frightened**<
GATTI Cats	→	>chi/cosa **spaventa** qualcuno< >(s)he who/what **frightens** someone<

Above, meaning is represented by means of semantic roles called ‘coradicals’ or, better, *cognates* (following the model of *Cognate objects*, see [3]). A cognate semantic role (CSR) is expressed with a notation using (where possible, otherwise suppletive forms are needed) the content morpheme of the predicate which licenses arguments (henceforth, the licenser)[1]. CSRs originate from the valence of predicates and should be defined in a relational way, i.e. as those units of sentence meaning which *any* valid paraphrase must preserve.

Section 2 of this paper introduces the software and shows the parsing of sentences (1) and (2). Section 3 illustrates how the tool distinguishes support verb constructions and ordinary verb constructions. Section 4 regards the Conclusion.

2 The Software: NLP^{YTALY}

The outcomes of the tool we propose, nominated NLP^{YTALY} are available on a website, which can be accessed once a free account has been opened. The software processes Italian sentences (unannotated text), functioning by means of a rule-based system implemented with Python 3.7.

Basic operations are performed with TreeTagger, a statistical parser which only provides words as they appear in the text, their POS and their lemmas. NLP^{YTALY} permits the modification, where necessary, of such outcomes, whether that be the POS or the lemma (numerous errors occur in both). Having inputted the text, the following

[1] We have deployed Cognate Semantic Roles in e.g. [3, 4]. Points of contact can be observed with the *rolesets* from PropBank. We are aware that rule-based systems are considered by many as being rather dated, and we know that the importance of statistical-based approaches cannot be underestimated. No references will be provided of the latter because in our opinion they hardly compare with the former.

will be obtained: (a) an identification of inflected verbal phrases and proclitics (if any); (b) an identification of the diathesis on a morphological basis (i.e. without lists of verbs, unless they are necessary). The four diathesis types are: passive (*Fui graffiato* 'I was scratched'), middle reflexive (*Mi graffiai* 'I scratched myself') and middle non-reflexive (*Tornammo* 'We came back'), and, by exclusion, active (*I gatti graffiano* 'Cats scratch'); (c) the detection of proper names and of noun phrases introduced by a determiner; (d) the identifying of Subject and Direct object; (e) the disambiguation of some function words (e.g. the use of *del, della, degli* etc. as either prepositions or partitive articles, see below); and lastly, (f) the automatic extraction of CSRs.

2.1 The Software: Sentence (1)

When NLP^{YTALY} analyses a sentence, the software yields three kinds of outcome for as many corresponding areas. The first, named *Results*, contains information which is either syntactic (e.g. chunking) or semantic in nature; the second and third outcomes display tags. Table 3 illustrates the Results of NLP^{YTALY} with sentence (1):

Table 3. The parsing of sentence (1)

Results

The sentence analysed is: "I cani spaventano dei gatti"

Diathesis	**Active**	**Passive**	**Middle +RFL**	**Middle -RFL**
spaventano	✓			

I. Candidates for subjects/direct objects

1. *[(0, 1)]*
2. *[(3, 4)]*

II. Syntactic functions (putative)

1. *The subject of 'SPAVENTANO' is 'I CANI'.*
2. *The direct object of 'SPAVENTANO' is 'DEI GATTI'.*

III. Semantic Roles from ordinary verbs:

1. *The subject of SPAVENTANO is >chi|cosa spaventa<*
2. *The direct object of SPAVENTANO is >chi|cosa è spaventato<*

IV. Function of the lemma 'DEL'

1. *DEI is a partitive article*

Below, we provide a description of the box showing the diathesis and of each of the four parts indicated with Roman numerals:

- the verb phrase (i.e. *spaventano* 'frighten.3.PL') is isolated and its diathesis is calculated; in (1) it is active;
- Part I, i.e. Candidates for Subjects/Direct Objects, provides the indexes of the noun phrases occurring in the text;
- Part II identifies the syntactic function for each candidate in Part I;
- Part III associates a semantic role to each of the syntactic functions. Since the licenser may be non-verbal (as in support verb constructions), an automatic procedure will

become necessary in order to identify it (see Sect. 3). Above, the Subject is matched with the CSR >the one who frightens someone<, whilst the referent of the Direct Object is >the one who is frightened<;

- Part IV displays the results of a disambiguation script for the series *<del, dello, della, dell', dei, degli, delle>*, that is words which in Italian can function either as prepositions coalesced with a definite article (e.g. *Parlò dei professori* 'She spoke about the professors') or as partitive articles (e.g. *Intervennero dei professori* 'Some professors intervened'); in (1), *dei* works as a partitive article and *dei gatti* 'some cats' is the Direct object.

The *New Tags*, with at least eight columns, provide information related to e.g. constituency and morphosyntactic features such as number, person, and gender. The second column in Table 4, i.e. *Block*, shows the constituency for sentence (1):

Table 4. The New Tags for sentence (1) in NLP^{YTALY}

New tags							
Index	Block	Occurrence	POS	Lemma	Gender and number	Person	Notes
0	1	i	DET:def	il	mp	3a	–
1	1	cani	NOM	cane	mp	3a	–
2	2	spaventano	VER.pres	spaventare	−p	3a	–
3	3	dei	DET:part	del	mp	3a	–
4	3	gatti	NOM	gatto	mp	3a	–

This obtains by isolating the VP and by assigning a progressive number to each block. The following three columns in Table 4 are those determined by TreeTagger (with an important difference, discussed below), whilst the others provide the above-mentioned features and some notes, when necessary.

Finally, the *Original Tags*, with three columns only, display those tags as produced by TreeTagger, as illustrated in Table 5:

Table 5. The Original tags for sentence (1) in NLP^{YTALY}

Original tags		
Occurrence	POS	Lemma
i	DET:def	il
cani	NOM	cane
spaventano	VER.pres	spaventare
dei	PRE:det	del
gatti	NOM	gatto

Noteworthy is the POS difference concerning *dei*, which TreeTagger invariably tags as PRE:det. By comparing Tables 4 and 5, it can be seen that the POS in the former is DET:part, namely a partitive article. Such a change is made possible by a function of NLP^{YTALY} which disambiguates the use of the words in the series and modifies the POS accordingly.

2.2 The Software: Sentence (2)

The *Results* area for sentence (2) is shown in Table 6 below. A number of differences between the *Results* boxes for sentence (1) and sentence (2) immediately stand out. First of all the diathesis:

Table 6. The Results for sentence (2)

Results

The sentence analysed is: "I cani si spaventano dei gatti"

Diathesis	**Active**	**Passive**	**Middle +RFL**	**Middle -RFL**
si spaventano			✓	

I. **Candidates for subjects/direct objects**
   1. *[(0, 1)]*
   2. *[]*

II. **Syntactic functions (putative)**
   1. *The subject of 'SI SPAVENTANO' is 'I CANI'.*

III. **Semantic Roles from ordinary verbs:**
   1. *The subject of SI SPAVENTANO is >chi|cosa è spaventato<*

IV. **Function of the lemma 'DEL'**
   1. *DEI introduces a genitive licensed by SPAVENTARSI*

The verb is third person plural and preceded by a clitic which is also third person, which means that (a) the diathesis is Middle+RFL; (b) the sentence is intransitive; and (c) the verb *spaventare* still licenses two arguments, but the structure is finally unaccusative. Turning now to the Roman numerals, Part I shows the candidate detected on the left of the VP and no candidate (the empty list in I.2) for Direct objects. In Part II the Subject of the sentence is found, but no Direct object, in accordance with the candidates shown in Part I. Part III illustrates the CSR assigned to the Subject, which is the same CSR assigned to the *Direct object* of sentence (1). Part IV shows the function of *dei*, i.e. the preposition *di* 'of' coalesced with the article *i* 'the'.

The Tables below display the tags for sentence (2), obtained first with NLP^{YTALY} (Table 7) and then with TreeTagger (Table 8):

Table 7. The New Tags for sentence (2)

New tags							
Index	Block	Occurrence	POS	Lemma	Gender and number	Person	Notes
0	1	i	DET:def	il	mp	3a	–
1	1	cani	NOM	cane	mp	3a	–
2	2	si	CLIT:refl	si	−slp	3a	–
3	2	spaventano	VER.pres	spaventare	−p	3a	–
4	3	dei	PRE:det	del	mp	3a	–
5	3	gatti	NOM	gatto	mp	3a	–

Table 8. The Original tags for sentence (2)

Original tags		
Occurrence	POS	Lemma
i	DET:def	il
cani	NOM	cane
si	PRO:refl	si
spaventano	VER.pres	spaventare
dei	PRE:det	del
gatti	NOM	gatto

3 Support vs. Ordinary Verbs

It is unquestionably true that the sentences below convey distinct meanings due to the different verbs ((4) is ambiguous):

(3) *Max ha mosso alcune obiezioni al capo*
Max has moved some objections to the boss
'Max objected to the boss'

(4) *Max ha riferito alcune obiezioni al capo*
a. 'Max reported his objections to his boss' (*riferire* is a three-place verb)
b. 'Max reported on some objections to the boss' (*riferire* is a two-place verb)

However, this is not in an obvious way. The meaning of (4) can be expressed as with (1) and (2) because its CSRs depend on the verb. One line of research couched in well-known theoretical frameworks (see [1]) has investigated dozens of languages over a 50-year period and it has established that the semantic roles in clauses such as (3) depend on the post-verbal *obiezioni* 'objections'. Expressed with CSRs, such differences can be summarised as follows:

a) whilst the verb in (4) is *riferire* 'to report' and the Subject Max is >he-who-reports(on-something)<, the verb in (3) is *muovere* 'to move', but the Subject Max is *not* >he-who-moves(something)<; in (3) the verb *muovere* has no literal meaning, as is often the case in SVCs; on the other hand, the meaning which *riferire* conveys in (4) is literal;
b) on a par with (a) above, the Direct object of (4), i.e. *alcune obiezioni*, is paired with >what-is-reported<, a semantic role which derives from the verb, whilst the Direct object of (3) is not paired with >what-is-moved<;
c) in (3) Max conveys the semantic role >he-who-objects-to-something<; this CSR formation makes use of the verb *obiettare*, which is cognate to *obiezioni* 'objections', the head of the Direct object;
d) unlike (3), the noun *obiezioni* in (4) plays no role in determining the CSR which the Subject Max conveys.

In other words, (3) guarantees *unambiguous knowledge* of the person who made the objection, whilst (4) does not. As already noted, whilst Max, the referent of the Subject in (3), undoubtedly is >he-who-objects<, in (4) >he-who-objects< could be *anyone*. Clearly, in no case can this semantic difference pass unobserved. On the one hand, the two clauses display the same linear succession of constituents and POS. On the other hand, the semantic roles of (3) cannot be obtained in the same way as those in (4). If CL aims to provide accurate semantic representations of clauses such as (3) and (4), the depicted similarities and the same sequence 'Subject + Verb + Direct object + Indirect object' will not assist in achieving this goal. In order to correctly obtain paraphrases, machine translations etc., the tool used must contain and exploit *additional information*. An examination of Tables 9 and 10 will reveal which type of additional information will differentiate the two structures and obtain appropriate semantic representations by exploiting CSRs. In Table 9, the summary of syntax and semantics (referring to the meaning in (4a)) is obtained as in Tables 3 and 6: (a) the verb phrase has been isolated; (b) the diathesis has been calculated; (c) the candidates for Subject and Direct object have been identified (Part I); (d) syntactic functions have been assigned to such candidates (Part II); and, finally, (e) CSRs have been associated to the nuclear syntactic functions (Part III).

Table 9. The Results for sentence (4)

Results

The sentence analysed is: "Max ha riferito alcune obiezioni al capo"

Diathesis	Active	Passive	Middle +RFL	Middle -RFL
ha riferito	✓			

I. **Candidates for subjects/direct objects**

1. *[(0,)]*
2. *[(3, 4)]*

II. **Syntactic functions (putative)**

1. *The subject of 'HA RIFERITO' is 'MAX'.*
2. *The direct object of 'HA RIFERITO' is 'ALCUNE OBIEZIONI'.*

III. **Semantic Roles from ordinary verbs:**

1. *The subject of HA RIFERITO is >chi|cosa riferisce<*
2. *The direct object of HA RIFERITO is >chi|cosa è riferito<*

The summary for sentence (3), shown in Table 10, is identical to that in Table 9 with regard to the following operations: the identification of the verb phrase, the assigning of the diathesis and the recognition of the syntactic functions. However, the semantic roles in this summary prove to be markedly diverse.

Table 10. The Results for sentence (3)

Results

The sentence analysed is: "Max ha mosso alcune obiezioni al capo"

Diathesis	**Active**	**Passive**	**Middle +RFL**	**Middle -RFL**
ha mosso	✓			

I. **Candidates for subjects/direct objects**
   1. *[(0,)]*
   2. *[(3, 4)]*

II. **Syntactic functions (putative)**
   1. *The subject of 'HA MOSSO' is 'MAX'.*
   2. *The direct object of 'HA MOSSO' is 'ALCUNE OBIEZIONI'.*

III. **Semantic Roles**
   1. *First semantic role: the subject of 'HA MOSSO' is >chi obietta qualcosa<*
   2. *Second semantic role: CAPO is >a chi|cosa si obietta qualcosa<*

The heading of Part III in Table 9 makes it explicit that CSRs derive from an *ordinary verb construction*, just as occurs in Tables 3 and 6 for (1) and (2) respectively. This means that the predicate licensing arguments is a verb and that its content morpheme (in bold in Tables 1 and 2) is used for CSR formation. However, the same heading in Table 10 is not explicit, which means that the licenser is not a verb. The reasons for this outcome can be observed in the *New Tags* area for sentence (3) in Table 11, to be compared with the *New Tags* for sentence (4) in Table 12:

Crucial is the different tag at index 4 in the *Notes* column, regarding *obiezioni*: only in Table 11 is the tag PRED. This is because in sentence (3) the predicate licensing syntactic functions is the noun *obiezioni*, since it combines with *muovere*, whose valence is blank. The verb does not determine the CSRs, which instead depend on the predicate noun *obiezioni*. The difference in the Tables is made possible by a dictionary which lists the nouns that, combined with certain verbs, function as sentence-level predicates, giving rise to support verb constructions. The dictionary also specifies which prepositions will be used for additional arguments (if any). Notice that the predicate noun itself does *not* receive a CSR, but rather *assigns* semantic roles. It mandatorily assigns a CSR to the Subject, whereas the presence of other CSRs depends on the noun's valence.

Table 11. The New Tags for sentence (3)

New tags									
Index	Block	Occurrence	POS	Lemma	Gender and number	Person	Notes	Diathesis	RSyn
0	1	max	NPR	max	–s	3a	–	None	SOGG
1	2	ha	VER:pres	avere	–s	3a	AUX	ACTIVE	None
2	2	mosso	VER:pper	muovere	ms	–	–	ACTIVE	None
3	3	alcune	PRO:indef	alcun|alcuno	fp	3a	–	None	OD
4	3	obiezioni	NOM	obiezione	fp	3a	PRED	None	OD
5	4	al	PRE:det	al	ms	3a	–	None	None
6	4	capo	NOM	capo	ms	3a	–	None	None

Table 12. The New Tags for sentence (4)

New tags									
Index	Block	Occurrence	POS	Lemma	Gender and number	Person	Notes	Diathesis	RSyn
0	1	max	NPR	max	–s	3a	–	None	SOGG
1	2	ha	VER:pres	avere	–s	3a	AUX	ACTIVE	None
2	2	riferito	VER:pper	riferire	ms	–	–	ACTIVE	None
3	3	alcune	PRO:indef	alcun|alcuno	fp	3a	–	None	OD
4	3	obiezioni	NOM	obiezione	fp	3a	–	None	OD
5	3	al	PRE:det	al	ms	3a	–	None	None
6	3	capo	NOM	capo	ms	3a	–	None	None

4 Conclusion

In relation to the extraction of meaning, we maintain that: (a) a semantic role must match a syntactic function; (b) the semantic core of a sentence can be broken down into CSRs. The basic meaning of a sentence can be expressed as the sum of the CSRs assigned by the predicate[s]. Above, the sentences (1) and (2) are represented with two units because their licenser is a two-place predicate. Sentence (4), whose licenser is a ditransitive predicate, will be broken down into three units of meaning. On the other hand, sentence (3) expresses two units because its licenser, a noun, is a two-place predicate. The Tables 1 and 2 illustrate how CSRs can be used with reference to MT. Once CSRs have been extracted, a suitable content morpheme of the target language will be found (e.g. ***spaventare*** - to ***frighten***). The translation will then rest on identifying clauses in the target language which convey the same CSRs and have an appropriate syntactic finish (with regard to diathesis and the packaging of information). CSRs are useful in capturing entailments, and therefore paraphrases, e.g. in: diathesis alternations (see (1) and (2)), OVC-SVC pairs (see (3) and (4)), double valence of the same verbal root, as in *The government disbanded the secret society/The secret society disbanded*, or in pairs involving mail order unaccusatives (as Carol Rosen facetiously called them[2]), for example *She assembled the parts/The parts assemble in minutes.* Other such pairs, traditionally analysed as transformations, involve passives, clefts, pseudoclefts, nominal phrases, etc. CSRs appear in strict correlation with a 55-year old observation by Z. S. Harris: «To what extent, and in what sense, transformations hold meaning constant is a matter for investigation» ([2]: 203). The proposal advanced in this paper is that, regarding transformations (an old fashioned term with too many interpretations), CSRs are what remains constant. We would contend that, in addition to Fillmore's Agent, Patient etc., there are advantages if semantic roles are expressed by using the content morpheme of the predicates licensing arguments.

References

1. Gross, M.: Les bases empiriques de la notion de prédicat sémantique. *Langages* **63**, 7–52 (1981)
2. Harris, Z.S.: Transformations in linguistic structures. Proc. Am. Philos. Soc. **108**(5), 418–422 (1964). In: Hiż, H. (ed.) *Papers on Syntax*. D. Reidel Publishing Company, London (1981)
3. Mirto, I.M.: Oggetti interni e *reaction objects* come nomi predicativi di costrutti a verbo supporto. *Écho des Études Romanes* **7**(1), 22–47 (2011)
4. Mirto, I.M.: Dream a little dream of me. Cognate predicates in English. In: Camugli, C., Constant, M., Dister, A. (eds.) Actes du 26e Colloque International Lexique-Grammaire, Bonifacio, Corse, 2–6 October 2007, pp. 121–128 (2007). http://infolingu.univ-mlv.fr/Colloques/Bonifacio/proceedings/mirto.pdf

[2] "Because mail order sales catalogs abound in examples of it" (C. G. Rosen, p.c., 2012).

Comparative Analysis Between Different Automatic Learning Environments for Sentiment Analysis

Amelec Viloria[1(✉)], Noel Varela[1], Jesús Vargas[1], and Omar Bonerge Pineda Lezama[2]

[1] Universidad de la Costa, Barranquilla, Colombia
{aviloria7,nvarela2,jvargas41}@cuc.edu.co
[2] Universidad Tecnológica Centroamericana (UNITEC), San Pedro Sula, Honduras
omarpineda@unitec.edu

Abstract. Sentiment Analysis is a branch of Natural Language Processing in which an emotion is identified through a sentence, phrase or written expression on the Internet, allowing the monitoring of opinions on different topics discussed on the Web. The study discussed in this paper analyzed phrases or sentences written in Spanish and English expressing opinions about the service of Restaurants and opinions written in the English language about Laptops. Experiments were carried out using 3 automatic classifiers: Support Vector Machine (SVM), Naïve Bayes and Multinomial Naïve Bayes, each one being tested with the three data sets in the Weka automatic learning software and in Python, in order to make a comparison of results between these two tools.

Keywords: Comparative analysis · Automatic learning · Sentiment analysis

1 Introduction

There are different methods for the study of Sentiment Analysis, however, the most studied is Automatic Learning and there are studies that apply different ways of using such learning. Below, some participations in the research of Opinion Mining are shown [1–11].

For the development of this research, three supervised machine learning classifiers are considered. Some of them are used in the literature mentioned above and show good results in Sentiment Analysis. Support Vector Machine (SVM), Naïve Bayes and Multinomial Naïve Bayes are the ones studied in this research. A unigram is used for each one as a lexical characteristic and a weighing scheme as TF-IDF, three data sets with two domains and two languages are used. Additionally, a comparison is made between two types of machine learning languages such as Python and Weka.

Y. Dong et al. (Eds.): DCAI 2020, AISC 1237, pp. 134–141, 2021.
https://doi.org/10.1007/978-3-030-53036-5_14

2 Proposed Approach

The following stages or phases are carried out for the proposed objectives: Pre-processing, Characteristic Extraction, Training Phase, Test Phase and Evaluation. The main feature is the use of the Term Frequency - Inverse Document Frequency (TF-IDF), which represents the number of documents in which a given term is calculated. To classify the polarities in the opinions, supervised automatic learning is used with the classifiers: Support Vector machine, Naïve Bayes and Multinomial Naïve Bayes [12–14]. The three proposed phases are described in detail below.

2.1 Pre-processing

- Collection of opinions. Extract only opinions from documents in XML format.
- Purification of opinions. Process in which opinions are obtained free of empty words, punctuation marks, accents and isolated characters [15].
- Tokenization. Tokenize every opinion by word.
- Stemming. Process in which the words of each opinion are reduced, often eliminating derived affixes.
- Opinion filtering. In this part, the training opinions are classified by the possible polarities: positive, negative, neutral and conflict.

2.2 Characteristic Extraction

The lexical characteristic of the Term Frequency - Inverse Document Frequency (TF-IDF) [16] is a technique that indicates the relevance of a word with respect to the selected document and the corpus in general, which allows the qualification of the documents of the corpus based on the key words, that is, if the words have more weight, then it means that the document is more related to the words than one with the same words but with less weight.

2.3 Training Phase

With the characteristics obtained in the previous phase, the training process is carried out. According to the classification algorithm, the model is built and then used in the test phase. The classes considered are four: positive, negative, neutral and conflict. The supervised classification algorithms used in this phase, both in WEKA and in Python, are: Support Vector Machine, Naïve Bayes and Multinomial Naïve Bayes [14].

2.4 Test Phase

The test data are classified according to the model proposed by the classifier. The polarities proposed to each opinion of the test data are compared with the data of the standard gold.

2.5 Evaluation

To measure the results of the experiments, Precision and Recall are used as evaluation measures to evaluate the results of the algorithm using the SVM classifier. The results of the Multinomial Naïve Bayes classifier, and the measure of Precision are used for the results of the algorithm with the Naïve Bayes classifier [15].

2.6 Accuracy and Recall

These are measures based on the comparison of an expected outcome and the actual outcome of the system being evaluated [4]. These measures were adapted for the evaluation of classification in sentiment analysis. Precision is measured with Eq. (1) and recall with Eq. (2) [5]:

$$Precision = \frac{t_p}{t_p + f_p}, \tag{1}$$

$$Recall = \frac{t_p}{t_p + f_n} \tag{2}$$

Where:

tp: true positives,
tn: true negatives,
fp: false positives,
fn: false negatives.

The harmonic measurement between the recall and the precision is function F1 (see Eq. 3) [5]:

$$\mathrm{F}_1 = \frac{2\mathrm{pr}}{(\mathrm{p}+\mathrm{r})}, \tag{3}$$

Where:

p: Precision,
r: Recall.

The Precision measure refers to the assessment of prediction bias, i.e. it answers the question, what is the average of correct predictions? The accuracy equation is shown in formula (4) [8]:

$$\mathrm{Accurrancy} = \frac{\mathrm{t_p} + \mathrm{t_n}}{\mathrm{t_p} + \mathrm{t_n} + \mathrm{f_p} + \mathrm{f_n}} \tag{4}$$

3 Obtained Results

This section describes the data used to test the proposed approach, and also shows the results obtained by applying the above to each data set.

3.1 Data Set

The data used for this research were taken from the data sets provided by Semeval 2019 for the solution of Task 5, subtask 2 [9]. 3 different sets were used: Spanish-language restaurant opinion set, English-language restaurant opinion set, and English-language laptop opinion set.

Table 1 shows the number of opinions in the training data and in the test data for each domain. Table 2 shows the total opinions by type of polarity: positive (Pos), negative (Neg), neutral (Neu) and conflict (Con), in each domain of the training data.

Table 1. Total, of opinions by dataset and domain.

Domain	Training	Try	Gold
Restaurants (English)	3452	1020	1035
Restaurants (English)	2410	452	498
Laptops (English)	2154	741	721

Table 2. Polarity training data set.

Domain	Pos	Neg	Neu	With	Total
Restaurants (Spanish)	2750	402	248	52	3452
Restaurants (English)	1754	407	145	104	2410
Laptops (English)	1352	685	210	93	2154

3.2 Experimental Results

The aim of this article is to compare the use of classifiers in different environments, Weka and Python, testing the same data sets in each.

To measure the results in these experiments, Precision and Recall are used as evaluation measures to measure the results of SVM and Multinomial Naïve Bayes, and the measure of Accuracy used only for the results obtained with the Naïve Bayes classifier.

Results in Python. For the case of the automatic classifier Support Vector Machine (SVM) in Python, the best result is achieved with the data set of the Spanish Restaurants domain obtaining 70% accuracy, in the same way as in the case of the Multinomial Naïve Bayes classifier obtaining 69% accuracy. In the case of Naïve Bayes, 74% accuracy is achieved.

Tables 3, 4 and 5 show in detail the results obtained in each classifier tested and in each data set.

Table 3. Results obtained in Python and SVM.

Domain	Precision	Recall	F1
Restaurants (Spanish)	0.70	0.79	0.73
Restaurants (English)	0.70	0.77	0.70
Laptops (English)	0.65	0.70	0.67

Table 4. Results obtained in Python and Multinomial Naïve Bayes.

Domain	Precision	Recall	F1
Restaurants (Spanish)	0.69	0.72	0.69
Restaurants (English)	0.65	0.74	0.70
Laptops (English)	0.66	0.74	0.69

Table 5. Results obtained in Python and Naïve Bayes.

Domain	Accuracy
Restaurants (Spanish)	0.74
Restaurants (English)	0.70
Laptops (English)	0.56

Results in Weka. By testing the data in the Weka environment, the results obtained are a lower percentage of Accuracy in the English language laptop domain. However, for the set of Spanish-language Restaurants, 0.70% accuracy is achieved with the classifier Naïve Bayes Multinomial and SVM. Tables 6, 7 and 8 with the results obtained for each data set are shown below.

Table 6. Results obtained in Weka and SVM.

Domain	Precision	Recall	F1
Restaurants (Spanish)	0.70	0.74	0.70
Restaurants (English)	0.69	0.74	0.70
Laptops (English)	0.61	0.65	0.67

Table 7. Results obtained in Weka and Multinomial Naïve Bayes.

Domain	Precision	Recall	F1
Restaurants (Spanish)	0.70	0.76	0.72
Restaurants (English)	0.70	0.75	0.71
Laptops (English)	0.69	0.65	0.64

Table 8. Results obtained in Weka and Naïve Bayes.

Domain	Accuracy
Restaurants (Spanish)	0.73
Restaurants (English)	0.60
Laptops (English)	0.44

4 Comparison of Results Between Python and Weka

Having the results with each environment and classifier, based on the results of the F1 harmonic measurement and accuracy, it can be noted that SVM and Naïve Bayes have a better behavior in Python, and with Multinomial Naïve Bayes the algorithm behaves better in Weka (see Table 9). A better automatic learning behavior is obtained using Python, since feature extractions are performed automatically using the data analysis and natural language processing tools scikit-learn and NLTK.

Table 9. Comparison between Python and Weka.

Domain	SVM		NB Multinomial		Naïve Bayes	
	Weka	Python	Weka	Python	Weka	Python
	F1	F1	F1	F1	Accuracy	Accuracy
Restaurants (Spanish)	0.70	0.73	0.72	0.69	0.73	0.74
Restaurants (English)	0.70	0.70	0.71	0.70	0.60	0.70
Laptops (English)	0.67	0.67	0.64	0.69	0.44	0.56

5 Conclusions

This research presents the results obtained in the analysis of sentiments using three supervised learning classifiers: Support Vector machine, Naïve Bayes and Multinomial Naïve Bayes. These algorithms are used to classify opinions from the domains of Restaurants and Laptops. For each opinion, one of the four possible polarities is detected: positive, negative, neutral and conflict. Python with nltk and scikit learn, and Weka are also used. Based on the results obtained, it is shown that the best results are from the classifiers designed in Python. In the case of accuracy in all three domains, the results are better in Python. This behavior is due to the fact that feature extraction, in the case of Python, is with tools built in the language. It is also concluded that with the tests made in this research, the best classifier is SVM with Python, Multimonial Naïve Bayes with Weka and Naïve Bayes with Python. Future research will consider the use of other types of features as well as the use of tools to measure the polarity of words such as SentiWordNet.

References

1. Zhang, Z., Ye, Q., Zhang, Z., Li, Y.: Sentiment classification of internet restaurant reviews written in cantonese. Expert Syst. Appl. **38**(6), 7674–7682 (2011)
2. Billyan, B., Sarno, R., Sungkono, K.R., Tangkawarow, I.R.: Fuzzy K-nearest neighbor for restaurants business sentiment analysis on TripAdvisor. In: 2019 International Conference on Information and Communications Technology (ICOIACT), pp. 543–548. IEEE, July 2019
3. Buitinck, L., Louppe, G., Blondel, M., Pedregosa, F., Mueller, A., Grisel, O., Niculae, V., Prettenhofer, P., Gramfort, A., Grobler, J., Layton, R., VanderPlas, J., Joly, A., Holt, B., Varoquaux, G.: API design for machine learning software: experiences from the scikit-learn project. In: ECML PKDD Workshop: Languages for Data Mining and Machine Learning, pp. 108–122 (2013)
4. Laksono, R.A., Sungkono, K.R., Sarno, R., Wahyuni, C.S.: Sentiment analysis of restaurant customer reviews on TripAdvisor using Naïve Bayes. In: 2019 12th International Conference on Information & Communication Technology and System (ICTS), pp. 49–54. IEEE, July 2019
5. Singh, S., Saikia, L.P.: A comparative analysis of text classification algorithms for ambiguity detection in requirement engineering document using WEKA. In: ICT Analysis and Applications, pp. 345–354. Springer, Singapore (2020)
6. Kumar, A., Jaiswal, A.: Systematic literature review of sentiment analysis on Twitter using soft computing techniques. Concurr. Computa.: Pract. Exp. **32**(1), e5107 (2020)
7. Mulay, S.A., Joshi, S.J., Shaha, M.R., Vibhute, H.V., Panaskar, M.P.: Sentiment analysis and opinion mining with social networking for predicting box office collection of movie. Int. J. Emerg. Res. Manag. Technol. **5**(1), 74–79 (2016)
8. Liu, S., Lee, I.: Email sentiment analysis through k-means labeling and support vector machine classification. Cybern. Syst. **49**(3), 181–199 (2018)
9. Pontiki, M., Galanis, D., Papageorgiou, H., Androutsopoulos, I., Manandhar, S., AL-Smadi, M., Al-Ayyoub, M., Zhao, Y., Qin, B., De Clercq, O., Hoste, V., Apidianaki, M., Tannier, X., Loukachevitch, N., Kotelnikov, E., Bel, N., Jimenez-Zafra, S.M., Eryigit, G.: Semeval-2016 task 5: aspect based sentiment analysis. In: Proceedings of the 10th International Workshop on Semantic Evaluation (SemEval-2016), pp. 19–30. Association for Computational Linguistics, San Diego, June 2016. http://www.aclweb.org/anthology/S16-1002
10. Ahmad, M., Aftab, S., Bashir, M.S., Hameed, N., Ali, I., Nawaz, Z.: SVM optimization for sentiment analysis. Int. J. Adv. Comput. Sci. Appl. **9**(4), 393–398 (2018)
11. Rennie, J.D., Shih, L., Teevan, J., Karger, D.R., et al.: Tackling the poor assumptions of Naive Bayes text classifiers. In: Proceedings of the Twentieth International Conference on Machine Learning, Washington DC, vol. 3, pp. 616–623 (2003)
12. Ahmad, M., Aftab, S., Ali, I.: Sentiment analysis of tweets using SVM. Int. J. Comput. Appl. **177**(5), 25–29 (2017)

13. Ducange, P., Fazzolari, M., Petrocchi, M., Vecchio, M.: An effective decision support system for social media listening based on cross-source sentiment analysis models. Eng. Appl. Artif. Intell. **78**, 71–85 (2019)
14. Iqbal, F., Hashmi, J.M., Fung, B.C., Batool, R., Khattak, A.M., Aleem, S., Hung, P.C.: A hybrid framework for sentiment analysis using genetic algorithm based feature reduction. IEEE Access **7**, 14637–14652 (2019)
15. Silva, J., Varela, N., Ovallos-Gazabon, D., Palma, H.H., Cazallo-Antunez, A., Bilbao, O.R., Llinás, N.O., Lezama, O.B.P.: Data mining and social network analysis on Twitter. In: International Conference on Communication, Computing and Electronics Systems, pp. 401–408. Springer, Singapore (2020)
16. Silva, J., Naveda, A.S., Suarez, R.G., Palma, H.H., Núñez, W.N.: Method for collecting relevant topics from Twitter supported by big data. In: Journal of Physics: Conference Series, vol. 1432, no. 1, p. 012094. IOP Publishing, January 2020

Context-Aware Music Recommender System Based on Automatic Detection of the User's Physical Activity

Alejandra Ospina-Bohórquez, Ana B. Gil-González(✉), María N. Moreno-García, and Ana de Luis-Reboredo

Department of Computer Science and Automation, Science Faculty, University of Salamanca, Plaza de los Caídos. s/n, 37008 Salamanca, Spain
{ale.ospina15,abg,mmg,adeluis}@usal.es

Abstract. The large amount of music that can be accessed in streaming nowadays has led to the development of more reliable music recommendation systems. To this end, context-aware music recommendation systems, capable of suggesting music taking into account contextual information, have emerged. Studies have shown that music helps to improve mood while can change the focus of attention of users during the performance of some activity, helping to make this activity more bearable. This work presents a music Context Aware Recommender System in order to motivate users in their daily activities. Its main purpose is to suggest to the user the most appropriate music to improve the performance of the physical activity at recommending time. The conducted experiments along a case study prove that this system is useful and satisfactory when the activity does not require a great deal of concentration. During activities that required movement, most users indicated that the perceived effort decreases when using the recommendation system proposed. They also indicated that their mood had improved after using this system. This demonstrates the usefulness of this recommender system while doing physical activities.

Keywords: Context-aware recommender systems · Music · Physical activities · Entrainment · Emotional state

1 Introduction

There are studies that show that listening to music improves performance when doing some type of exercise, as well as motivating and distracting users from fatigue [1]. It is impractical to choose these songs that are suitable for listening while doing exercise. In this scenario, the technological solution would be to implement an application with an intelligent an intelligent music recommender system to minimize the effort spent by the user in the search for music customized according his/her activities while minimizing the interaction effort.

Y. Dong et al. (Eds.): DCAI 2020, AISC 1237, pp. 142–151, 2021.
https://doi.org/10.1007/978-3-030-53036-5_15

Recommender systems are designed to provide users with personalized products, being music an important field [2]. In the last years, new recommender systems have emerged taking into account not only the user's tastes but also her/his environment, the so-called Context-Aware Recommender Systems (CARS) [3] as described in more detail in the following sections.

A context-aware music recommender system is proposed in this work, which takes into account the user's activity in order to recommend music that may motivate her/him to continue such activity. To strengthen the knowledge about this proposal a mobile application that detects the posture and movement of listeners has been developed to make music recommendations.

The reminder of this paper is organized as follow: previous research on CARS in music are stated in Sect. 2. Section 3 elaborates the proposal of the physical activity detection, including the recommendation strategy along the data-driven architecture and shows the experimental development while provides some experimental results. Finally, the conclusions are presented in Sect. 4.

2 Context-Aware Recommendation Systems in Music

Prior to the recommendation process, a set of ratings given to items by users, are obtained and stored in the matrix R then, this matrix is used by the recommendation method to predict the ratings that the user would give to the items not rated by him/her, and finally the k items with the highest predicted ratings are recommended to the user.

$$R : User \times Item \rightarrow Rating \tag{1}$$

There are different ways to make the recommendation to the user, one of them is just to make a prediction of ratings for a specific item that a user would assign to that item. On the other hand, there are also the *top-N* recommendations, which recommends an ordered list of N items. Classical recommender systems are classified into three categories according to the methods they use: content-based, collaborative filtering and knowledge-based. In many cases, recommendation systems not only analyse users' preferences, but also take into account the context in which they find themselves. With the help of current mobile devices, different personal, social and contextual factors can be integrated into the recommendation process, so that a more appropriate recommendation can be given to a specific user at the right time based on their current activity or some other factor that can be deduced from the data obtained by the device [4]. The importance of incorporating contextual information into these systems, has led to the emergence of Context-Aware Recommender Systems (CARS). This specialisation on recommenders has begun to be studied by extending the dimensional model ($User \times Item$) and proposing a new multidimensional rating function:

$$R : User \times Item \times Context \rightarrow rating \tag{2}$$

They can be classified into three categories according to how they incorporate contextual information. They are explained below:

Contextual Pre-filtering. With the pre-filtering approach, contextual information is incorporated before the calculation of recommendations by eliminating data that do not correspond to the context being recommended. Context information is used during the selection and construction of data sets. Ratings are then predicted using the two-dimensional recommender systems for the selected data.

Contextual Post-filtering. At these approaches contextual information is ignored when the recommendation is being generated and then the list of recommendations is adjusted in the way that those considered irrelevant for a given context are discarded.

Contextual Modeling. The contextual modeling approach is the one that uses contextual information directly in the multidimensional function as part of the prediction of ratings.

There are a big amount of music recommender systems. Nowadays with the increased capabilities of mobile devices and the immediate availability of different types of information on the Web, the opportunities to include additional contextual information has steadily increased [5]. Some approaches are detailed below.

2.1 Related Work

In this section, we give a description of some approaches to music recommender systems with context feed, their advantages and limitations.

Greenhalgh *et al.* [6] propose the recommendation of songs (*geotracks*) and playlist (*geolists*) aligned and adapted to specific routes, such as the route to the workplace. Reproduction aspects of *geotracks* and the transition between them may be influenced by the user's progress, activity and context. Geotracks are chosen because they are somehow appropriate for the place where they are to be listened to.

Braunhofer *et al.* [7] proposed a location-based recommendation system. The idea is to recommend music that fits with places of interest. For this, they used tags about emotions that users assigned to both songs and places of interest. With this goal in mind, they developed a mobile application that suggested a route and played the music recommended for each place of interest visited.

Moens *et al.* [8] proposed a musical system called *D-Jogger* that uses the movement of the body to select music dynamically and adapt its tempo to the user's pace. They conducted a pilot experiment that suggested that when most users synchronize the musical tempo and the user's step. In addition, a user questionnaire indicated that participants experienced this as both stimulating and motivating.

Oliver *et al.* [9] proposed a mobile personal system (hardware and software) that users could use during physical activity. The hardware of this system includes a heart rate monitor and acceleration wirelessly connected to the mobile phone that the user would carry. The software allows the user to enter an exercise routine, then help the user achieve the objectives of the routine by constantly monitoring heart rate and movement as well as selecting and playing music with specific characteristics that would guide him to achieve the objectives of the routine.

Sen *et al.* [4] proposed a musician recommender system aimed to recommend to novel music users based on contextual information obtained from sensors.

Based on these works, we have selected four characteristics that summarizes the requirements considered interesting for a music recommendation system on the context of physical activity: minimal user interaction, entrainment (BPM), need for a smartphone only and recommendation during physical exercise.

2.2 Adding the Context: Influence of Music on Activity Performance

Establishing contextual information on the use of music will give us the opportunity to develop a tailored and accurate recommendation system. Music as a natural form of expression is fundamental to human beings. Also has been shown music to have a motivational effect that encourages people to exercise more vigorously or for longer periods of time [11].

There are few defining properties of rhythmicity [12]. We point out three of them for the proposal, which are relevant in the task of studying the linkage of music in physical activity. First of this elements is the phenomenon of *entrainment* [8], which talks about how two or more independent rhythmic processes synchronize with each other. For example, during dances in which people perform rhythmic movements in synchrony with the perceived musical pulse or in the case of users who listen to music while performing some activity. It is said that they try to synchronize their activity with the musical tempo, improving the experience.

The second property is the called *Rate of Perceived Exertion scale* (RPE) [13], that recently has been proposed to explain the performance during the exercise. RPE measures the entire range of effort that an individual perceives The RPE scale runs from 0–10. This individual perception is really important, allowing us to implement ways to personalization. It has been noted that some psychological manipulation techniques alter the RPE response during constant exercise.

We ended up with the third and final, the called *Extended parallel process model* (EPPM) [14] which is a model that suggests that when music is listened to during exercise, there is competition from cognitive information, where this information comes from different sources (external conditions, such as music, and internal conditions of the body, such as respiratory rate, ventilation, among others) that compete for attention.

Several works and research suggests that music has very positive effects when you engage in physical activity [9]. Some of these works are referenced below.

Simpson and Karageorghis [15], examined the effect of music on performance during 400-m speed races while controlling mood before the race. Runners who listened to music were shown to perform better during the race. Styns [16] observed that participants in his study walked faster with music than with metronome ticks. In addition to the motivational factor, it is believed that exercises that are repetitive in nature benefit more from music that is synchronized with the rhythm of the exercise movement, with it the endurance of those who exercise can increase and can be exercised with greater intensity when they move in synchrony with the musical stimulus. It has been suggested that the effect of using synchronized music during exercise is because it has the ability to reduce the metabolic cost of exercise by improving neuromuscular or metabolic efficiency [17]. It is a fact that music is directly associated with emotions. Although the field of emotions is quite subjective, different investigations conclude that music influences people's emotional reactions [10]. Music affects mood in a positive way, increases confidence and self-esteem. All these elements analized so far motivate our proposal that we will detail in the following point.

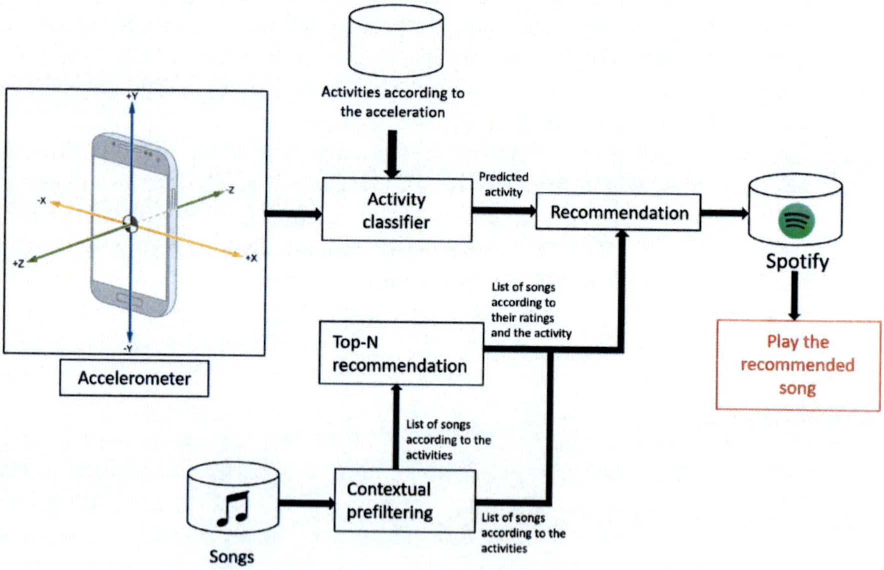

Fig. 1. Architecture of the proposed system

3 Case Study: CARS in Music Streaming Services

A CARS in music is proposed taken in account the daily activities of the users in order to motivate them to continue with that activity. Different studies indicate that listening to music while doing some type of activity may be beneficial. With this in mind, the proposal is a mobile application that implements a music recommendation system that can predict the user's activity and to make a recommendation while taking into account the user taste. As the idea is to recommend music while the user is doing some activity, it's required the system a low level of interaction. The purpose is to use a data set that relates acceleration data to the physical activities that a user performs based on it. By applying an automatic learning algorithm to this data set, it is possible to obtain as output the activity that the user is performing from the described contextual data. The model obtained will be used within a mobile application where physical activity first will be predicted with the help of this model by giving it as input the values returned by the accelerometer in the mobile phone [18].

It would also be important to recommend songs that had an appropriate BPM, and would also please the user. Therefore, we will also use a data set that contains ratings of the songs where applying a top-N technique. Finally, the song recommended to the user will be played from the Spotify repository. Our proposal, shown at Fig. 1, has all characteristics that were considered desirable in previous studies, detailed in Sect. 2.1.

3.1 Classification of Physical Activity

In order to achieve the objective of this system, it is necessary to know the activity that a user is carrying out. This activity can be inferred from the mobile accelerometer that measures the acceleration in the three spatial dimensions.

Table 1. Results for automatic learning algorithms

	Accuracy	Precision	Recall
Nearest centroid	0.28	0,27	0,42
Bayesian classifier	0.23	0,235	0,51
Multiperceptron neural network	0.66	0,65	0,52
Decision tree	0.98	0,99	0,98
LTMS neural network	0.95	0,96	0,97

For the materialization of this application, a data set provided by the laboratory of *Wireless Sensor Data Mining (WISDM)* [19] will be used. This dataset without unknown values, has 1098207 tuples and 6 attributes: *user*, *activity*, *timestamp*, *x-axis*, *y-axis* and *z-axis*. The class attribute in this case would be the activity to be predicted. The attribute *timestamp* indicates the time when the data was taken. On the other hand, *x-axis*, *y-axis* and *z-axis* are attributes that show the accelerations in each of the spatial axes (x, y and z).

Several algorithms were trained to build the activity classifier Table 1. The results were good for decision Tree and LSTM (Long Short Term Memory) neural network. Due the library *TensorFlow* was used to implementation, LTSM was the algorithm selected to export it to the Android application with a recall of 0.97, a precision of 0.96 and an accuracy of 0.95. This network has cyclic connections between the nodes, this leads to them being able to use their internal states to process input sequences. Table 2 shows the progress during training for LTSM.

Table 2. Progress during training

Epoch	Accuracy	Recall
1	0,77	0,99
10	0,94	0,56
20	0,96	0,39
30	0,97	0,29
40	0,97	0,25

3.2 Classification of the Songs According to the Activity

For the classification of the songs the data-set provided by Gomes *et al.* [20] was used, which relates the songs with a series of attributes. The attributes used were:

- *N*: indicates the numbering of the dataset tuples.
- *artist*: the names of the artists of each song are shown. This will be used to display the names of the artists in the interface.
- *bpm*: indicates the BPM of the songs, that as it has been said before, is the property that will be used to choose the songs based on the activity that the users are carrying out.
- *song_id*: indicates the song identifier in this dataset.
- *image*: this attribute refers to the image that is associated with each song. These images are taken from a Spotify repository; it used to display that image during song playback.
- *preview_url*: this attribute are URLs to fragments of songs also taken from the Spotify repository. These will be the ones that will be played in the application.

For the incorporation of contextual information, prefiltering approach is used. BPM associated with the entrainment is linked to the intensity and rhythm in which an activity is carried out. For this purpose the song data set is divided into different ranges according to the BPM. As many divisions are made as activities are able to predict, as follows: Sitting: 0–80 bpm; Standing: 80–100 bpm; Walking: 100–120 bpm; Downstairs: 120–140 bpm; Upstairs: 140–155 bpm and Run: more than 155 bpm. Based on this, the system play songs with a more intense musical tempo (a higher BPM) when the user is performing activities that require movement (walking, going down and up stairs and running), and it will be higher depending on the intensity of the activity while on activities where user needs relax, concentration or less activity, plays songs with a lower BPM, calmer songs.

3.3 Recommendation to the User

The ratings were obtained from the same data-set provided by Gomes *et al.* [20] along the songs associated with. The main attributes used at this procedure where *Song_id*, *Rating* and two new attributes were added to this data set, *Count* and *Mean*:

- *Count*: attribute that indicates the number of times a song has been rated. It was used both to get the average rating and to calculate the top-N table as will be seen below.
- *Mean*: attribute that indicates the average rating of each song. This was used to calculate the top-N table.

Because the ratings are not enough, at this prototype a simple recommender system was made. The Weighted Rating (WR) formula was used to take into account the average of the ratings and the number of votes of the songs:

$$WR = (\frac{v}{v+m} \cdot R) + (\frac{m}{v+m} \cdot C) \tag{3}$$

where v is the number of votes for the songs, m is the minimum number of votes required to be included in the table, R is the average of the song ratings and C is the average of all ratings. Once ratings are computed, the system provides the top-N list of songs.

Fig. 2. Application interface

The Android application, shown at Fig. 2, obtains data from the smartphone accelerometer, as inputs to the neural network that predicts the activity of the user. Once the classifier shows the user activity as an output, works for song selection. The recommendation is made by choosing the song that is within the top-N list and fits with the activity that has been predicted. Then the song selected is *previewed* and reproduced directly from Internet through the url associated in the dataset.

The simple interface makes possible for the user had the minimum interaction with the system so that she/he didn't have to interrupt his activity. It is based on a well-known music player interface and also includes a way to provide ratings to the songs (using the stars).

3.4 User Testing Evaluation

The application was tested with a small number of users at this stage. These users carried out activities with and without the use of the application in order to make a comparison. The RPE model was applied, through which the users were asked to indicate the perceived effort during the activities that required movement (walking, running, climbing and descending stairs). However, for activities that do not require movement, such as sitting or standing, the user was first asked to indicate what he/she was doing while the activity (relaxing or some activity that required concentration). With this in mind, he or she was asked whether they considered that they was able to concentrate or relax adequately on a scale of 1 to 5. Finally, for all cases, we asked about their mood after doing the activity (improved, the same or worsened). The perceived effort rate decreased using the recommendation system in most cases, and the mood improved in the majority of cases. This demonstrates the usefulness of this recommendation system for this type of activities. In the non-moving activities, where the activity carried out required concentration, users indicated that they achieved greater concentration without music. On the other hand, for the activities that had to do with relaxation, user opinion was split. But all those who performed a relaxing activity while using the recommendation system indicated that their mood improved. Based on this, it could be concluded that this recommender system is more suitable for activities that require movement.

4 Conclusions and Future Lines

In this paper, we proposed a CARS applied to music recommendation while performing physical activities. The proposal is able to detect the user activity and make the music recommendation according to it. Some parameters are considered, such as RPE, entrainment and PPM in order to contribute to the improvement of the physcical activity. Along the user's activity detection, LSTM type neural network proved to give good results as a classifier of activities, taking into account both the results of its quality metrics and the tests of the experimental study. The case study allowed to demonstrate that the phenomenon of *entrainment* is very useful to link the activities of the users with the music while improving their mood. The research would have to continue to implement the model with another learning algorithm and to extend functionalities. A new method for predicting ratings in order to obtain user implicit feedback. At a higher level, a playlist generation technique could be applied aligned with kinds of activity.

Acknowledgments. This research has been supported by the project RTI2018-095390-B-C32 (MCIU/AEI/FEDER, UE).

References

1. Karageorghis, C.I., Priest, D.L.: Music in the exercise domain: a review and synthesis (Part I). Int. Rev. Sport Exerc. Psychol. **5**(1), 44–66 (2012)
2. Song, Y., Dixon, S., Pearce, M.: A survey of music recommendation systems and future perspectives. In: 9th International Symposium on Computer Music Modeling and Retrieval, vol. 4, pp. 395–410, June 2012
3. Adomavicius, G., Tuzhilin, A.: Context-aware recommender systems. In: Recommender Systems Handbook, pp. 217–253. Springer, Boston (2011)
4. Sen, A., Larson, M.: From sensor to songs: a learning-free novel music recommendation system using contextual sensor data (2015)
5. Bonnin, G., Jannach, D.: Automated generation of music playlists: survey and experiments. ACM Comput. Surv. (CSUR) **47**(2), 1–35 (2015)
6. Greenhalgh, C., Hazzard, A., McGrath, S., Benford, S.: GeoTracks: adpative music for everyday journeys (2016)
7. Braunhofer, M., Kaminskas, M., Ricci, F.: Location-aware music recommendation (2013)
8. Moens, B., Van Noorden, L., Leman, M.: D-Jogger: sycing music with walking (2010)
9. Oliver, N., Kreger-Stickle, L.: Enhancing exercise performance through real-time physiological monitoring and music: a user study (2007)
10. García Vicente, J., Gil, A.B., Reboredo, A.L., Sánchez-Moreno, D., Moreno-García, M.N.: Moodsically. Personal Music Management Tool with Automatic Classification of Emotions (2019)
11. Fang, J., Grunberg, D., Lui, S., Wang, Y.: Development of a music recommendation system for motivating exercise (2017)
12. Priest, D.L., Karageorghis, C.I., Sharp, N.C.: The characteristics and effects of motivational music in exercise settings: the possible influence of gender, age, frequency of attendance, and time of attendance (2004)
13. Lopes-Silva, J.P., Lima-Silva, A.E., Bertuzzi, R., Silva-Cavalcante, M.D.: Influence of music on performance and psychophysiological responses during moderate-intensity exercise preceded by fatigue. Physiol. Behav. **139**, 274–280 (2013)
14. Popova, L.: The extended parallel process model: illuminating the gaps in research. Health Educ. Behav. **39**(4), 455–473 (2012)
15. Simpson, S.D., Karageorghis, C.I.: The effects of synchronous music on 400-m sprint performance. J. Sports Sci. **24**(10), 1095–1102 (2006)
16. Styns, F., van Noorden, L., Moelants, D., Leman, M.: Walking on music. Hum. Mov. Sci. **26**, 769–85 (2007)
17. Van Dyck, E., Moens, B., Buhmann, J., Demey, M., Coorevits, E., Bella, S.D., Leman, M.: Spontaneous entrainment of running cadence to music tempo. Sports Med.-Open **1**(1), 15 (2015)
18. Sassi, I.B., Mellouli, S., Yahia, S.B.: Context-aware recommender systems in mobile environment: on the road of future research. Inf. Syst. **72**, 27–61 (2017)
19. Dash, Y., Kumar, S., Patle, V.K.: A novel data mining scheme for smartphone activity recognition by accelerometer sensor. In: Proceedings of the 4th International Conference on Frontiers in Intelligent Computing: Theory and Applications (FICTA) 2015, pp. 131–140. Springer, New Delhi (2016)
20. Gomes, C., Moreno, M.N., Gil, A.B.: Sistema de Recomendación de Música Sensible al Contexto, pp. 65–80. Avances en Informática y Automática, Duodécimo Workshop (2018)

Classification of Chest Diseases Using Deep Learning

Jesús Silva[1(✉)], Jack Zilberman[1], Yisel Pinillos Patiño[2], Noel Varela[3], and Omar Bonerge Pineda Lezama[4]

[1] Universidad Peruana de Ciencias Aplicadas, Lima, Peru
jesussilvaUPC@gmail.com, jack.zilberman@upc.pe
[2] Universidad Simón Bolívar, Barranquilla, Colombia
ypinillos@unisimonbolivar.edu.co
[3] Universidad de la Costa, Barranquilla, Colombia
nvarela2@cuc.edu.co
[4] Universidad Tecnológica Centroamericana (UNITEC), San Pedro Sula, Honduras
omarpineda@unitec.edu

Abstract. The field of computer vision has had exponential progress in a wide range of applications due to the use of deep learning and especially the existence of large annotated image data sets [1]. Significant improvements have been shown in the performance of problems previously considered difficult, such as object recognition, detection and segmentation over approaches based on obtaining the characteristics of the image by hand [2]. This article presents a novel method for the classification of chest diseases in the standard and widely used data set ChestX-ray8, which contains more than 100,000 front view images with 8 diseases.

Keywords: Classification of chest diseases · Deep learning · ChestX-ray8

1 Introduction

In recent years, deep learning has shown to be effective for producing similar performance increases in the domain of medical image analysis for object detection and segmentation tasks [3]. Recent notable studies include important medical applications, for example in the domain of pneumology (classification of lung diseases [4] and detection of lung nodules on CT images [5]). Compared to other applications in computer vision, the main limitation in medical machine learning applications is that most of the proposed methods are evaluated using a rather small dataset with hundreds of patients at most, but progress has been made in recent years with the introduction of publicly available datasets [6, 7]. This paper focuses on the identification and classification of chest diseases, a problem presented in the following sections.

Compared to previous studies [8–12], this approach differs in three main ways: 1. It uses data augmentation to increase the size of the ChestX-ray8 dataset; 2. It implements a data filtering scheme to eliminate images that could negatively impact the training process; and 3. It designs a much smaller and reduced convolutional network that achieves better results than those presented by [13].

Y. Dong et al. (Eds.): DCAI 2020, AISC 1237, pp. 152–158, 2021.
https://doi.org/10.1007/978-3-030-53036-5_16

2 Problem Statement

The approach presented in this paper is a multiple classification problem, since 8 possible classes or diseases were identified and classified by the X-ray image. The data was extracted from the ChestX-ray8 set and trained with a convolutional network to make the classification, and obtain a model after some fine adjustments such as dropout or weight removal and L2 regularization, among other strategies. However, one of the biggest problems of deep learning, when working with data sets like the one used in this study, is the problem of overtraining or overfitting [14].

One widely used solution is data augmentation (D.A.), that is, generating more training data from existing samples, applying a variety of random transformations to these samples to produce credible images. The data augmentation process can significantly reduce the loss of validation [15], as illustrated in Fig. 1, which shows the graph with training and validation data for two identical convolution networks, one without using data augmentation and the other using this strategy. Using data augmentation clearly addresses the problem of overtraining that can be seen on the left side of the figure, where it is observed that the value of the loss function with the validation data initially decreases and then increases again. The right side of the figure shows how the loss of validation can be regularized using the increase of data, among other strategies.

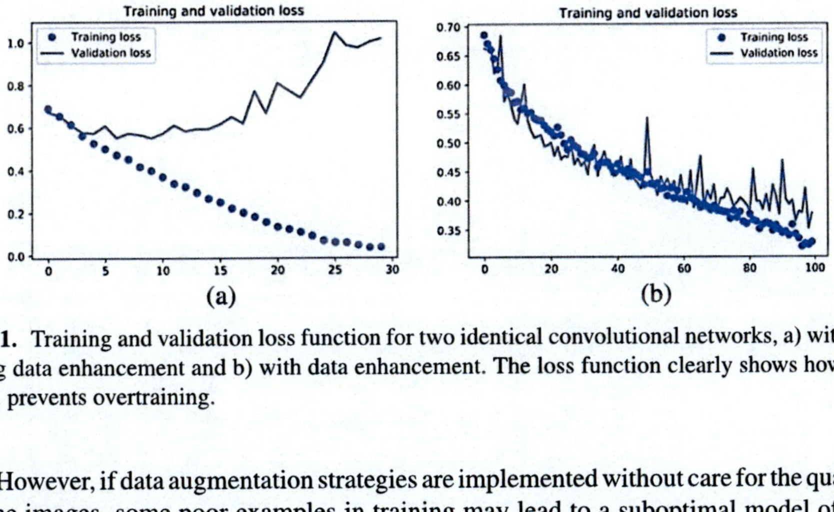

(a) (b)

Fig. 1. Training and validation loss function for two identical convolutional networks, a) without using data enhancement and b) with data enhancement. The loss function clearly shows how the A.D. prevents overtraining.

However, if data augmentation strategies are implemented without care for the quality of the images, some poor examples in training may lead to a suboptimal model of the convolution network [16].

3 Proposed Solution

Compared to other classification datasets such as COCO and ImageNet, the spatial extent of the diseases on ChestX-ray8 occupy a rather small region of the image, which can affect the performance of the convolutional network when limited computing power is available, as is the case here. The authors of [11] extracted the X-ray images from the

DICOM file and adjusted them to a size of 1,024 × 1,024 pixels (from their original size of 3,000 × 2,000) without losing significant detail. In this study, the images of the modified data set were further reduced to a size of 224 × 224 pixels, due to limitations of the available hardware, but satisfactory results were obtained, as shown below.

The following sections show the depth of the convolutional network that required the researchers to adjust the hardware used in the study. However, Sect. 4 shows the results reported, which in some cases exceeded the expectations [11].

3.1 Proposed Architecture

The convolutional network implemented in the study can be seen at the top of Fig. 2. It is a rather minimal architecture, as it is made up of 5 convolutional layers and 3 fully connected layers, with a low parameter count of just 254,138.

The architecture is based on AlexNet which has a total of 63.2 million parameters and was gradually simplified until the model matched the available hardware, a Nvidia GeForce 1050 GTX GPU. It is compared to the architecture used in [11] that was tested on several models but had the best results with ResNet-50, shown at the bottom of Fig. 2, and which has 25.6 million parameters.

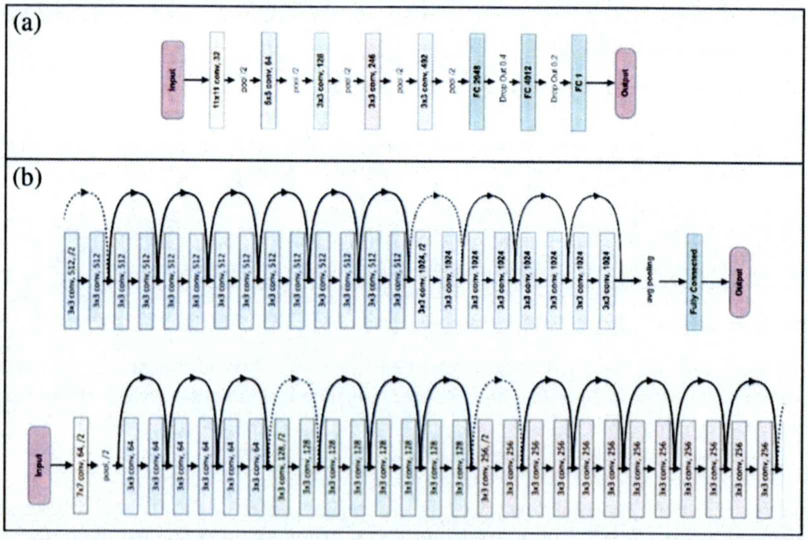

Fig. 2. Architectures for a) the proposed convolutional network, and b) the system in [11] based on RestNet50

3.2 Training Process

Due to the image size and the memory limit of the GPU, it was necessary to reduce the image and batch sizes to load the entire network onto the GPU and increase the number of iterations to accumulate the gradients. The combination of both can vary in different convolutional networks, but in this case the batch size was kept constant at 32.

The network was trained from start to finish using Adam's optimizer with the standard parameters ($\beta 1 = 0.9$ and $\beta 2 = 0.999$). An initial value of 0.001 was used for the learning rate that was decreased by a factor of 10 each time the validation loss stagnated after an epoch and the model with the lowest validation loss was chosen.

In total, 253,147 front view X-ray images were included in the database, of which 89,320 images contained one or more pathologies. The remaining 63,832 images were normal cases. For the pathology classification task, the complete data set was randomly divided into three groups: training (70%), validation (10%), and testing (20%). In training and validation, the model was fine-tuned with Stochastic Gradient Descent. In these experiments, just the results obtained in the classification of the 8 diseases on the test data were reported.

4 Results and Discussion

This section presents the results obtained from the experimentation and compare them with similar approaches, specifically, the models presented by [11].

4.1 Training and Validation Results

The CNN was trained to classify the 8 chest diseases presented in the ChestX-ray8 dataset. In most cases, the models converged quickly, due to the low complexity of the model of the CNN architecture, requiring less than 3,000 iterations or epochs and obtaining better results than the base research reported in [11]. The graphs in Fig. 3 show the accuracy obtained with the validation data in 4 of the 8 diseases; the red line represents the testing of the model using the original, unaugmented data, while the blue line shows the results in the modified, augmented data set.

The accuracy obtained with the validation data for all 8 classes is shown in Table 1, for both the original unaugmented data and the data augmented with quality images in the ChestX-ray data set8. As mentioned above, there is a significant improvement in all 8 classes, which is especially surprising given that the convolutional network is simpler.

4.2 Comparison with the Original ChestX-Ray8 Model

Very good results were obtained using a relatively simple convolutional network inspired by AlexNet, as shown in Table 2. The authors of [11] tested 4 different networks with ChestX-ray8: AlexNet, GoogLeNet, VGG-Net-16 and RestNet50. The best results were obtained with ResNet50, except for "Mass" where AlexNet obtained the best results. When comparing the model with the results reported for AlexNet and ResNet50, a significant improvement was evident for all cases of the first one, except "Effusion" and better results for "Cardiomegaly", "Infiltration", "Mass" (a particularly difficult class) and "Pneumonia" with the second one. Increasing data quality played a significant role in achieving these results, and for future research it is desired to investigate whether the classes that did not obtain the best results can be improved with other network models or by modifying the filtering process [17].

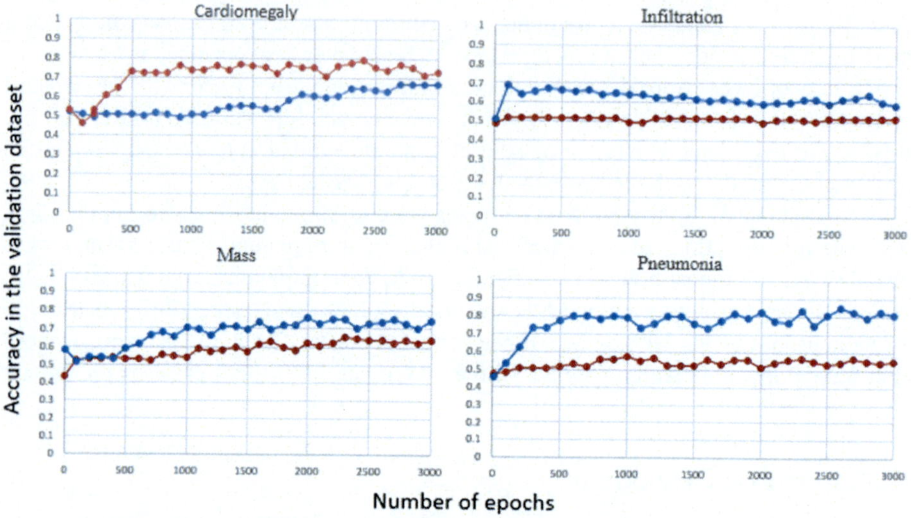

Fig. 3. Accuracy test results using ChestX-Ray validation data, without using data enhancement (red line) and with data enhancement and filtering (blue line), using the proposed network

Table 1. Results for accuracy at best for multiple classification in different DCNN model settings.

	Cardiomegaly	Effusion	Infiltrate	Mass	Nodule	Pneumonia
No DA	0.7475	0.6254	0.5698	0.6962	0.6523	0.6047
W/DA	0.8123	0.62014	0.7045	0.79247	0.7012	0.88521

Table 2. Comparison in terms of accuracy of the models with other state-of-the-art models

Network-class	Cardiomegaly	Effusion	Infiltrated	Mass	Node	Pneumonia
AlexNet	0.6574	0.7025	0.6014	0.57123	0.6541	0.55123
ResNet	0.8142	0.7347	0.6078	0.57023	0.7247	0.64021
Our	0.8213	0.6521	0.7025	0.7963	0.7023	0.87412

4.3 Discussion on Limitations and Extensions of the Proposed Approach

As discussed in this paper, the presented approach of data enhancement based on image quality produced very interesting results with ChestX-ray8 data. Up to date, this approach has not been reported in the literature and this may be the main contribution of this study. As mentioned, the research used a relatively simple convolutional network architecture, mainly due to the availability of hardware. This limitation also led to use smaller images than those reported in the literature (224 × 224, in contrast to 1024 × 1024 images as mentioned in [11] and other papers).

However, the results reported in the previous subsection are very promising and the approach to data augmentation can be tested using other, more complex convolutional networks, such as ResNet-50, a line that can be explored in the future using tools such as Google Colab or Kaggle Kernels.

5 Conclusions

Medical diagnosis has become a more interesting and viable domain for machine learning. In particular, chest x-rays are the most common type of radiological study in the world and a particularly challenging example of multiple classification for medical diagnosis. Accumulating about 45% of all radiological studies, the chest plate has achieved global ubiquity as a low-cost screening tool for a wide range of pathologies including lung cancer, tuberculosis and pneumonia.

However, the general shortage of publicly available medical data sets has inhibited the development of deep learning solutions in this field. In addition, pre-processing on datasets such as ImageNet or COCO may introduce unintended biases which may be unacceptable in clinical settings, making the transfer of learning more difficult to achieve in this context.

Likewise, most clinical environments demand models that can accurately predict multiple diagnoses. This transforms many medical problems into multiple classification problems, where a single image can have a large number of results, which can be ambiguous or poorly defined and the image is likely to be labeled inconsistently. These problems are exposed due to the introduction of the ChestX-ray8 dataset in 2018, since the previously available datasets such as Open-I were considered too small.

Since then, several papers have progressively improved the results by diagnosing thoracic diseases based on this dataset. However, these studies do not consider the image quality of the training samples, which has produced poor results in previous studies, as demonstrated by experimentation using the data enhancement approach with quality images. The results of these experiments were able to match those achieved in [11] for 4 diseases and surpass them in the other 4, using a smaller network. A more comprehensive study is being planned for evaluating other more complex networks such as ResNet-50 to further validate these results.

References

1. Song, Q., Zhao, L., Luo, X., Dou, X.: Using deep learning for classification of lung nodules on computed tomography images. J. Healthc. Eng. (2017)
2. Pouyanfar, S., Sadiq, S., Yan, Y., Tian, H., Tao, Y., Presa Reyes, M., Shyu, M.-L., Chen, S.-C., Iyengar, S.S.: A survey on deep learning: algorithms, techniques, and applications. ACM Comput. Surv. **51**(5), 36 (2018). Article 92
3. Litjens, G., et al.: A survey on deep learning in medical image analysis. Med. Image Anal. **42**, 60–88 (2017)
4. Wang, H., Jia, H., Lu, L., Xia, Y.: Thorax-Net: an attention regularized deep neural network for classification of Thoracic diseases on chest radiography. IEEE J. Biomed. Health Inform. **24**(2), 475–485 (2019)

5. Shadeed, G.A., Tawfeeq, M.A., Mahmoud, S.M.: Deep learning model for thorax diseases detection. Telkomnika **18**(1), 441–449 (2020)
6. Viloria, A., Bucci, N., Luna, M., Lis-Gutiérrez, J.P., Parody, A., Bent, D.E.S., López, L.A.B.: Determination of dimensionality of the psychosocial risk assessment of internal, individual, double presence and external factors in work environments. In: International Conference on Data Mining and Big Data, pp. 304–313. Springer, Cham, June 2018
7. Mao, K.P., Xie, S.P., Shao, W.Z.: Automatic Segmentation of Thorax CT Images with Fully Convolutional Networks. In: Current Trends in Computer Science and Mechanical Automation vol. 1, pp. 402–412. Sciendo Migration (2017)
8. Krishna, R., Zhu, Y., Groth, O., Johnson, J., Hata, K., Kravitz, J., Chen, S., Kalantidis, Y., Li, L.J., Shamma, D. A., Bernstein, M., Fei-Fei, L.: Visual genome: Connecting language and vision using crowdsourced dense image annotations (2016)
9. Wang, X., Peng, Y., Lu, L., Lu, Z., Summers, R. M.: Automatic classification and reporting of multiple common thorax diseases using chest radiographs. In: Deep Learning and Convolutional Neural Networks for Medical Imaging and Clinical Informatics, pp. 393–412. Springer, Cham (2019)
10. Trullo, R., Petitjean, C., Ruan, S., Dubray, B., Nie, D., Shen, D.: Segmentation of organs at risk in thoracic CT images using a sharpmask architecture and conditional random fields. In: 2017 IEEE 14th International Symposium on Biomedical Imaging (ISBI 2017), pp. 1003–1006. IEEE, April 2017
11. Wang, X., Peng, Y., Lu, L., Lu, Z., Bagheri, M., Summers, R.M.: Chestx-ray8: Hospital- scale chest x-ray database and benchmarks on weakly-supervised classification and localization of common thorax diseases. arXiv preprint arXiv:1705.02315 (2017)
12. Ming, J.T.C., Noor, N.M., Rijal, O.M., Kassim, R.M., Yunus, A.: Lung disease classification using different deep learning architectures and principal component analysis. In: 2018 2nd International Conference on BioSignal Analysis, Processing and Systems (ICBAPS), pp. 187–190. IEEE, July 2018
13. Yao, L., Poblenz, E., Dagunts, D., Covington, B., Bernard, D., Lyman, K.: Learning to diagnose from scratch by exploiting dependencies among labels. arXiv preprint arXiv:1710.10501 (2017)
14. Dodge, S., Karam, L.: Understanding how image quality affects deep neural networks. In: Eighth International Conference on Quality of Multimedia Experience (QoMEX), IEEE, pp. 1–6. arXiv:1604.04004v2 (2016)
15. Wang, X., Peng, Y., Lu, L., Lu, Z., Bagheri, M., Summers, R.M.: Chestx-ray8: Hospital-scale chest x-ray database and benchmarks on weakly-supervised classification and localization of common thorax diseases. In: Proceedings of the IEEE Conference on Computer Vision and Pattern Recognition, pp. 2097–2106 (2017)
16. Liu, Z., Chen, H., Liu, H.: Deep Learning Based Framework for Direct Reconstruction of PET Images. In International Conference on Medical Image Computing and Computer-Assisted Intervention, pp. 48–56. Springer, Cham, October 2019
17. Gamero, W.M., Agudelo-Castañeda, D., Ramirez, M. C., Hernandez, M. M., Mendoza, H. P., Parody, A., Viloria, A.: Hospital admission and risk assessment associated to exposure of fungal bioaerosols at a municipal landfill using statistical models. In: International Conference on Intelligent Data Engineering and Automated Learning, pp. 210–218. Springer, Cham, November 2018

Mobile Device-Based Speech Enhancement System Using Lip-Reading

Tomonori Nakahara[1], Kohei Fukuyama[1], Mitsuru Hamada[1], Kenji Matsui[1(✉)], Yoshihisa Nakatoh[2], Yumiko O. Kato[3], Alberto Rivas[4], and Juan Manuel Corchado[4]

[1] Osaka Institute of Technology, Osaka, Japan
m1m19r24@st.oit.ac.jp, kenji.matsui@oit.ac.jp
[2] Kyushu Institute of Technology, Fukuoka, Japan
[3] St. Marianna University School of Medicine, Kawasaki, Japan
[4] BISITE Digital Innovation Hub, University of Salamanca, Salamanca, Spain

Abstract. The lip-reading based speech enhancement method for laryngectomees is proposed to improve their communication in an inconspicuous way. First, we developed a simple lip-reading mobile phone application for Japanese using the YOLOv3-Tiny, which can recognize Japanese vowel sequences. Four laryngectomees tested the application, and we confirmed that the system design concept is along the line of the user needs. Second, the user-dependent lip-reading algorithm with very small training data set was developed. Each 36 viseme images were converted into very small data using VAE(Variational Autoencoder), then the training data for the word recognition model was generated. Viseme is a group of phonemes with identical appearances on the lips. Our viseme sequence representation with the VAE was used to be able to adapt users with very small amount of training data set. Word recognition experiment using the VAE encoder and CNN was performed with 20 Japanese words. The experimental result showed 65% recognition accuracy, and 100% including 1st and 2nd candidates. The lip-reading type speech enhancement seems appropriate for embedding mobile devices in consideration of both usability and small vocabulary recognition accuracy.

Keywords: Lip-reading · Laryngectomy · Viseme

1 Introduction

People who have had laryngectomies have several options for restoration of speech, but currently available devices are not satisfactory. The electrolarynx (EL), typically a hand-held device which introduces a source vibration into the vocal tract by vibrating the external walls, has been used for decades by laryngectomees for speech communication. It is easy to master with relatively short-term practice period regardless of the post-operative changes in the neck. However, it has a couple of disadvantages. Firstly, it does not produce airflow, so the intelligibility of consonants is diminished and the speech is very mechanical tone that does not sound natural. Secondly, it is far from normal appearance. Alternatively, esophageal speech does not require any special equipment,

Y. Dong et al. (Eds.): DCAI 2020, AISC 1237, pp. 159–167, 2021.
https://doi.org/10.1007/978-3-030-53036-5_17

but requires speakers to insufflate, or inject air into the esophagus. It takes long time to master that method, especially, elderly laryngectomees face difficulty in mastering the esophageal speech or keep using it because of the waning strength. To understand the user needs more precisely, questionnaires were used to understand implicit user needs with 121 laryngectomees (87% male, 13% female), including 65% esophageal talkers, 12% EL users, 7% both, and 21% used writing messages to communicate.

We extracted primary needs of laryngectomees from the result as shown in Fig. 1. Then, this time, we focused on three of them, i.e. "Use of existing devices", "Ordinary-looking", "Easy to Use" [1].

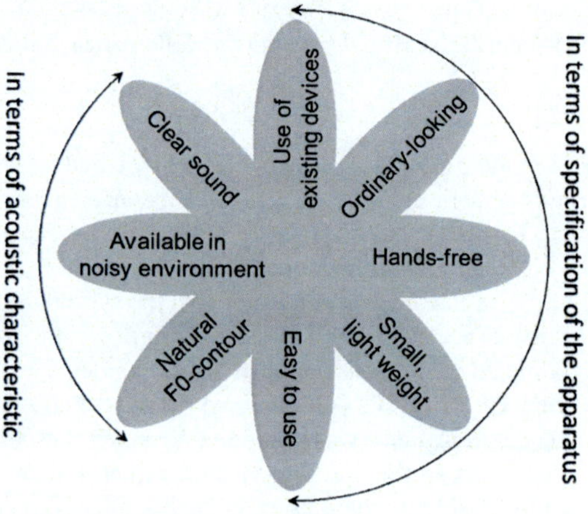

Fig. 1. The primary needs of laryngectomees

1) "Use of existing devices": Mobile phone, especially smartphone is becoming very popular and it has a lot of computational power. Therefore, we use the smartphone as the central unit of speech enhancement system.
2) "Ordinary-looking": Again, mobile devices are widely used and no one thinks it is strange even if you are talking to your mobile phone. We plan to develop a system which can recognize user's lip motion and generate the corresponding synthesized speech.
3) "Easy to use": By combining lip-reading and speech synthesis, people can communicate without using either electrolarynx or esophageal speech. That makes users much easier to communicate.

Recently, image processing technique (lip-reading) for handicapped people has been developed by several researchers [2] such as visual only speech recognition (VSR). However, most of those technologies are developed for ordinary people and implemented on PC [3]. The present study aims to develop a speech enhancement tool using technique of lip-reading on mobile device for laryngectomies to meet the essential user needs.

2 Preliminary UI Testing of the Speech Enhancement System

Based on the user needs, a mobile phone based speech enhancement system was proposed. Figure 2 shows the concept image of the proposed system. Users just need to silently talk to the smart phone, then the system captures the lip images via the tiny camera, and recognizes each phoneme using lip-reading capability. The recognized phrases are converted into speech output by speech synthesis application. We can expect to obtain relatively higher lip-reading accuracy by utilizing the feedback from the display.

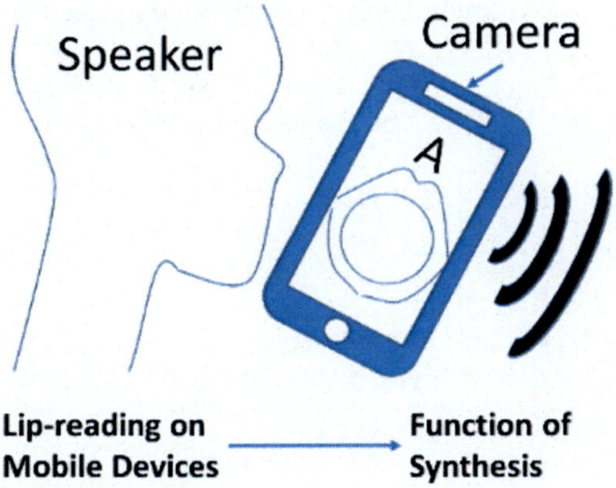

Fig. 2. A concept image of the proposed system

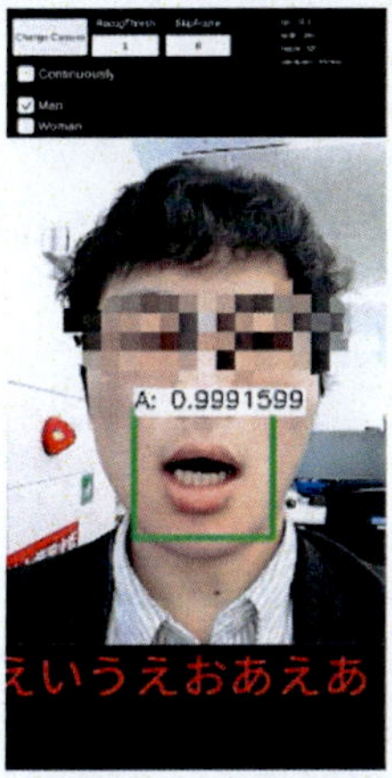

Fig. 3. Real-time monitoring program

Figure 3 shows the screen shot of the real-time monitoring program. The red colored symbols are the recognition results. YOLOv3-tiny and Unity was used for the early prototype. The overall concept and the UI design was tested with users. Four laryngectomees (three males and one female) tested the system. After short training, users were able to use the device. Their feedbacks are;

1) Although the experimental device can only recognize Japanese vowels, the basic concept of the speech enhancement application using lip-reading was well accepted.
2) Lip-reading with some kind of solution to be able to recognize consonants would be needed.
3) If the system can recognize phrases, the delayed response is acceptable.
4) It is very nice to be able to choose the female/male speech output.
5) Additional loudspeaker could be necessary in case of very noisy environment.

Based on the user test in terms of the basic concept of system design, we are currently working on short phrase recognition.

3 Word Recognition Algorithm

As for the lip-reading device for laryngectomies, we could assume a customized system where we can use user's lip images for the machine learning. Our tentative goal is to develop speaker dependent, easy to train, small foot print word recognition algorithm [4–6].

3.1 Data Collection and Preprocessing of the Lip-Reading System

Table 1 shows the two syllable patterns of visemes including closure X. In this preliminary study, those 36 viseme movies were captured only once and used for the training data.

Table 1. 36 patterns of visemes consist of five Japanese vowels and closure (X)

X	A	I	U	E	O
XA	AX	IX	UX	EX	OX
XI	AI	IA	UA	EA	OA
XU	AU	IU	UI	EI	OI
XE	AE	IE	UE	EU	OU
XO	AO	IO	UO	EO	OE

Figure 4 shows the frontend process of the lip-reading system. Firstly, in order to extract the face images, we used HOG (Histogram of Oriented Gradients) detector and SVM based algorithm [7]. Next, the face image inputs a GBDT (Gradient Boosting Decision Tree) based algorithm to extract lip images [8]. Finally, the mouth region of interest is normalized by histogram, and the image is resized 64 × 64 pixels without changing the aspect ratio. Figure 5 shows the viseme images of closure **X** and five Japanese vowels.

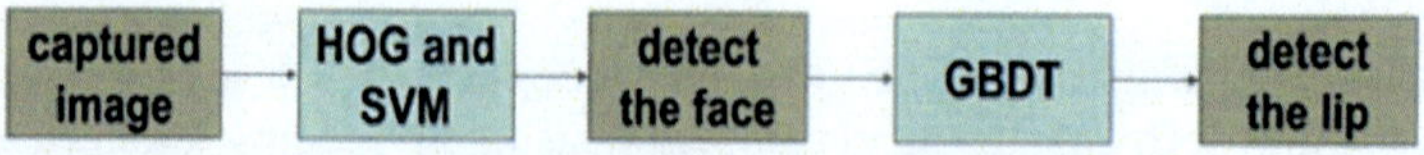

Fig. 4. A block diagram of proposed system

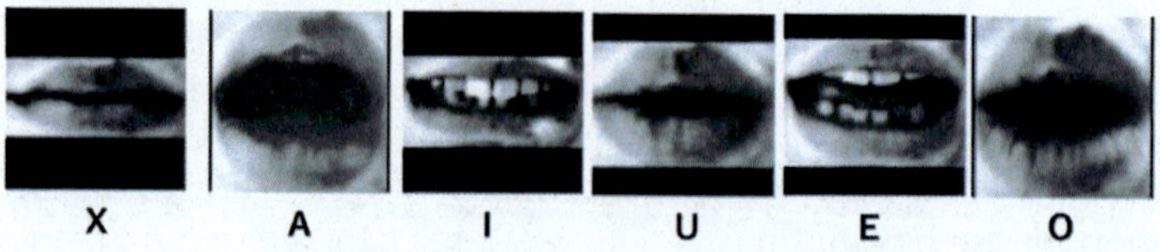

Fig. 5. Preprocessed viseme images

3.2 Feature Extraction of Viseme Images

We took VAE as the feature extraction model of viseme images. Figure 6 shows the VAE model. VAE consists of an encoder, a decoder, and a loss function. The encoder 'encodes' the input data into latent (hidden) representation space z, which is much less dimensions. Normally the z is specified as a standard normal distribution, Therefore the variational parameters would be the mean and variance of the latent variables for each data. In our study, the convolution layer of the VAE encoder is shown in Fig. 7, and Fig. 8 shows the VAE encoder model. Also the decoder model is shown in Fig. 9.

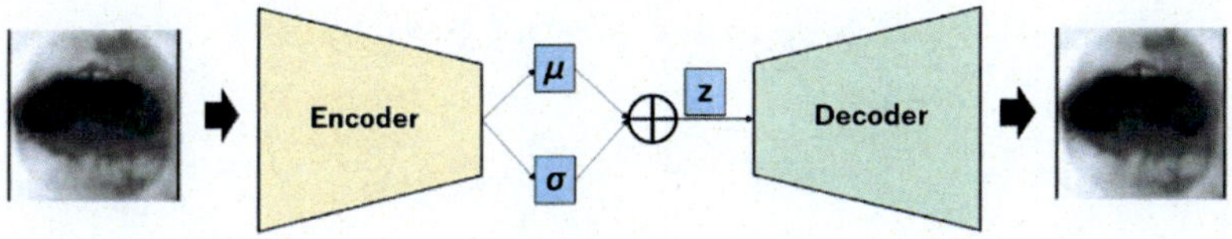

Fig. 6. VAE model

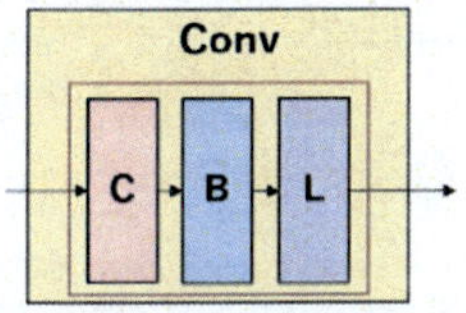

Fig. 7. Convolution layer of the encoder

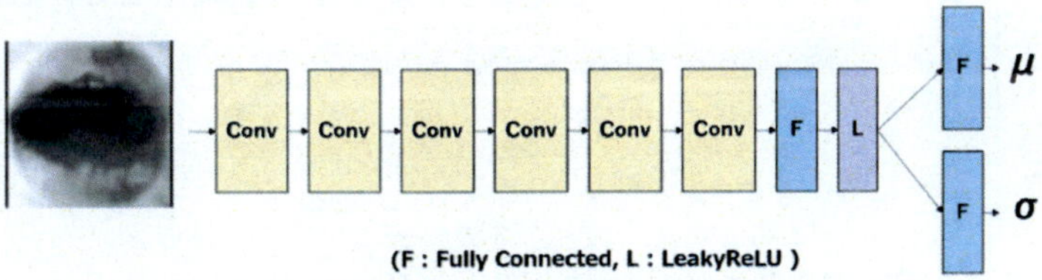

Fig. 8. VAE encoder model

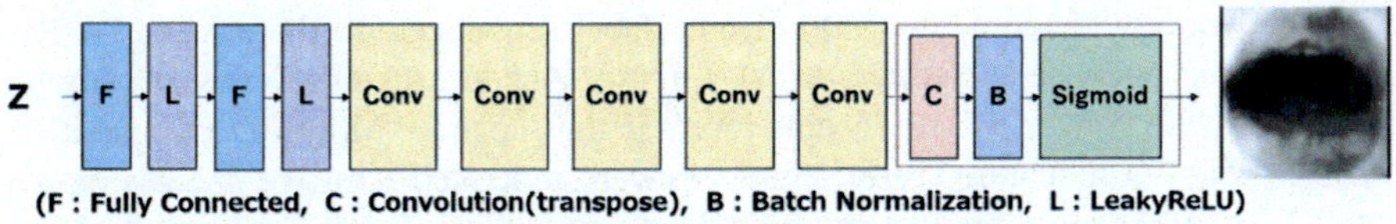

Fig. 9. VAE decoder model

3.3 Generation of the Feature Vector Sequences

Those 36 viseme images were recorded from one male speaker. VAE was trained using those images. For each of those 36 image data, five consecutive frames were extracted around the highest frame difference point. Then, using those processed data, feature vector sequences were generated. Figure 10 shows the optical flow and generation of the vector sequences. The largest part of the optical flow is the center of the sequence.

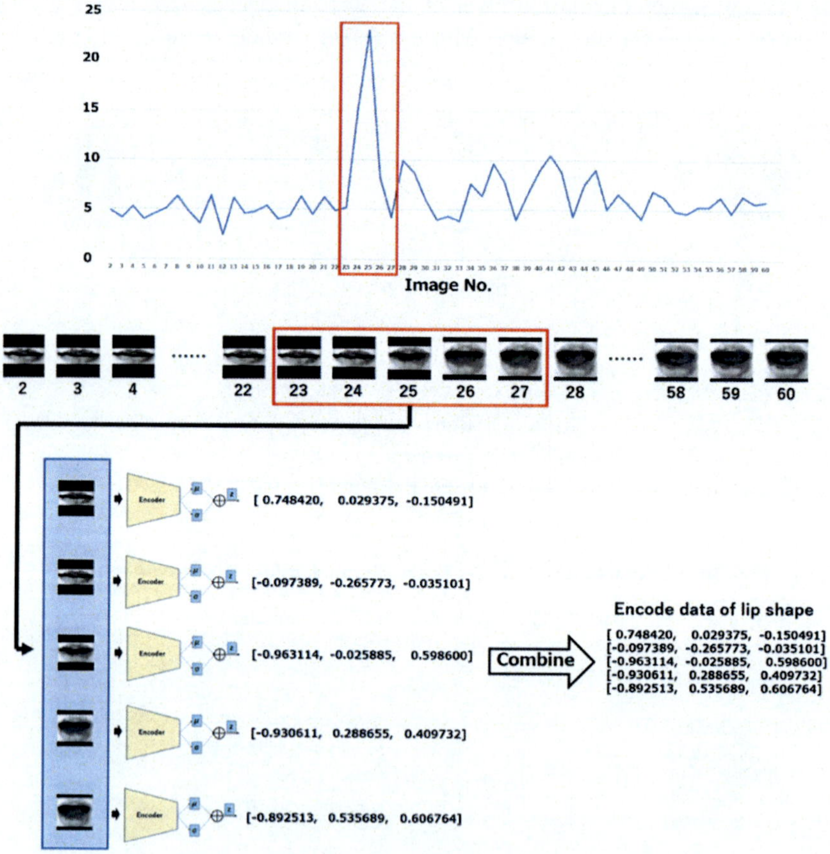

Fig. 10. Generation of the Feature Vector Sequence

3.4 Generation of the Training Data for Word Recognition Model

Vocabulary words were converted into viseme label sequences. In the case of "A-RI-GA-TO-U", the sequence is "X, XA, A, AI, I, IA, A, AU, UO, O, OX, X", for example. Those words were converted into viseme labels using Japanese Morph Decomposition Library "MeCab". The feature vector sequences were generated using the process shown in Fig. 10. Then, the word recognition model was trained using those training data set. Figure 11 shows the word recognition model used for our experiment. Normally, RNN is used for the time series data analysis, however, by using CNN and learning the changes of time series data, the network is able to perform like RNN.

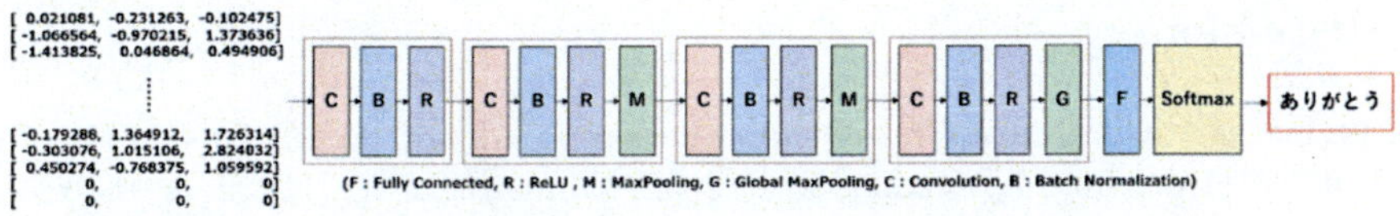

Fig. 11. Word Recognition Model

3.5 Word Recognition Experiment Using VAE Encoder and CNN

LipNet [9–11] is the first end-to-end sentence-level lip-reading model, which achieves more than 95.2% accuracy in English sentence-level. Asami, et al. tested LipNet with Japanese 20 words [12]. We used the same 20 words for our experiment. Table 2 shows the recognition result [13. 14]. The same subject was asked to record the 20 word lip-images, and checked the recognition accuracy. Table 2 shows the recognition result.

The green colored words were recognized correctly. The recognition accuracy is 63%, and 100% including 1st and 2nd candidates. The recognition accuracy is relatively high by comparing with the Asami's result, which was 36% [12].

Table 2. 15 frequently used word recognition result

	True	Result	
		First	Second
1	A-RI-GA-TO	A-RI-GA-TO	O-HA-YO
2	I-I-E	I-I-E	KI-I-RO
3	O-HA-YO	O-HA-YO	A-O
4	O-ME-DE-TO	A-RI-GA-TO	O-ME-DE-TO
5	GO-ME-N-NA-SA-I	GO-ME-N-NA-SA-I	O-ME-DE-TO
6	SU-MI-MA-SE-N	SU-MI-MA-SE-N	I-I-E
7	DO-I-TA-SHI-MA-SHI-TE	DO-I-TA-SHI-MA-SHI-TE	GO-ME-N-NA-SA-I
8	HA-I	SA-N	HA-I
9	MO-SHI-MO-SHI	MO-SHI-MO-SHI	SA-N
10	A-KA	SA-N	A-KA
11	A-O	O-HA-YO	A-O
12	KI-I-RO	KI-I-RO	SI-RO
13	MI-DO-RI	MI-DO-RI	I-CHI
14	SHI-RO	KI-I-RO	SHI-RO
15	ZE-RO	ZE-RO	SA-N
16	I-TI	I-TI	NI
17	NI	I-TI	NI
18	SA-N	SA-N	HA-I
19	YO-N	GO	YO-N
20	GO	GO	YO-N

4 Discussion

Our preliminary experimental result shows the proposed method is relatively effective for word level machine lip-reading with small training data set. However, the representation using viseme sequence makes some of the word classification difficult. For example, the word "A-KA" was mis-recognized as "SA-N". The possible reason is that the representation of "A-KA" is "X, XA, A, A, AX, X", and "SA-N" is very similar sequence, "X, XA, A, AX, X". Therefore, more carefully exploring the training data padding method is required.

When 36 viseme images were recorded, the subject was asked to move lips articulately. To be able to increase the word vocabulary with natural lip movement, we need to obtain not only the lip images, but some other features related to some of the consonants.

In this research, we tested the system with one subject, and obtained promising result. We plan to evaluate if the proposed method is effective as the speaker-dependent system.

5 Conclusions

We performed preliminary study of machine lip-reading for laryngectomies. A mobile phone based user interface design was tested and confirmed the effectiveness. As for the lip-reading algorithm, viseme sequence representation with VAE were used to be able to adapt users with very small amount of training data set. The experimental result showed 63% recognition accuracy under the condition of 20 word vocabulary size with one subject. For future study, we plan to test with various subjects to see the recognition accuracy and the error pattern. Also we need to test with larger size of word list with mobile device platform.

Acknowledgment. This work was supported by JSPS KAKENHI Grant-in-Aid for Scientific Research(C) Grant Number 19K012905.

References

1. Kimura, K., et al.: Development of wearable speech enhancement system for laryngectomees. In: NCSP2016, pp. 339–342, March (2016)
2. Denby, B., Schultz, T., Honda, K., Hueber, T., Gilbert, J.M., et al.: Silent speech interfaces. Speech Commun. **52**(4), 270 (2010)
3. Kapur, A., Kapur, S., Maes, P.: AlterEgo: a personalized wearable silent speech interface. In: IUI 2018, Tokyo, Japan, 7–11 March 2018
4. Goodfellow, Ian, Bengio, Yoshua: Aaron Courville, Deep Leaning. MIT Press, Cambridge (2016)
5. Saito, Y.: Deep Learning from Scratch. O'Reilly, Japan (2016)
6. Hideki, A., et al.: Deep Leaning. Kindai Kagakusya, Tokyo (2015)
7. King, D.E.: Max-margin object detection. arXiv:1502.00046v1 31 Jan 2015
8. Kazemi, V., Sullivan, J.: One millisecond face alignment with an ensemble of regression trees. In: IEEE Conference on Computer Vision and Pattern Recognition, pp. 1867–1874 (2014)
9. Assael, Y.M., Shillingford, B., Whiteson, S., de Freitas, N.: LipNet: end-to-end sentence-level lipreading. In: GPU Technology Conference (2017)

10. Ito, D., Takiguchi, T., Ariki Y.: Lip image to speech conversion using LipNet. Acoustic Society of Japan articles, March 2018
11. Kawahara, H.: STRAIGHT, exploitation of the other aspect of vocoder: perceptually isomorphic decomposition of speech sounds. Acoust. Sci. Technol. **27**(6), 349–353 (2006)
12. Asami, et al.: Basic study on lip reading for Japanese speaker by machine learning. In: 33rd, Picture Coding Symposium (PCSJ/IMPS2018), P–3–08, November 2018
13. Saitoh, T., Kubokawa, M.: SSSD: Japanese speech scene database by smart device for visual speech recognition. In: IEICE, vol. 117, no. 513, pp. 163–168 (2018)
14. Saitoh, T., Kubokawa, M.: SSSD: speech scene database by smart device for visual speech recognition. In: Proceedings of ICPR2018, pp. 3228–3232 (2018)

Mobile Networks and Internet of Things: Contributions to Smart Human Mobility

Luís Rosa[1(✉)], Fábio Silva[1,2(✉)], and Cesar Analide[1(✉)]

[1] ALGORITMI Center, Department of Informatics, University of Minho, Braga, Portugal
id8123@alunos.uminho.pt, analide@di.uminho.pt
[2] CIICESI, ESTG, Politécnico do Porto, Felgueiras, Portugal
fas@estg.ipp.pt

Abstract. Nowadays, our society can be considered as a "connected society" thanks to a heterogeneous network and the growth of mobile technologies. This growth has meant new devices are now supporting Internet of Things (IoT) architecture. Consequently, a new look at the current design of wireless communication systems is needed. Smart mobility concerns the massive movement of people and requires a complex infra-structure that produces a lot of data, creating new interesting challenges in terms of network management and data processing. In this paper, we address classic generations of mobile technology until the latest 5G implementation and its alternatives. This analysis is contextualized for the problem of smart mobility services and people-centric services for the internet of things that have a wide range of application scenarios within smart cities.

Keywords: Mobile generation technology · Internet of Things · Networked society · Smart human mobility

1 Introduction

The rise of the digital world has placed technology at the heart of how our society is run, from working to banking to human mobility. The digital age is not just about creating new trends in technology, it is also about impacting society positively. Following on from technology trends based on the digitalization of society, the vital step is the fusion of the digital and physical worlds into cyber-physical environments, namely, an ecosystem of smart applications and services based on the interconnectivity of heterogeneous sensors. In other words, the digital transformation of society has been accelerated with the contributions of new network technologies and the Internet of Things (IoT).

In the last decades, the IoT has seen an increase in density because its real value comes from analyzing the information provided by available sensors. Information fusion improves information completeness/quality and, hence, enhances estimation about the state of things. This trust in activities of the information

Y. Dong et al. (Eds.): DCAI 2020, AISC 1237, pp. 168–178, 2021.
https://doi.org/10.1007/978-3-030-53036-5_18

fusion process and hence IoT smart applications has lead to an increase in the number of connected users. Therefore, there is a need for simultaneous connection with a spectral efficiency, great data rate, security and signaling efficiency, and reduced latency to the final users and the deployed sensors. To answer this need a migration to a new wireless communication should be implemented. In other words, the next mobile generation technology such as 5G should address many challenges for better connectivity and applications of mobile users.

The challenges in each technology lead to the creation of a newer one. Therefore, the tremendous growth wireless communications including mobile generation technologies had in the past 50 years is not surprising [1]. Over the years, each one of them has had an impact on "mobile society" and IoT device users. For example, 1G came when salespersons suddenly had a phone in their car so they could call orders while they were traveling. 2G moved phones into people's pockets, and we were able to start sending text messages to each other. 3G phones began working everywhere, not being limited to a certain region. In its turn, 4G offered inter-operability between various networks. And to increase coverage and to suite both consumers, businesses, and authorities 5G came along.

5G Wireless Communication provides interoperability between all networks and uses packet switching rather than circuit switching. For example, data transfer takes place directly without a pre-connection or no resource reservation because bandwidth is shared among users. Indeed, its flexibility meets different performance requirements of real smart city services such as human mobility. No matter where we are or what we are trying to achieve, the first step to build a human mobility service is to have a robust, reliable, and high speed wireless network. This is the backbone of all potential smart human mobility applications.

In this article, we discuss how important wireless network technologies are to the success of smart human mobility applications. Fortunately, 5G is going to get implemented in various projects but we need further research to understand the potential of connectivity in human mobility services. The aim of this article is to raise awareness to the opportunities generated by the constant evolution of IoT architecture and mobile connectivity. This study intends to survey the level of coverage of IoT architectures in human activity monitoring. Finally, we discuss future directions towards new mobile technologies as an important step to optimize and present new solutions for smart human mobility.

2 A Survey on Connectivity Platforms in IoT

The evolution of Information and Communication Technology (ICT) generates more and more devices which have the ability to communicate with other devices, which is transforming the physical world itself into an information and knowledge system. This digital transformation, although having started in a modest way in the 80's (e.g. with analog devices), for increased capacity and intelligence within networks, has led to the development of new IT infrastructure paradigms. For example, in the newest Internet of Things (IoT) architecture whereby capacity, computing, connectivity are moving closer to the devices themselves, closer to the edge of the network, Cloud technologies are becoming important. As a matter of fact, Cisco Systems, Inc. is already a driver of evolution and development in networks that involve Fog Computing, when until recently we only talked about Cloud Computing [10]. However, to better understand where we are in terms of connectivity and computing approaches we need to understand the past and its evolution over time. Table 1 summarizes the properties of each connectivity system and their effects on IoT architectures in the last 40 years.

Table 1. Evolution of mobile networks and internet of things architectures.

Mobile networks	1G	2G	2.5G	3G	3.5G	4G	5G
Year	1984	1990	1995	2000	2003	2010	2020
Frequency	800–900 MHz	850–1900 MHz	850–1900 MHz	1.6–2.5 GHz	1.6–2.5 GHz	2–8 GHz	≥6 GHz
Data capacity	24 Kbps	56 Kbps	56–115 Kbps	2 Mbps	30 Mbps	200 Mbps–1 GBps	≥1 GBps
Services	Mobile telephony	Mobile telephony and SMS	E-mail, Information and Entertainment	E-productivity, E-commerce, ...	Skype, Facebook, YouTube, ...	Identification, Tacking, ...	Monitoring, Metering, ...
Generation internet	Human to Human	WWW	WWW	Web 2.0	Social Media	Cloud Computing	Machine to Machine
Internet of Things	Network	Internet	Mobile and Internet	Mobile, People and Internet	Mobile, People and Internet	Objects, Devices, Tags and Internet	Data, Ambient Context and Internet

The evolution of Mobile Networks and Internet of Things had 7 phases. Because 0G generation/pre-cellular technology did not support roaming and lead to the to expansion of next generation mobile technology, it all began with 1G technology. This generation had a basic voice analog phone system using circuited technology for the transmission of radio signals. Moreover, it connected two computers together, creating the first network. It used an architecture called Advance Mobile Phone Service (AMPS). This architecture makes a bridge connection between a base transceiver station with a car or mobile, and also links with a mobile telecommunications switching office. This office services to customers from public telecommunications switching networks.

The next generation allowed for the combination of data and voice services such as Short Message Services (SMS), Multimedia Message Services (MMS), Global System For Mobile (GSM), and E-mail with capacity. At this stage, the World Wide Web was invented by connecting a large number of computers, but, although this was innovative, it did not support complex data such as videos. Moreover, it relied on techniques in digital multiple accessing called time division multiple access (TDMA) and code division multiple access (CDMA) standards. In 2.5G, the mobile-Internet emerged by connecting mobile devices to the Internet and also provided support for Wireless Application Protocol (WAP) and Access Multimedia Messaging Services (AMMS), and Mobile games.

To be 3 times better than GSM technology (also called 3GSM) a third generation was created. It is usually referred to as Universal Mobile Telecommunication Standard (UMTS). Beyond UMTS, the 3G was also supported by CDMA2000 technology where, to frame the International standard for 3G cellular networks, the International Telecommunication Union (ITU) signed the International Mobile Telecommunications (IMT) standard in 1999. Thereby, this generation supported 3GPP and 3GPP2 [3]. Its main disadvantage is that it requires higher bandwith, needs more network capacity that does not fail in real time applications. This called for a new connectivity technology and this is where the fourth generation comes in.

4G incorporates major requirements such as Quality of Service (QoS). After people joined the Internet via social networks on the previous connectivity platform, then, inter-operability between various networks, monitoring, tracking, and moving bigger data volume began to be accepted. This is done by small applications such as MMS, Video Chat, HDTV, Digital Video Broadcasting (DVB) and Wireless Mobile Broadcast Access (WMBA). Besides, the OFDM technique divides the channel in narrowband to transmit data pockets with greater efficiency. It is provided by a combination of CDMA and Interim Standard 95 (IS-95) (first CDMA-based digital cellular technology) due volume of data implemented. However, there are some major issues in this mobile network generation including data modification, scrambling attacks, expensive hardware and higher dependency on battery.

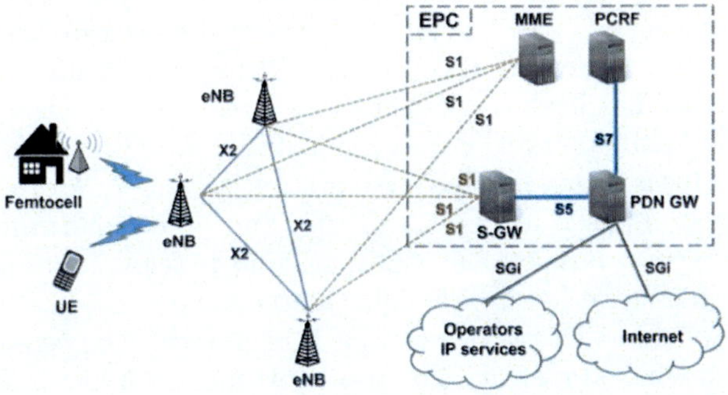

Fig. 1. Architecture of LTE-Advanced network with Evolved Packet Core (EPC), Mobility Management Entity (MME), Public Data Network Gateway (PDN GW), Serving Gateway (S-GW), evolved Node B (eNB), Serving Gateway interface (SGi), Policy and Charging Rules Function (PCRF), and User Equipment (UE) (based on [12]).

Once again, these issues called for a new connectivity platform, and that is 5G. The technology in this new mobile generation is called LTE-Advanced (LTE-A), and it is configured to support, just like 3G, Frequency Division Duplexing (FDD) and Time Division Duplexing (TDD) [8]. The LTE-A architecture is presented in Fig. 1. 5G improves communication coverage, data rate, reduces latency, and creates a true wireless world experience. Although it is not yet totally deployed, IoT is one of the main benefits of the 5G cellular network. This would make the most of the higher speed connectivity to allow for seamless integration of devices on a scale never achieved before. In other sections, we will provide more details about the 5G cellular network, the IoT, and contributions both offer to implement smart human mobility services.

3 Mobile Networks Applications

Evolution from the first generation (1G) to the nascent, fifth generation (5G) networks has been gradual. As mentioned in Sect. 2, every successive mobile generation technology has introduced advances in data throughput capacity, growth interconnectivity between devices, and decreases in latency, and 6G and 7G are no exception. Although formal 6G standards are yet to be set, 6G is expected to be at least faster than current five generation (5G) standards. To truly understand how we got here, it is useful to understand what changes and benefits 6G will bring in supporting smart human mobility services.

3.1 5G Architecture

Thanks to 5G, we are entering a new IoT era, with even more pervasive connectivity, where a very wide variety of appliances will be connected to the web. Besides, different authors have proposed architectures to meet the continuous increase in the density of cells. There is no universal consensus on 5G architecture for IoT, but Fig. 2 presents an overview of the components.

This architecture has been implemented in three levels: transport layer, processing layer, and business layer. For example, in [2], a self-configuring and scalable architecture for a large-scale IoT network was proposed. Moreover, with the integration of a cloud-based IoT platform, also known as cloud computing, data processing is done in a largely centralized fashion by cloud computers. Such a cloud centric architecture keeps the cloud at the center, applications above it, and the network of smart things below it [5]. This information center networking is combined by all wireless transmission protocols spectrum bands and standards under a single network control plane, introducing a high number of small cells as pico-cells and femtocells, as well as Wi-Fi hotspots in order to increase bandwidth per cell and provide throughput for end users. For short-range data transmission, which benefit the 5G-IoT with lower power consumption, and better Quality of Service (QoS) for users and load balancing, this architecture seems like a promising solution to enhance network capabilities by allowing nearby devices to establish direct links between themselves. Part of human mobility tracking is transmitted between devices instead of being transmitted through the base station. The main advantages are flexibility and efficiency, thanks to the use of dynamic small cells instead of fixed cells. Therefore, the 5G architecture should be included in smart city solutions.

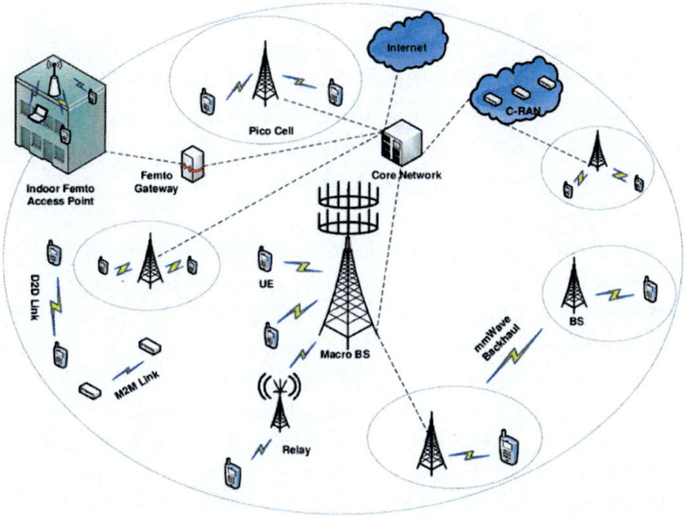

Fig. 2. Architecture of 5G cellular network with Device-to-Device (D2D Link), Base Station (BS), Macro Base Station (Macro BS), Cloud Radio Access Network (C-RAN), User Equipment (UE), and Machine-to-Machine (M2M Link) (based on [7]).

3.2 Impact on Cities

5G networks are predicted to carry 45% of global mobile traffic by 2025 - a 10% increase from what was forecasted in June 2019. Basically, 5G will support nearly half of the world's mobile data traffic, but has the potential to cover up to 65% of the world's population [4]. Moreover, we are in the final stage of the next technological revolution: the development of a ubiquitous wireless network that will marry data collection and computation with billions of devices. This will push the number of mobile devices to the extreme, with 10^7 devices per km^2 in dense areas (up from 10^6 in 5G) and more than 125 billion devices worldwide by 2030.

Service providers and consumers have shown enthusiasm in different mobile generations. This reflects they are improving people's lives in many different aspects. Unlike cell phones in 1995, cell phones can do much more than just make a phone call. Nowadays, 5G is definitely taking off, while 6G and 7G will be next, and having a positive impact on the development of new services and solutions for smart cities. If we review the characteristics and foreseen requirements of use cases that represent 5G services, we provide a comprehensive view in many new applications that are already viable today, particularly in urban areas and cities.

Autonomous Vehicles. This new application requires a response in fractions of a second, and the use of Ultra Reliable Low Latency Communication (uRLLC). A main goal is a vehicle-to-everything communication network. This enables vehicles to automatically respond to objects and changes around them almost instantaneously. A vehicle must be able to send and receive messages in milliseconds in order to brake or shift directions in response to road signs, hazards and people crossing the street. Sure, autonomous cars may be fitted with cameras capable of face recognition and/or LIDAR scanners that can pick out moving objects in the vicinity of the vehicle. Eventually all devices should alert passing vehicles to the presence of a pedestrian.

Drone Management. Drones have a vast and growing set of use cases today beyond the consumer use for filming and photography. For example, utilities are using drones today for equipment inspection. Logistics and retail companies are looking for drone delivery services. The trend is to push the limits of drones that exist today, especially in terms of range and interactivity. With 5G we are able to see beyond current limits with low latency and high resolution video. It also extends the reach of controllers beyond a few kilometers or miles. Moreover, these developments have implications for use cases in search and rescue, border security, or surveillance.

Human Activity Trackers. 5G has a very gradual and initially negligible impact on consumer wearables such as smartwatches, mobile phones. Where, for example, the maker of the smartwatch provides secure cloud services and enhanced big data analytics to truly become the personal trainer of the wearer. An ultra-fast connection to massively powerful compute power on the cloud will come in useful for keeping wearables small, light and battery-efficient. 5G-enabled wearables play a major part in 'smart roads' of the future. While the smartwatch or Fitbit may not require high-speed connectivity with ultra-reliable, low-latency communications, it is used to alert autonomous and assisted vehicles to the presence of its wearer close to the road, or crossing the road, a scenario called "vehicle to pedestrian" communications.

Although 5G enables the connection of thousands of wearables and sensors per square kilometre, supporting both the high-speed, high-bandwidth mode and the low-power, long-life, small-messaging model, this use-case won't support much data per device, so it will be hard to apply AI to it. Basically, what this mobile phone generation will do is to provide a permanent connection where wearables could swap between them as and when they need to.

4 Emergent Mobile Technologies and Infrastructures

5G continues to be at the center of the discussion around networking technologies, as well as Internet of Things use cases. Nevertheless, Low Power Wide-Area Networks (LPWANs), Wireless Local Area Networks (WLANs), Bluetooth, NB-IoT to CAT-M1, and LoRa to SigFox, are some of the emergent alternatives able to produce solid applications with their own unique use case scenarios. In fact, an IoT connectivity option that is ideal for one application may be awful for another. No single technology is able to cover all applications. This opens the door for a much broader ecosystem of "best fit" technologies.

4.1 Impact on IoT

For years, the prospect of reaching billions of connected devices is coming from applications that promise yielding revolutionary results. The successful deployment of many of them requires applying transformative enabling technologies and new connectivity models. However, industries such as agriculture, healthcare, activity trackers and smart buildings or homes that may take advantage of short-range technologies such as Wi-Fi, Bluetooth or Radio Frequency Identification (RFID), or long-range communications technology like cellular or satellite, do not need to implement a long-range connectivity solution. Nowadays, there are alternatives with low-power consumption and more advantageous customer engagement models to meet their digital transformation goals.

To this end, LPWANs and WLANs are fundamentally changing the IoT landscape [11]. LPWANs are designed for connecting devices that send and receive small data packets over long distances. Currently, Sigfox and LoRa (beyond NB-IoT, LTE-M, Weightless or InGenu) are the major competitors in the LPWAN space. And while the business models and technologies behind the companies are quite different, the end goals of both Sigfox and the LoRa Alliance are very similar: that mobile network operators adopt their technology for IoT deployments over both city and nationwide low power, wide-area networks (LPWANs), using unlicensed communication technologies. The Sigfox business model takes a top-down approach. The company owns all of its technology, from the backend data and cloud server to the endpoints software. But the differentiator is that SigFox is essentially an open market for the endpoints where manufacturers like STMicroelectronics, Atmel, and Texas Instruments make Sigfox radios. Additionally, only one Sigfox network can be deployed in an area because the company has exclusive arrangements with network operators when they work together. While LoRa or LoRaWAN is differentiated by its open ecosystem, strong security specifications, bi-directional communication, optimization for mobility, and scalability for capacity. The LoRaWAN architecture is a high availability, fault tolerant and redundant platform. It is usually used on fuel tanks, fill level and other valuable data is sent over LoRaWAN networks to tank monitoring solutions used by supply companies [6].

However, LPWANs are not available in many places of the planet, for example, inside a division of a house. Fortunately, there is another alternative to turn to: WLAN (used by Bluetooth 5 and Bluetooth Low Energy, Thread, Wi-Fi, ZigBee or Z-Wave). Although, LPWAN and Wi-Fi 6 use similar technology (e.g. methods for encoding data into radio waves), in building infrastructure, and the way we interact with systems are different. Because they are designed to ensure that people operating from a workstation or in their home can connect on a variety of different devices by using the internet and, consequently, provide different coverage levels. Moreover, in this type of set up, there are no wires connecting the devices to the network,ensuring that they can be distributed across a significant distance. Thus, the major advances coming with 5G and its alternatives will open up ever more opportunities and surprising applications for all of us.

4.2 Impact on Asset Tracking Applications

In a smart city, asset tracking services are expected to overcome different social challenges: be safe, efficient, reliable and provide custom responses. Besides, WLAN and LPWAN networks offer advanced real-time communication services and with a simple density of network nodes, covering indoor and outdoor places where 5G Wireless Communication does not reach efficiently. In this context, these 5G Infrastructure Alternatives are regarded as the ultimate platform to track human mobility in smart buildings or use cases where the frequency/size of messages require low bandwidth.

However, the challenges that exist in the asset tracking space (i.e tracking of human mobility) and indoor positioning include installation, scale and availability. The truth is that it is hard to install a correct amount of hardware required for a given space. The addition of unnecessary hardware leads to redundancy and expenses. On the other hand, not enough hardware can result in dead zones and a failed solution. In addition, as the spaces we wish to use indoor positioning and asset tracking on expands, the complexity, security vulnerability and cost increases as well [9]. What may work well in a small 400 square foot area can cost dramatically different compared to a 16,400 square foot warehouse.

5 Conclusions

According to the evolution and use cases from mobile technologies, we can consider that these technologies resulted from a strong relationship between humans and devices associated to the Internet of Things. As we have seen, the growth of IoT correlates with the direct growth in device users via the Internet. This rapid growth and increase in the interconnectedness between devices and people means there is a wide range of requirements that needs to be met. Thus, the advent of 5G network is expected to massively improve or expand IoT security, increase speed, boost cellular operation with increased bandwidth while also overcoming several network challenges faced in the previous generations of mobile networks.

On the other hand, thanks to the Internet of Things the information we are looking for will help us to know more about our activities at anytime, anywhere and at any place of our life which is only possible thanks to the evolution of connectivity platforms in IoT. For example, it can provide location information from massive device connectivity and improve coverage for everybody. Once people use a set of devices (e.g. smartphones, smartwatches, etc) that take advantage of Global Positioning System (GPS) and Bluetooth Low Energy (BLE) technologies, tracking them is already possible, allowing up-to-date location information.

In its turn, although generations - 2G, 3G, 4G and 5G - have succeeded, different alternatives arise from the design of their infrastructure and their use cases in a internet of things scenario. Using these infrastructure alternatives, for example, we can easily find and locate areas most visited by people or even identify patterns of human mobility in indoor places. But, despite all this, to make all of them available to customers there are costs and challenges. One of the challenges is the cost of rolling out to different proportions of the population and the investment cost for rolling out a high coverage solution, mainly in relation to urban-rural settlement patterns. In addition, there are a wide range of use cases which fall within fast mobile broadband and massive machine communications. All this is built to be available and used to track human mobility.

In the future, we hope to keep the balance between these different network technologies, expanding IoT architectures and enabling the smart composition of smart mobility services that can track user and device actions, with a connection suited to the rquirements of each use case, and the capability to travel between different network endpoints. To this end, we aim to extend frameworks that use multiple connectivity technologies to build services in the context of the internet of things and smart cities.

Acknowledgements. This work has been supported by FCT – Fundação para a Ciência e Tecnologia within the R&D Units Project Scope: UIDB/00319/2020. It has also been supported by national funds through FCT – Fundação para a Ciência e Tecnologia through project UIDB/04728/2020.

References

1. Benisha, M., Prabu, R.T., Bai, T.: Evolution of mobile generation technology. Int. J. Recent Technol. Eng. **7**(5), 449–454 (2019)
2. Cirani, S., Davoli, L., Ferrari, G., Leone, R., Medagliani, P., Picone, M., Veltri, L.: A scalable and self-configuring architecture for service discovery in the internet of things. IEEE Internet Things J. **1**, 508–521 (2014)
3. Del Peral-Rosado, J.A., Raulefs, R., López-Salcedo, J.A., Seco-Granados, G.: Survey of cellular mobile radio localization methods: from 1G to 5G (2018). https://doi.org/10.1109/COMST.2017.2785181
4. Ericsson: Ericsson Mobility Report, June 2019 (2019)
5. Gubbi, J., Buyya, R., Marusic, S., Palaniswami, M.: Internet of Things (IoT): a vision, architectural elements, and future directions. Future Gener. Comput. Syst. (2013). https://doi.org/10.1016/j.future.2013.01.010
6. Ibrahim, D.M.: Internet of things technology based on LoRaWAN revolution. In: 2019 10th International Conference on Information and Communication Systems, ICICS 2019, pp. 234–237 (2019). https://doi.org/10.1109/IACS.2019.8809176
7. Kar, U.N., Sanyal, D.K.: A sneak peek into 5G communications. Resonance **23**, 555–572 (2018). https://doi.org/10.1007/s12045-018-0649-4
8. Levanen, T.A., Pirskanen, J., Koskela, T., Talvitie, J., Valkama, M.: Radio interface evolution towards 5G and enhanced local area communications. IEEE Access (2014). https://doi.org/10.1109/ACCESS.2014.2355415
9. Malik, H., Khan, S.Z., Sarmiento, J.L.R., Kuusik, A., Alam, M.M., Le Moullec, Y., Parand, S.: NB-IoT network field trial: Indoor, outdoor and underground coverage campaign. In: 2019 15th International Wireless Communications and Mobile Computing Conference, IWCMC 2019, pp. 537–542 (2019)
10. Moffett, M., Dodson, K.: Understanding the value of edge and fog computing to your community (2018). https://bit.ly/2JvDRql

11. Nikoukar, A., Raza, S., Poole, A., Gunes, M., Dezfouli, B.: Low-power wireless for the internet of things: standards and applications. IEEE Access (2018). https://doi.org/10.1109/ACCESS.2018.2879189
12. Tran, T.T., Shin, Y., Shin, O.S.: Overview of enabling technologies for 3GPP LTE-advanced. Eurasip J. Wirel. Commun. Netw. **2012**, 54 (2012). https://doi.org/10.1186/1687-1499-2012-54

Design and Implementation of a System to Determine Property Tax Through the Processing and Analysis of Satellite Images

Jesús Silva[1] (✉), Darwin Solano[2], Roberto Jimenez[2], and Omar Bonerge Pineda Lezama[3]

[1] Universidad Peruana de Ciencias Aplicadas, Lima, Peru
jesussilvaUPC@gmail.com
[2] Universidad de la Costa, Barranquilla, Colombia
{dsolano1,rjimenez12}@cuc.edu.co
[3] Universidad Tecnológica Centroamericana (UNITEC), San Pedro Sula, Honduras
omarpineda@unitec.edu

Abstract. One of the main objectives when implementing metaheuristics in engineering problems, is to solve complex situations and look for feasible solutions within a defined interval by the design dimensions. With the support of heuristic techniques such as neural networks, it was possible to find the sections that allow to obtain the characteristics of interest to carry out the study of the important regions of an image. The analysis and digital processing of images allows to smooth the file and to section the area of analysis in regions defined as rows and columns, results in a matrix of pixels, this way carrying out the measurement of the coordinates of the beginning and end of the region under analysis, taking it as a starting point for the creation of a frame of references to be examined. Once this requirement is completed, it is possible to return to the smoothed image with which the high edges of the image will be determined by means of the Gaussian function, thus finding the edges generated for the structures of interest.

Keywords: Property tax · Processing and analysis of satellite images · Gaussian function

1 Introduction

In [1], a prototype system was designed to use different types of computer techniques to perform image analysis using neural network models. This system uses numerical information (at the pixel level) to perform image interpretation tasks, identifying the presence of certain ocean structures in Africa and the Canary Islands.

While in [2], a computer tool for medical support is developed to identify whether one person has a malignant or benign tumor, making use of digital image processing as well as artificial neural networks.

Y. Dong et al. (Eds.): DCAI 2020, AISC 1237, pp. 179–186, 2021.
https://doi.org/10.1007/978-3-030-53036-5_19

On the other hand, in [3], the authors designed a geospatial tool that can automatically detect objects in satellite photographs using image processing and artificial vision techniques.

Additionally, a multibiometric system was designed and built in [4] for the identification of personnel by means of iris and fingerprint artificial vision. The image obtained is manipulated to extract information through the Haar Wavelet transform and/or bifurcations and adaptive correlation in 2D, using neural networks for its classification.

Similarly, in [5] an attempt is made to survey forest areas using satellite image processing for subsequent classification and management of forest resources in Tierra del Fuego. Likewise, in [6], the analysis and detection of buildings is carried out automatically/semi-automatically on aerial and satellite images by obtaining thresholds using computer tools.

Also, in [7], the design of an automatic inspection system that can improve this quality control process in textile companies, using image analysis to eliminate imperfections in the fabrics is discussed.

Furthermore, in [8], an image processing software is developed for metallographic analysis for the determination and automation of the metric estimation of the grain and ferritic-perlitic phases in low carbon steels.

Finally, in [9], a digital classification method was developed to generate cartography of the Valdivia National Reserve using satellite images, using a supervised classification method and applying a maximum probability algorithm.

With respect to the mentioned background, the disadvantages are that the results obtained are very idealistic and difficult to put into practice. However, this research intends to contribute to correct this situation, through the implementation of restrictions that allow obtaining feasible results and that lead to using this method for diverse designs with image processing and neural networks in engineering.

2 Methodology

According to the problem posed, it is possible to adapt the use of heuristics to find the necessary regions in the system, since they are optimization and analysis methods that can be used to find solutions to non-linear or analytically complex problems that are difficult to specify in a non-intelligent system [10].

Based on this, the artificial neural networks are applied to image analysis, thus two types of training are used for the network used in this study, supervised learning and learning by reinforcement [11], allowing the network to be trained in such a way that it is able to carry out the analysis required by the system and train itself constantly with each use. In this case, the system is analyzed according to Fig. 1 [8, 10, 12, 13].

2.1 Neural Networks

For the implementation of an artificial neural network it is necessary to understand that they are based on a biological neural network, so its basic element is the neuron, and the research seeks to generate a structure that emulates this condition. A biological neuronal system is composed of millions of neurons organized in layers. In the emulation of this

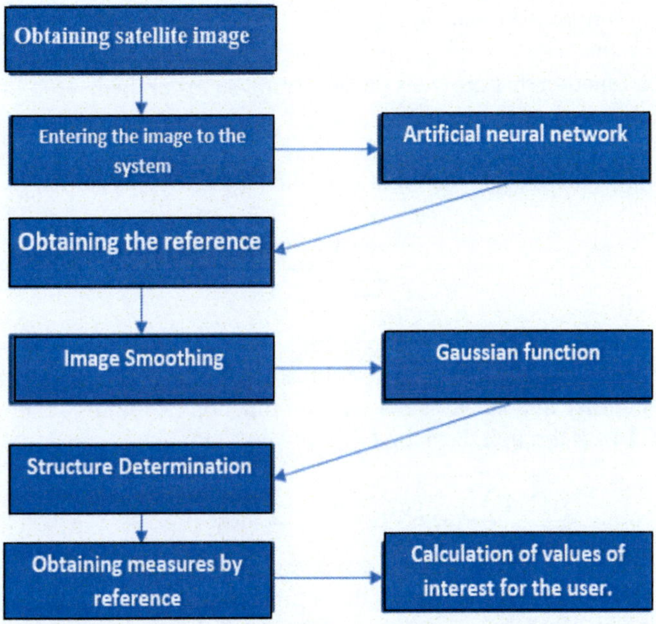

Fig. 1. Process considered for the design of the system

biological neuronal system, by means of an artificial neuronal system, a hierarchical structure can be established, similar to that existing in the brain [14].

To carry out the design of the system it is necessary to consider the functioning of an artificial neural network, so it is important to analyze the structure and architecture of the neurons. This research is based on a McCulloch-Pitts neuron architecture [8], which works in a very similar way to a biological neuron. To be more precise, it is a calculation unit that attempts to emulate the behavior of a biological neuron. In Fig. 2, the architecture of the artificial neuron can be seen.

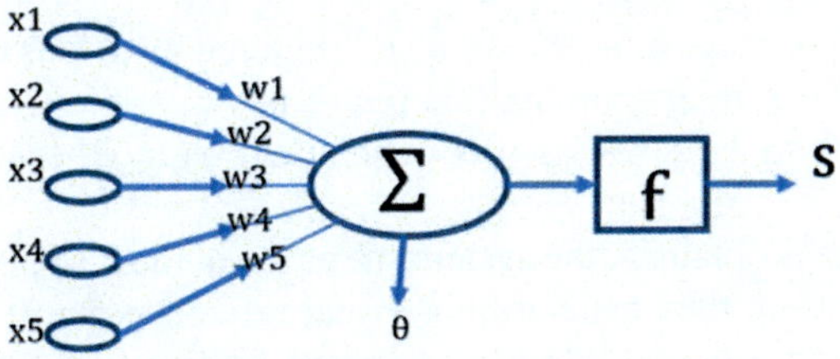

Fig. 2. Architecture of the artificial neuron [13].

This artificial neuron consists of the following elements [2]:

X: A set of n components called a system input vector.
W: Set of synaptic weights of the connection between the input and the neuron.

Σ: Processing sum of the input values.
θ: threshold value.
F: Activation function that provides the activation status of the neuron in function.
S: neuron output

Within the neuron, the calculation of the following Eq. (1) is carried out [7]:

$$\sum_{i=1}^{n} w_i x_i. \quad (1)$$

It should be noted that the McCulloch-Pitts architecture allows two possible outcomes 1 and 0, thus making it necessary to compare the result obtained with the threshold of the neuron in order to detonate the activation function (Eq. 2) by sending the result of the neural network through the axon [12]:

$$s_1\{1\, si \sum\nolimits_{i=1}^{n} w_i x_i \geq \theta\; 0\, si \sum\nolimits_{i=1}^{n} w_i x_i < \theta. \quad (2)$$

Neurons function as simple information processors. Simplistically, dendrites are the input channel of information, the soma is the computational organ, the axon corresponds to the output channel, and at the same time it sends information to other neurons. Each neuron receives information from approximately 10,000 neurons and sends impulses to hundreds of them.

2.2 Obtaining the Reference

In this part of the system, the satellite image will be searched for and used as a reference, in this case for analysis and as a system test. This section takes the Tecnológico de Estudios Superiores de Ecatepec as a reference, specifically the computer division building as an example, as shown in Figs. 3 and 4 respectively. Once the image is obtained, it is possible to notice that in the lower right part it contains an approximate reference measurement, which will be taken as a reference in the system to approximate a unit of measurement (see Fig. 5).

Taking the obtained image as an example, it is entered by means of the system's file scanning assistant, where by processing this image it is possible to detect the position of the reference that will be taken as part of the data required for the calculation of results [8].

Once the reference is obtained, the system stores the number of pixels obtained from the metric approximation, thus generating a visual reference that it uses so that once the objects are detected it is possible to compare the dimensions of each object with the value, generating a scale between pixel per meter or number of pixels in a meter in the same way that the reference of the satellite image is obtained, in this step this information is added by means of the image processing. The result will be something similar to what is shown in Fig. 6, which shows the file in the processing stage, where the visual reference is added for the analysis of the structure dimensions detected in the photograph.

Fig. 3. Google Maps platform.

Fig. 4. Test satellite image

Fig. 5. Reference of measures to be taken by the system.

Fig. 6. Satellite image with measurement reference

The figure above helps to understand the importance of the measurement reference, since the system generates the reference through the measurement estimate provided by the platform from which the image was obtained. Based on this, the measurement units required for the system will be obtained.

2.3 Edge Detection

Edge detection is an image analysis and processing technique that allows objects to be identified and isolated. Once the projections that make up the edges of a figure have been obtained, it is possible to count the objects shown in the image that is being used as a study object, in order to store the positions in coordinates of each point found as an edge by means of matrices [13].

This research uses the so-called artificial vision, which is a subfield of mathematics that encompasses many techniques that allow digital image processing. With this technique, specific objects can be counted using the OpenCV tool. During image analysis, it is preferable to have a clear contrast between the objects being searched for and the background. This will allow the algorithm to perform the analysis in a more efficient way.

The process is divided into 5 phases [8]:

1. Convert image to grayscale
2. Filter the image to remove noise
3. Apply the edge detector
4. Searching for contours within the detected edges
5. Drawing these contours.

3 Results and Discussion

The found set is painted on the original image, which will allow users to see the edges detected by the system represented by red lines on the original image. This color is used with the aim of making them as visible as possible and highlighting them for both the system and the user. At the end, the results can be shown in pixels and the amount of objects that were detected within the image, thus generating the calculation of the measurements in meters, as shown in Fig. 7.

We have found 74 objects.

Perimeter	=	1845
Side	=	450.77
Area	=	167452.1

Fig. 7. Pixel measurements and object counting.

Fig. 8. Object measurement

Finally, as a result, the measurements obtained in meters superimposed on the image are shown in Fig. 8, which will allow the corresponding calculations to be made to find the required values.

4 Conclusions

It is possible to develop the design of a system for the determination of property tax through the processing and analysis of satellite images, using methods such as artificial neural networks, image analysis and processing, determination of measurements with respect to matrices generated from images. It is possible to emphasize that the use of heuristics in the implementation of the designed system has been very useful due to the complexity of the detection of certain components required by the system, within both the images and the bit matrices used during the structure detection.

References

1. Dorai, C., Jerome, W.F., Stern, E.H., Winegust, F.S.: U.S. Patent No. 7,809,587. U.S. Patent and Trademark Office, Washington, DC (2010)
2. Vargas, R., Torres-Samuel, M., Luna, M., Viloria, A., Fernández, O.S.: Formulation of strategies for efficient cadastral management. In: International Conference on Data Mining and Big Data, pp. 523–532. Springer, Cham (2018)
3. Ali, D.A., Deininger, K., Wild, M.: Using satellite imagery to revolutionize creation of tax maps and local revenue collection. The World Bank (2018)
4. Ali, D.A., Deininger, K., Wild, M.: Using satellite imagery to create tax maps and enhance local revenue collection. Appl. Econ. **52**(4), 415–429 (2020)
5. Vishnoi, D.A., Padaliya, S., Garg, P.K.: Various segmentation techniques for extraction of buildings using high resolution satellite images. In: On a Sustainable Future of the Earth's Natural Resources, pp. 251–261. Springer, Heidelberg (2013)
6. Brimble, P., McSharry, P., Bachofer, F., Bower, J., Braun, A.: Using machine learning and remote sensing to value property in Kigali (2020)
7. Llulluna, L., Fredy, R.: Image processing using free software python for metallographic analysis in low-carbon steels Quito, pp. 60–79 (2014)
8. Yildirim, V., Ural, H.: A geographic information system for prevention of property tax evasion. In: Proceedings of the Institution of Civil Engineers-Municipal Engineer, vol. 173, no. 1, pp. 25–35. Thomas Telford Ltd., March 2020
9. Awasthi, R., Nagarajan, M., Deininger, K.W.: Property taxation in India: Issues impacting revenue performance and suggestions for reform. Land Use Policy, 104539 (2020)
10. McCluskey, W., Huang, C.Y.: The role of ICT in property tax administration: lessons from Tanzania. CMI Brief **2019**(06) (2019)
11. Duncan, M., Horner, M. W., Chapin, T., Crute, J., Finch, K., Sharmin, N., Stansbury, C.: Assessing the property value and tax revenue impacts of SunRail stations in Orlando, Florida. Case Stud. Transp. Policy (2020)
12. Canaz, S., Aliefendioğlu, Y., Tanrıvermiş, H.: Change detection using Landsat images and an analysis of the linkages between the change and property tax values in the Istanbul Province of Turkey. J. Environ. Manage. **200**, 446–455 (2017)

13. Collier, P., Glaeser, E., Venables, T., Manwaring, P., Blake, M: Land and property taxes: exploiting untapped municipal revenues. Policy brief (2017)
14. Gaitán-Angulo, M., Viloria, A., Lis-Gutiérrez, J.P., Neira, D., López, E., Sanabria, E.J.S., Castro, C.P.F.: Influence of the management of the innovation in the business performance of the family business: application to the printing sector in Colombia. In: International Conference on Data Mining and Big Data, pp. 349–359. Springer, Cham (2018)

Multi-step Ultraviolet Index Forecasting Using Long Short-Term Memory Networks

Pedro Oliveira(✉), Bruno Fernandes, Cesar Analide, and Paulo Novais

Department of Informatics, ALGORITMI Centre, University of Minho, Braga, Portugal
poliveira199208@gmail.com, bruno.fmf.8@gmail.com, {analide,pjon}@di.uminho.pt

Abstract. The ultraviolet index is an international standard metric for measuring the strength of the ultraviolet radiation reaching Earth's surface at a particular time, at a particular place. Major health problems may arise from an overexposure to such radiation, including skin cancer or premature ageing, just to name a few. Hence, the goal of this work is to make use of Deep Learning models to forecast the ultraviolet index at a certain area for future timesteps. With the problem framed as a time series one, candidate models are based on Recurring Neural Networks, a particular class of Artificial Neural Networks that have been shown to produce promising results when handling time series. In particular, candidate models implement Long Short-Term Memory networks, with the models' input ranging from uni to multi-variate. The used dataset was collected by the authors of this work. On the other hand, the models' output follows a recursive multi-step approach to forecast several future timesteps. The obtained results strengthen the use of Long Short-Term Memory networks to handle time series problems, with the best candidate model achieving high performance and accuracy for ultraviolet index forecasting.

Keywords: Deep Learning · Ultraviolet index · Long Short-Term Memory networks · Time series forecasting

1 Introduction

Ultraviolet (UV) index is a standard metric used to express the magnitude of UV radiation reaching Earth's surface at a particular time, at a given region. Ozone in the stratosphere, also known as "good" ozone, protects life from harmful UV radiation. However, due to the burning of fossil fuels, which releases carbon into the atmosphere, the ozone layer has become thinner, leading to dangerous UV radiation reaching Earth's surface [1].

Over the years, the increase of UV radiation reaching Earth's surface has been associated with increased rates of skin cancers, particularly melanomas [2].

Y. Dong et al. (Eds.): DCAI 2020, AISC 1237, pp. 187–197, 2021.
https://doi.org/10.1007/978-3-030-53036-5_20

Indeed, information regarding UV index variations can be essential for the human being. Despite harmful in high concentrations, UV radiation is also essential for humanity. As an example, exposure to UV radiation activates Vitamin D, a key regulator in the calcium and phosphate homeostasis with implications in several human body systems [2]. For a better perception of which concentrations lead to harmful UV radiation, the World Health Organisation and the World Meteorological Organisation proposed a standardised global UV Index scale. UV index values between 0 and 2 are of low risk; between 3 and 5 are of moderate risk; between 6 and 7 start carrying some risk; between 8 and 11 are very dangerous to the human being; and values higher than 11 are of extreme danger [3].

Knowing, beforehand, when the UV index will achieve high or extreme values is of the utmost importance as it allows one to adjust his behaviour and avoid risky moves. Hence, the goal of this work is to make use of Deep Learning models to forecast the UV index at a certain area for several future timesteps, in particular for the next three days. With the use of Deep Learning models, it becomes possible to forecast future time points in a given scope. Being this a time series problem, uni and multi-variate Long Short-Term Memory networks (LSTMs), a subset of Recurrent Neural Networks (RNNs), were conceived and evaluated, with the goal being to forecast the UV index.

The remainder of this paper is structured as follows: the next section summarises the state of the art on the subject of UV forecasting. The third section aims to present the materials and methods, focusing on the collected dataset, its exploration and all the applied treatments. The fourth and fifth sections yield a description of the performed experiences and a discussion of the obtained results, respectively. The sixth and final section notes major conclusions taken from this work and presents future perspectives.

2 State of the Art

Across the years several studies have been carried out on the topic of UV forecasting [4–6]. Gómez et al. [4], used the Santa Barbara DISTORT Atmospheric Radiative Transfer (SBDART) algorithm, developed by Ricchiazzi et al. [7], to predict the UV index for Valencia, Spain, in a period of 3 days. The SBDART model calculates the radiative transfer parallel to the Earth's atmosphere [7]. In this study, this model had, as input, the Total Ozone Columns (TOC) data through the Global Forecast System (GFS). Hence, as the UV incidence forecast would be for the next 3 days, there was a gap of 4 days between the last available data, also limiting a forecast to the next day. The metrics used to evaluate the models were the Root Mean Square (RMS) and Mean Value (MV). With these metrics, the authors concluded that the TOC GFS obtained interesting results when forecasting the UV index for the next day.

Another study was conducted by Ravinesh et al. [5], with the authors focusing on a short term forecast of 10 min. These authors propose an Extreme Machine Learning (EML) model that is based on a Single Layer Feed Forward Network (SLFN), that use a FeedForward Back Propagation (FFBP). To evaluate the performance of the models, the authors used two metrics as criteria: Mean Absolute Error (MAE) and Root Mean Square Error (RMSE). In this study, the EML model outperformed the Multivariate Adaptive Regression Splines (MARS) model and a Hierarchical model (M5 Model Tree). The data was partitioned with 80% for training and the remaining 20% for testing and validation. The ELM model shows a decrease of 0.1, in both metrics, compared to the other models.

Puerta et al. [6] proposed a Deep Learning Model to forecast UV index to predict erythema formation in the human skin. Erythema corresponds to redness of the skin due to dilation of the superficial capillaries. The fit of the model is carried out using a Deep Belief Network (DNB). In this model, the backpropagation method is also applied, adjusting the weight values according to the Mean Squared Error (MSE). The conceived neural network has, as input, the average temperature, the clear sky index, the insolation, and the UV index of the day before the one it intends to predict. Regarding the validation of the model, it was carried out by comparison with the records extracted from the National Aeronautics And Space Administration (NASA) Prediction Of Worldwide Energy Resource. A value of 66.8% of correct classifications was obtained in the values predicted by the conceived model.

Interestingly, in the environmental sustainability domain, there is a clear lack of studies that use machine or deep learning to predict the UV index. There are however studies that focuses on related topics such as forecasting ozone values at ground level. Ghoneim et al. [8], conducted a study to predict ozone concentration in a smart city. This study aimed to compare a deep learning model to regression techniques, such as Support Vector Machines (SVM). In this study, data related to pollution and weather from the CityPulse project [9] were used. Grid search was used to optimise hyperparameters of the model, such as the number of hidden layers and the number of neurons in each layer. MSE, RMSE and MAE were the used metrics. With a focus on RMSE, the deep learning model had a lower RMSE when compared with the other models. In fact, the deep learning model outperformed all other models for all used metrics.

3 Materials and Methods

The materials and methods used in this study to conceive and evaluate the candidate models are detailed in the next lines. The used evaluation metrics are also described as well as the conceived deep learning models.

3.1 Data Collection

The UV index dataset used in this study was created from scratch using real-world data. For that purpose, a software was developed to collect data from a set of soft sensors. In this case, the Open Weather Map API, whose features are the type of pollutant, the name of the city, the value of the UV index, the source from which the record is taken, and the timestamp. Data collection started on July

24, 2018 and was maintained, uninterruptedly, until the present. The developed software makes API calls every hour using an HTTP Get request. It parses the received JSON object and saves the records on a database. The software was developed so that any other type of polluter can be easily added to the list. The present work analyses data collected until February 24, 2020.

3.2 Data Exploration

The collected dataset consists of a total of 16375 timesteps with data being physically gathered by three distinct hard sensors. Each observation, from now on designated as timestep, consists of a total of 8 features. Most features are of the string type, with the UV index being a double-typed feature. The remaining features are integers. The creation date feature consists of the date and time. Table 1 describes the features available in the collected dataset.

Table 1. Features of the collected dataset.

#	Features	Description
1	*id*	Record identifier
2	*pollution_type*	Pollutant type
3	*city_name*	Name of the city under analysis
4	*value*	UV index
5	*data_precision*	Precision of the value feature
6	*source_name*	Source from where the timestep was obtained
7	*last_update*	Last date when the value was updated
8	*creation_date*	Timestamp (YYYY-MM-DD HH24:MI:SS)

After an analysis of the dataset, it is possible to understand that the *last update* feature does not have any value assigned as well as the *data precision* feature which is always filled with the same value. Being this a time series problem, one must be aware of all missing timesteps. In total there were missing 102 timesteps. Some of which corresponded to periods of roughly 1 month, between December 13, 2018 to January 14, 2019, and between March 7, 2019 to April 9, 2019.

To understand the variation of the UV index over the course of a year, the monthly average was analysed. Figure 1 illustrates the average of the UV index for the years 2018 and 2019. It is possible to verify that the peak of the UV index is reached during July. From that month onwards, the index declines until December and starts to increase again in January until reaching its peak. Therefore, it is possible to highlight the existence of seasonality as well as cyclicality. The highest values are reached during the summer, reducing during the fall and the beginning of the winter.

3.3 Data Preparation

The available dataset includes observations from July 24th, 2018 to February 24th, 2020 made at one hour time intervals. The first step in the preparation of the dataset is to apply feature engineering and create the year, month, day, hour, and minutes as features to each observation. Daily forecast requires daily UV index measures. Thus, the average of observations was considered to create daily timesteps.

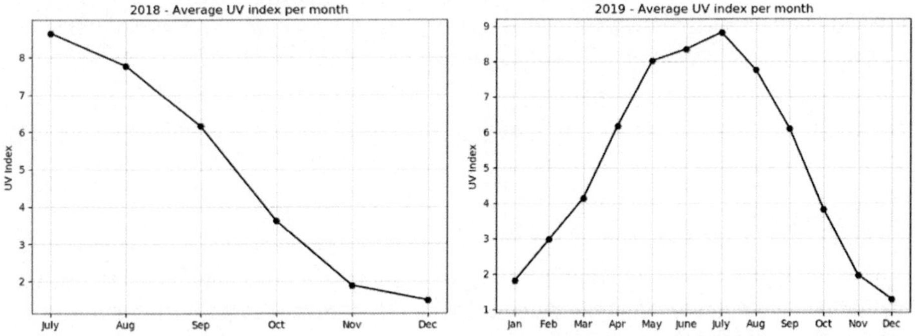

Fig. 1. Average UV index per month in the years of 2018 and 2019.

No missing values were found in the dataset. However, missing timesteps were present in the dataset, either due to certain limitations of the API or because it was unavailable (not recorded or not measured). The lack of timesteps can result in the development of incorrect standards, so it was necessary to fill in the missing values. For that purpose, the Open Weather Map Historical UV API was used. By the end of this step, the dataset consisted of 581 daily timesteps.

The next step consists in removing some informative features that will not be used by the conceived models such as the *id*, *pollution type*, *city name*, *data precision*, *source name* and *last update*. The *hour* and *minute* features were also disregarded since the dataset was grouped by day.

Since LSTMs work internally with the hyperbolic tangent, normalisation of the dataset was performed, with all features falling within the range [−1, 1], according to the following equation:

$$\frac{x_i - min(x)}{max(x) - min(x)} \tag{1}$$

At the end of the data preparation process, two datasets were created. Both datasets contain 581 timesteps, varying only in the number of features. One dataset (**Uni-variate**) contains only the value of the UV index for each timestep. The second dataset (**Multi-variate**) contains, beside the UV index for each timestep, the month of the year and the day.

3.4 Evaluation Metrics

To obtain the best combination of parameters of the candidate models, two error metrics were used. One, the RMSE, is a measure of accuracy, as it measures the difference between the values predicted by the model and the true values observed. RMSE equation is as follows:

$$\mathrm{RMSE} = \sqrt{\frac{\sum_{i=1}^{n}(y_i - \hat{y}_i)^2}{n}} \tag{2}$$

The second metric, the MAE, is the mean of the differences between predicted and observed values. Its use is mainly to complement and strengthen the confidence of the obtained values. Its equation is as follows:

$$\mathrm{MAE} = \frac{1}{n}\sum_{i=1}^{n}|y_i - \hat{y}_i| \tag{3}$$

3.5 LSTMs

RNNs constitute a class of artificial neural networks where the evolution of the state depends on the current input as well as the input at previous timesteps. This property makes it possible to carry out context-dependent processing, allowing long-term dependencies to be learned [10]. A recurrent network can have connections that return from the outgoing nodes to the incoming nodes, or even arbitrary connections between any nodes.

LSTMs are a type of RNN architecture. Unlike a traditional neural network, this architecture is used to learn from experience how to classify, process and predict time series, as is the case with this study. This type of network aims to help preserve the error, which can be propagated through time and layers. The technique of keeping the error constant, allows this type of networks to continue their learning process over many "steps" in time [10].

LSTMs contain information outside the normal flow of the recurring network, more specifically in a gated cell. Information can be stored, written or read from a given cell, in an approach similar to data in a computer's memory. These networks are used to process, predict and classify based on time series data.

4 Experiments

To achieve the objective of predicting the UV index, it was necessary to develop and tune several candidate LSTM models. The conceived models forecast the UV index for three consecutive days, following a recursive multi-step approach, i.e., predicting recursively each future timestep until it achieves the time window of three predicted days. With this multi-step approach, the prediction for timestep t is used as input to predict timestep $t + 1$.

Several experiments were carried out to find the best combination of hyperparameters for both the uni-variate and the multi-variate approaches. Performance was compared in terms of error-based accuracy, for both the uni-variate and multi-variate candidate LSTM models. Regarding the first, it uses only one feature as input, unlike the second which takes into account several features. In fact, the uni-variate models use only the UV index *value* feature to recursively predict the future values of this index. On the other hand, the multi-variate model uses the UV index *value* as well as the *month* and *day* features, giving a stronger temporal context to the network. Regarding the searched hyperparameter configuration, they were identical between both approaches, being described in Table 2 which summarises the parameter searching space considered for uni and multi-variate models.

Table 2. Uni-variate vs multi-variate hyperparameters' searching space.

Parameter	Uni-variate	Multi-variate	Rationale
Epochs	[150, 300]	[300, 500]	–
Timesteps	[7, 14]	[7, 14]	Input of 1 and 2 weeks
Batch size	[16, 23]	[16, 23]	Batch of 2 to 3 weeks
LSTM layers	[3, 4]	[3, 4]	Number of LSTM layers
Dense layers	1	1	Number of dense layers
Dense activation	[ReLU, tanh]	[ReLU, tanh]	Activation function
Neurons	[32, 64, 128]	[32, 64, 128]	For dense and LSTM layers
Dropout	[0.0, 0.5]	[0.0, 0.5]	For dense and LSTM layers
Learning rate	Callback	Callback	Keras callback
Multisteps	3	3	3 days forecasts
Features	1*	3**	Used features
CV Splits	3	3	Time series cross-validator

* Used features: *UV index*

**Used features: *UV index*, *month* and *day*

Knime was the platform used for data exploration. Python, version 3.7, was the used programming language for data preparation, model development and evaluation. Pandas, NumPy, scikit-learn and matplotlib were the used libraries. Tensorflow v2.0.0 was used to implement the deep learning models. Tesla T4 GPUs were used as well as CUDNNLSTM layers for optimized performance in a GPU environment. All hardware was made available by Google's Colaboratory, a free python environment that runs entirely in the cloud.

5 Results and Discussion

Being this a time series forecasting problem, one particular time series cross validator was used, being entitled as TimeSeriesSplit. For each prediction of each split of this cross validator, RMSE and MAE were calculated to be able to evaluate the best set of parameters. The experiments carried out for both uni and multi-variate candidate models made it clear that a stronger temporal context results in an overall decrease of both error metrics even though the best uni-variate models has a lowest MAE than the best multi-variate one. Table 3 depicts the top 3 results for both approaches.

Table 3. Uni-Variate vs Multi-Variate LSTMs top-three results.

#	Timesteps	Batch	Layers	Neurons	Dropout	Act.	RMSE	MAE
Recursive multi-step uni-variate								
116	7	16	4	64	0.5	tanh	**0.325**	**0.236**
108	7	16	3	128	0.5	tanh	0.349	0.271
8	7	16	3	64	0.5	tanh	0.354	0.26
Recursive multi-step multi-variate								
31	14	16	3	64	0.0	tanh	**0.306**	**0.249**
125	14	16	3	64	0.0	relu	0.339	0.284
73	14	23	3	32	0.0	relu	0.34	0.275

The best LSTM model concerning the uni-variate model had an RMSE of 0.325 and an MAE of 0.236. On the other hand, the RMSE was 0.306 and the MAE was 0.249 for the best multi-variate model. Since both metrics are in the same unit of measurement as the UV index, an error of 0.3 shows that it is possible to forecast, very closely, the expected UV index for the next three days. In the multi-variate model, the inclusion of the month and day yields more accurate predictions in comparison to the uni-variate one. Interestingly, the number of inputs of the model is directly proportional to the number of timesteps, i.e., more features as input lead to an increase of the number of timesteps that are required to build a sequence. This is shown by Table 3, with the best uni-variate models requiring sequences of 7 timesteps (a week), while the best multi-variate ones require sequences of 14 timesteps (two weeks). The number of epochs and batch size was 300 and 16, respectively, for both models, with a early stopping callback stopping training when the monitored loss stopped improving. Concerning the number of hidden layers, the best uni-variate model required a more complex architecture of 4 hidden layers to achieve similar performances to the multi-variate ones, who required a shallow architecture. In addition, this shallow architecture ruled out the use of dropout, which was required by the best uni-variate models. Figure 2 presents the architectures of the best multi-variate model. Moreover, Fig. 3 illustrates six predictions of three days for the best Multi-Variate LSTM model, showing very close results to the known UV index value.

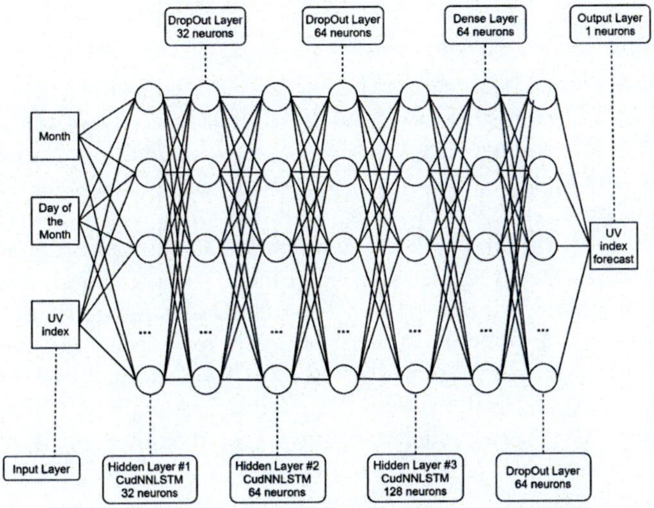

Fig. 2. Architecture of the best multi-variate model.

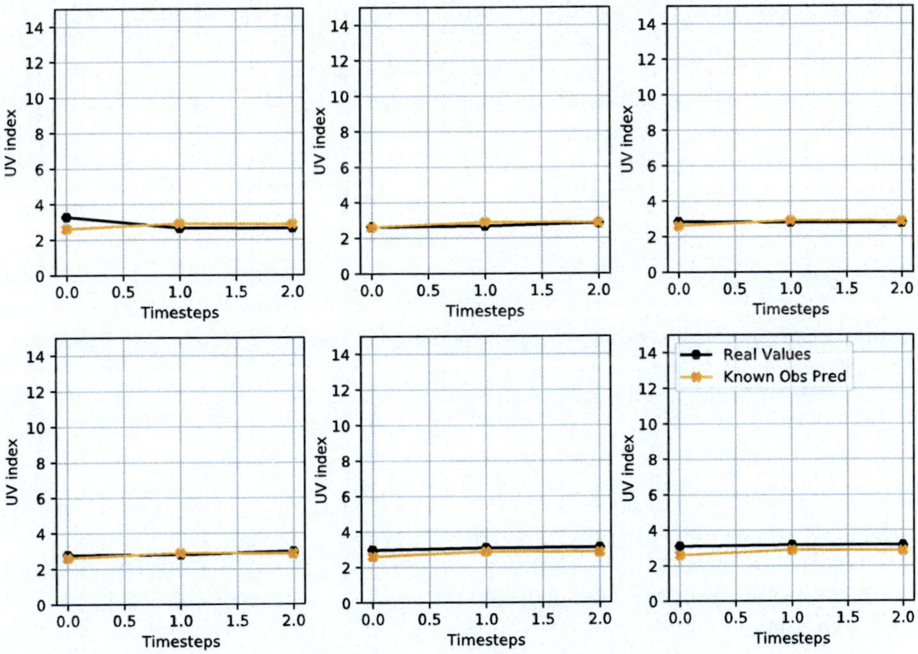

Fig. 3. Six random predictions of the best multi-variate LSTM model.

6 Conclusions

Over the past few years, skin cancer prevention campaigns have increased worldwide. Knowing that exposure to ultraviolet radiation is one of the main causes for such disease, forecasting the UV index assumes particular importance. Hence, this study focused on using deep learning models, in particular LSTMs, to forecast the UV index for the next three days. Multiple experiments were performed, using a wide combination of hyperparameters for all the candidate models. The model with the best accuracy in the prediction of the UV index was the Recursive Multi-Step Multi-Variate model with a RMSE of 0.306 and a MAE of 0.249, which depict that it is possible to forecast, with very accurate results, the UV index for the next few days. Nevertheless, the models that had only the UV index as input also presented interesting results. As expected, the number of input features impacts the models' accuracy. Yet it is interesting to note that the increase in input features led to an increase of the required timesteps as well as to shallow networks.

The obtained results show promising prospects for UV index forecasting. Hence, future work will consider the inclusion of more input features such as the temperature, ozone levels and the position of the sun expressed in terms of solar zenith angle. In addition, future work will also focus on different state-of-the-art recurrent networks such as the Gated Recurrent Unit network.

Acknowledgments. This work has been supported by FCT - Fundação para a Ciência e a Tecnologia within the R&D Units project scope UIDB/00319/2020 and DSAIPA/AI/0099/2019. The work of Bruno Fernandes is also supported by a Portuguese doctoral grant, SFRH/BD/130125/2017, issued by FCT in Portugal.

References

1. Lickley, M., Solomon, S., Fletcher, S., Velders, G., Daniel, J., Rigby, M., Montzka, S., Kuijpers, L., Stone, K.: Quantifying contributions of chlorofluorocarbon banks to emissions and impacts on the ozone layer and climate. Nat. Commun. **10**(1), 1–11 (2020)
2. Young, A., Narbutt, J., Harrison, G., Lawrence, K., Bell, M., O'Connor, C., Olsen, P., Grys, K., Baczynska, K., Rogowski-Tylman, M., et al.: Optimal sunscreen use, during a sun holiday with a very high ultraviolet index, allows vitamin D synthesis without sunburn. Br. J. Dermatol. **181**(5), 1052–1062 (2019)
3. World Health Organization and International Commission on Non-Ionizing Radiation Protection and others: Global solar UV index: a practical guide. World Health Organization (2002)
4. Gómez, I., Marín, M., Pastor, F., Estrela, M.: Improvement of the Valencia region ultraviolet index (UVI) forecasting system. Comput. Geosci. **41**, 72–82 (2012)
5. Deo, R., Downs, N., Parisi, A., Adamowski, J., Quilty, J.: Very short-term reactive forecasting of the solar ultraviolet index using an extreme learning machine integrated with the solar zenith angle. Environ. Res. **155**, 141–166 (2017)
6. Barrera, J., Hurtado, D., Moreno, R.: Prediction system of erythemas for phototypes i and ii, using deep-learning. Vitae **22**(3), 189–196 (2015)

7. Ricchiazzi, P., Yang, S., Gautier, C., Sowle, D.: SBDART: a research and teaching software tool for plane-parallel radiative transfer in the Earth's atmosphere. Bull. Am. Meteorol. Soc. **79**(10), 2101–2114 (1998)
8. Ghoneim, O., Manjunatha, B., et al.: Forecasting of ozone concentration in smart city using deep learning. In: 2017 International Conference on Advances in Computing, Communications and Informatics (ICACCI), pp. 1320–1326. IEEE (2017)
9. Barnaghi, P., Tönjes, R., Höller, J., Hauswirth, M., Sheth, A., Anantharam, P.: CityPulse: real-time IoT stream processing and large-scale data analytics for smart city applications. In: Europen Semantic Web Conference (ESWC) (2014)
10. Hochreiter, S., Schmidhuber, J.: Long short-term memory. Neural Comput. **9**(8), 1735–1780 (1997)

Multispectral Image Analysis for the Detection of Diseases in Coffee Production

Jesús Silva[1(✉)], Noel Varela[2], and Omar Bonerge Pineda Lezama[3]

[1] Universidad Peruana de Ciencias Aplicadas, Lima, Peru
jesussilvaUPC@gmail.com
[2] Universidad de la Costa, Barranquilla, Colombia
nvarela2@cuc.edu.co
[3] Universidad Tecnológica Centroamericana (UNITEC), San Pedro Sula, Honduras
omarpineda@unitec.edu

Abstract. Coffee is produced in Latin America, Africa and Asia, and is one of the most traded agricultural products in international markets. The coffee agribusiness has been diversified all over the world and constitutes an important source of employment, income and foreign exchange in many producing countries. In recent years, its global supply has been affected by adverse weather factors and pests such as rust, which has been reflected in a highly volatile international market for this product [1]. This paper shows a method for the detection of coffee crops and the presence of pests and diseases in the production of these crops, using multispectral images from the Landsat 8 satellite.

Keywords: Multispectral image analysis · Detection of diseases · Coffee production

1 Introduction

The implementation of this project is intended to help detect early infestation of coffee plantations by pests such as rust, providing farmers with information on the condition of their land and crops using a combination of multispectral imaging, pattern recognition and artificial intelligence.

Research will focus on the analysis and processing of multispectral images for the detection of diseases and pests in crops. This will be carried out on the basis that the analysis of satellite images provides a great help in the field of agriculture, due to the fact that significant results are obtained compared to the analysis of digital images showing only the visible range of the electromagnetic spectrum [2]. It is important to mention that in the analyzed researches not all of them are about pests and diseases in coffee crops, although most of them are focused on these. Those that are not focused on them are considered due to the techniques used in the analysis of multispectral images, however, they also consider the analysis of the quality of fruits.

Geographic information technologies (GIT) include remote sensing, which currently uses multispectral satellite images that are applied to estimate types of cover on the

Y. Dong et al. (Eds.): DCAI 2020, AISC 1237, pp. 198–205, 2021.
https://doi.org/10.1007/978-3-030-53036-5_21

Earth's surface and the condition of these covers [3]. In this way, this technology can be applied to identify areas with coffee plantations, differentiate their development conditions and discriminate those areas whose characteristics in the image suggest some restriction [4].

Since any phenomenon occurring on the surface of the earth: vegetation, crops, water bodies, soils, etc., can be detected and analyzed by remote sensing technology [5], it is feasible to infer the manifestation of the "health" of the crop or its development condition through the interpretation of multispectral satellite images.

Multispectral imaging is an arrangement of columns and lines that form a matrix of numerical data representing the intensity of electromagnetic energy reflected or emitted by objects on the Earth's surface. The images can be recorded in individual bands of the electromagnetic spectrum, as is the case with the Landsat 8 satellites which have 11 spectral bands, i.e. the same scene captured in different bands: band 1, band 2, band 3, etc. [6, 7].

The Landsat 8 satellite scene, acquired on February 4, 2018 was used; it was downloaded from the "Libra" satellite image viewer, carrying out Georeferencing processes. In order to obtain information from the multispectral images, an analysis is performed by means of digital processes applying image statistics (supervised classification) or by means of visual analysis through a color compound [8, 9].

The aim of this research was to analyze satellite multispectral images with the purpose of detecting coffee crops as well as pests and diseases in them.

2 Related Studies

The following studies make use of multispectral images with high spectral and spatial resolution, applying image processing and pattern recognition techniques (computer vision) in coffee crops with the intention of obtaining early specific detection of pests that may represent a risk for the production of coffee and the safety of their harvest [10].

In [11], they related foliar characteristics and spectral reflectance of sugar beet leaves with Cercospora leaf spot and leaf rust at different stages of development, using a hyperspectral imaging spectrometer (ImSpector V10E) with a 2.8 nm spectral resolution of 400 to 1000 nm and a 0.19 mm spatial resolution for the detection and continuous monitoring of disease symptoms during pathogenesis. They do not mention quantitative results, but concluded that spectral reflectance, in combination with spectral angle mapping classification, allowed differentiation of mature symptoms in areas showing all ontogenetic stages from young to mature symptoms [12].

In [13], they studied learning functions that could predict whether the value of a continuous target variable could be greater than a given threshold. The objective of their study was to warn about the high incidence of coffee rust, the main disease of coffee cultivation in the world. They made a comparison between the results of their confusion matrix, obtaining results where the costs of false negatives are higher than those of false positives, and both are higher than the cost of the warning predictions.

Other studies such as [14], detected that the wheat fields, in winter seasons, are affected by the disease called yellow rust, which harms the production of wheat, therefore, the objective of their study was to evaluate the accuracy of the optical spectrum, the

Photochemical Reflectance Index (PRI) for quantifying the index of this disease and its applicability in the detection of the disease through hyperspectral imaging. They tested the PRI over three seasons, showing that, in winter, with a 97% determination rate, the PRI shows a potential for quantifying levels of yellow rust in wheat and as a basis for the development of a proximal image sensor of yellow rust in wheat fields in winter.

In [15], they evaluated ten widely used vegetation indices, based on mathematical combinations of measurements from narrow-band optics reflectance in the visible and near-infrared wavelength range for their ability to discriminate leaves of one-month-old wheat plants infected with yellow stripes. They do not mention quantifiable results but conclude that no single index was able to discriminate the three rust species from each other, however, the sequential application of the Index of Anthocyanin Reflectance for separating the healthy, yellow and mixed classes of oxide and rust of the leaves followed by the absorption index of chlorophyll and the index of reflectance for separating the rust classes from leaves and stems, could form the basis of discrimination of rust species in wheat under field conditions [16].

3 Materials and Methods

3.1 Study Area

The area of influence of coffee production is of an interstate nature and the areas occupied by coffee are distributed in a dispersed way, although in certain localities it has a greater concentration. This research was carried out in the municipality of Tolima, an area of cultivation and production of Colombian coffee.

Figure 1 shows the polygon where an important concentration of coffee crops is located, for these municipalities.

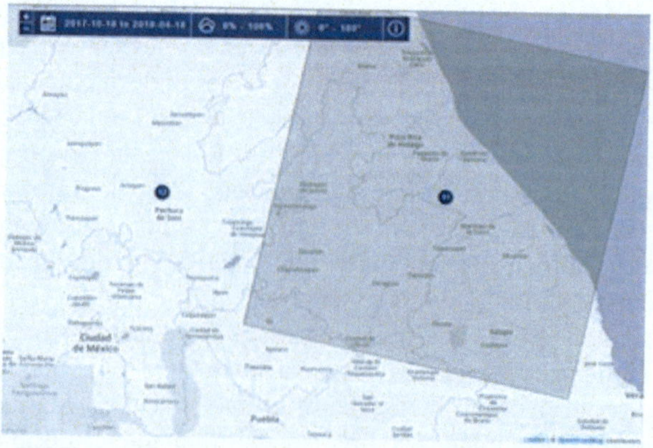

Fig. 1. Libra main interface once the map is placed over the Tolima area. Search filter is offered by date, cloud coverage or solar angle (on the map).

3.2 Used Images

For remote sensing recognition of factors restricting the development of coffee cultivation, a Landsat 8 satellite scene from January 6, 2019 was used, with a cloudiness of 17.19% and a sun inclination of 142.50. This scene was downloaded from the Landsat 8 satellite image viewer, "Libra" (see Fig. 1). The Landsat 8 satellite carries two instruments, OLI and TIRS, which stand for Operational Land Imager (OLI) and Thermal Infrared Sensor (TIRS). The OLI sensor provides access to nine spectral bands covering the spectrum from 0.434 μm to 1.400 μm, while TIRS records from 10.20 μm to 11.75 μm. Therefore, the scene is composed of 11 images which are bands of the scene with its corresponding spectral resolution. This makes it possible to capture radiation from the earth's surface in eleven spectral bands that record characteristics of objects on the surface: soils, vegetation, water, etc. [7].

3.3 Field Sampling for Training and Validation

The images were geo-referenced through Google Earth, using the values of the Landsat 8 image (see Fig. 2).

3.4 Band Combination

Different combinations of bands were made in order to analyze the most appropriate combination for crop detection and disease identification (see Fig. 3). The most relevant combinations performed in the processing of Landsat 8 satellite images are shown in Table 1, where the different uses of each band combination are identified.

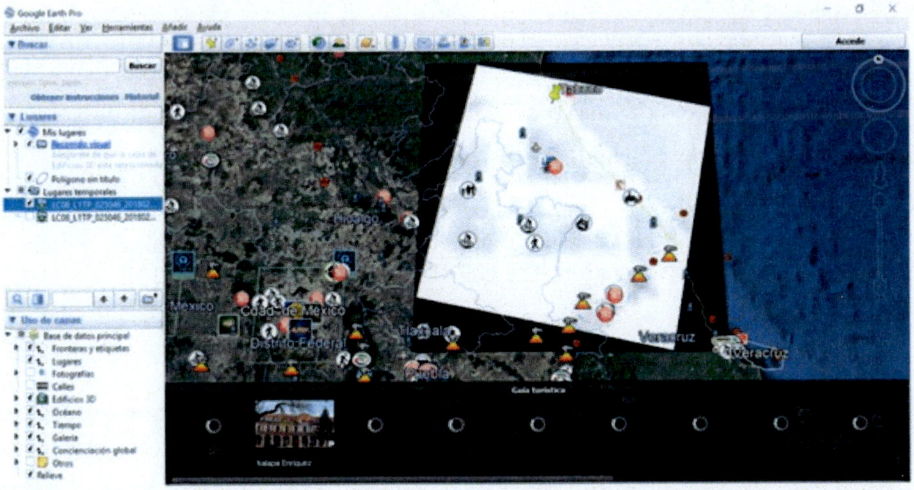

Fig. 2. The image is positioned on the geographical map, considering that it includes part of Tolima.

Fig. 3. The image is positioned on the geographical map, considering that it includes part of the regions that make up Tolima area.

Table 1. Landsat 8 satellite band combinations.

Use	Combination of bands
Natural color	5, 3, 2
False color (urban)	7, 5, 4
Infrared color (vegetation)	5, 2, 3
Agriculture	6, 4, 2
Atmospheric penetration	7, 6, 5
Healthy vegetation	5, 5, 2
Land/Water	6, 6, 2
Natural with atmospheric removal	6, 5, 4
Shortwave infrared	7, 4, 3
Vegetation analysis	8, 5, 4

3.5 Calculation of the Normalized Difference Vegetation Index (NDVI)

NDVI values vary between −1 and 1, where zero corresponds to an approximate value of non-vegetation [6]. Negative values represent areas with no vegetation, while values close to 1 contain dense vegetation. The NDVI is calculated by Eq. (1) [16]:

$$\mathrm{NDVI} = \frac{(\mathrm{NIR} \; - \mathrm{R})}{(\mathrm{NIR} \; - \mathrm{R})} \tag{1}$$

Where and refer to the reflectance values measured by the red () and near infrared () bands.

3.6 Segmentation and Classification of Images

The identification of the endmembers belonging to the georeferenced coffee crops in the scene required a Principal Component Analysis (PCA) [17] in order to eliminate the inherent redundancy of the used data. The multispectral or multidimensional nature of the images can be adjusted by reconstructing a vector space with a number of axes or dimensions equal to the number of components associated with each pixel [7].

This transformation generates a set of bands that correspond to each eigenvalue and are organized according to the estimation of the noise in the multispectral images [11]. For this reason, for obtaining a reliable result in the characterization of the spectral profiles of the crops, it is necessary to remove such noise from the image; however, it is necessary to avoid the loss of data as much as possible.

4 Experiments and Results

According to the NDVI values, photosynthetically active green vegetation is between 0.1 and 0.9 and crops tend to be between 0.3 and 0.8 depending on the leaf area index and soil layout. Likewise, NDVI is influenced by the percentage of soil cover, and the best correlation is found when the cover is between 26% and 83%. The low values presented by grasses are possibly due to low coverage, below 16%, in which case the NDVI does not accurately indicate the degree of vegetation biomass, since it is affected by the reflectance of bare soil [5]. Figure 4 shows how the reflectance values of each coverage type behave in the 11 bands that make up the Landsat 8 scene.

Three groups are clearly observed. Each pixel is represented by its reflectance value in each band, so each pixel is represented in 11 points (blue, green, yellow, purple, brown, orange, black, white, cyan, magenta, red, and wine), where each color represents a band.

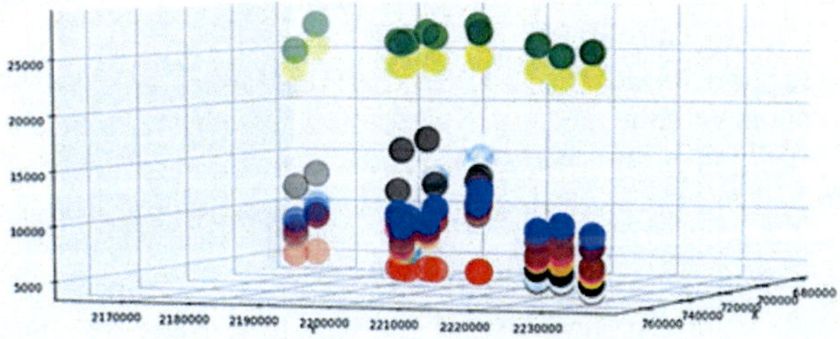

Fig. 4. Selection of 14 pixels chosen from the satellite image involving 3 types of coverage (vegetation, urban areas and water bodies). The behavior in each of the bands of the Landsat 8 satellite scene is observed.

5 Conclusions and Future Research

Using the multispectral images of the area that corresponds to an important coffee region in Tolima and performing the processing and analysis of the scene, the process marked by the proposed method is carried out as a reliable alternative for the identification of coffee crops, and consequently its usefulness in other crops and in various environmental conditions must be established. If the multispectral images are fully available, the use of those taken from different phenological states can be explored to better characterize the spectral behavior of the crop. It is important to mention that, considering the spatial resolution of the images, the results and information obtained are acceptable, although if it is possible to access satellite images with higher spatial and spectral resolution the results would be much better.

The advance of the research shows a part of the results that are expected to be obtained by implementing the whole process of the proposed method, although these preliminary results show high possibilities for the adequate detection of coffee crops and some restrictions in production, such as diseases and/or pests.

References

1. Chemura, A., Mutanga, O., Dube, T.: Separability of coffee leaf rust infection levels with machine learning methods at sentinel-2 MSI spectral resolutions. Precis. Agric. 23 (2016)
2. Landgrebe, D.: Hyperspectral image data analysis. IEEE Signal Process. Mag. **19**, 17–28 (2002)
3. Velásquez, D., Sánchez, A., Sarmiento, S., Toro, M., Maiza, M., Sierra, B.: A method for detecting coffee leaf rust through wireless sensor networks, remote sensing, and deep learning: case study of the caturra variety in Colombia. Appl. Sci. **10**(2), 697 (2020)
4. Mahlein, A.K., Steiner, U., Hillnhutter, C., Dehne, H.W., Oerke, E.C.: Hyperspectral imaging for small-scale analysis of symptoms caused by different sugar beet diseases. Plant Methods **8**, 3 (2012)
5. De Oliveira Pires, M.S., de Carvalho Alves, M., Pozza, E.A.: Multispectral radiometric characterization of coffee rust epidemic in different irrigation management systems. Int. J. Appl. Earth Obs. Geoinf. **86**, 102016 (2020)
6. Viloria, A.: Commercial strategies providers pharmaceutical chains for logistics cost reduction. Indian J. Sci. Technol. **8**(1), Q16 (2016)
7. Thomas, S., Wahabzada, M., Kuska, M.T., Rascher, U., Mahlein, A.K.: Observation of plant–pathogen interaction by simultaneous hyperspectral imaging reflection and transmission measurements. Funct. Plant Biol. **44**, 23–34 (2016)
8. Huang, W., Lamb, D.W., Niu, Z., Zhang, Y., Liu, L., Wang, J.: Identification of yellow rust in wheat using in-situ spectral reflectance measurements and airborne hyperspectral imaging. Precis. Agric. **8**(4–5), 187–197 (2007)
9. Da Rocha Miranda, J., de Carvalho Alves, M., Pozza, E.A., Neto, H.S.: Detection of coffee berry necrosis by digital image processing of landsat 8 oli satellite imagery. Int. J. Appl. Earth Obs. Geoinf. **85**, 101983 (2020)
10. Nzimande, N., Mutanga, O., Kiala, Z., Sibanda, M.: Mapping the spatial distribution of the yellowwood tree (Podocarpus henkelii) in the Weza-Ngele forest using the newly launched Sentinel-2 multispectral imager data. South Afr. Geogr. J. 1–19 (2020)
11. Marin, D.B., de Carvalho Alves, M., Pozza, E.A., Belan, L.L., de Oliveira Freitas, M.L.: Multispectral radiometric monitoring of bacterial blight of coffee. Precis. Agric. **20**(5), 959–982 (2019)
12. Oliveira, A.J., Assis, G.A., Guizilini, V., Faria, E.R., Souza, J.R.: Segmenting and detecting nematode in coffee crops using aerial images. In: International Conference on Computer Vision Systems, pp. 274–283. Springer, Cham (2019)
13. Folch-Fortuny, A., Prats-Montalbán, J.M., Cubero, S., Blasco, J., Ferrer, A.: VIS/NIR hyperspectral imaging and N-way PLS-DA models for detection of decay lesions in citrus fruits. Chemometr. Intell. Lab. Syst. **156**, 241–248 (2016)

14. Chemura, A., Mutanga, O., Sibanda, M., Chidoko, P.: Machine learning prediction of coffee rust severity on leaves using spectroradiometer data. Trop. Plant Pathol. **43**(2), 117–127 (2018)
15. Amelec, V.: Increased efficiency in a company of development of technological solutions in the areas commercial and of consultancy. Adv. Sci. Lett. **21**(5), 1406–1408 (2015)
16. Katsuhama, N., Imai, M., Naruse, N., Takahashi, Y.: Discrimination of areas infected with coffee leaf rust using a vegetation index. Remote Sens. Lett. **9**(12), 1186–1194 (2018)
17. Izquierdo, N.V., Lezama, O.B.P., Dorta, R.G., Viloria, A., Deras, I., Hernández-Fernández, L.: Fuzzy logic applied to the performance evaluation. Honduran coffee sector case. In: International Conference on Sensing and Imaging, pp. 164–173. Springer, Cham(2018)

Check for updates

Photograph Classification Based on Main Theme and Multiple Values by Deep Neural Networks

Toshinori Aoki(✉) and Miki Ueno

Osaka Institute of Technology, 5-16-1, Omiya, Asahi-ku, Osaka 535-8585, Japan
e1616002@st.oit.ac.jp, miki.ueno@oit.ac.jp

Abstract. Recently, numerous people take photographs easily by smartphones and post them on social network services. On the other hand, huge knowledge and technique are required in order to become skilled photographers. It is hard to acquire these information especially on real-time filming.

Thus, the aim of the research is to create a system to suggest advices for novices of photographers. To construct such a system, the method to evaluate photos precisely like a professional photographer is required.

In this paper, we have proposed original dataset of scenery photographs with detailed scores by professional photographers and construct Deep Convolutional Neural Networks to classify two class images with the labels evaluated by the author as a professional photographer.

Keywords: Photograph creation · Scenery photograph dataset · Professional knowledge · Deep Convolutional Neural Network · Visualization of image features

1 Introduction

Numerous people can easily aquire opportunities to take photographs with smartphone and post them to Social Network Services. Thus, the population of people get into the hobby of photographs as one of familiar creative works. Although the method of filming becomes easier, the beginners of photographers have lots of questions to win the photograph contests. Photographers have to aquire filming knowledge; selecting theme, structure, devices and so on.

The aim of research is 1) to construct of original scenery photograph dataset based on detailed scores by professional photographers, 2) consider the method of giving suggestions to several photographers based on machine learning.

In this paper, we construct original scenery dataset with scores by a professional photographer and consider classification result of each image by Deep Convolutional Neural Network.

2 Dataset

The dataset is consisted of 289 scenery photographs filmed by the first author. The main theme of all of the images is "sky".

Y. Dong et al. (Eds.): DCAI 2020, AISC 1237, pp. 206–210, 2021.
https://doi.org/10.1007/978-3-030-53036-5_22

Table 1. The guideline of photograph evaluation

Evaluation category	Detailed item	Score
Basic technique	Blurry	
	Bokeh	
	Exposures	5
	Color temperature	
	Blown out highlights. Blocked up shadows	
Content	Feelings	15
	Meaning	15
	Mainly opinion	15
	Imagery	15
Viewpoint	Selection of attractive target	10
	The shooting situation	10
Art technique	Filming technique	10
	The way of using functions on devices	10
	The way of retouch	10

2.1 Construction of Dataset

1. All of the images were trimmed subjectively and resized to 128 px. × 128 px. images.
2. Annotated point to each image based on the score by the author. The guideline of score was decided by the first author referred to existed research [1,2]. Table 1 shows the guideline of score.
3. Annotated "good" or "bad" label to each image based on step 2. Good or bad label is annotated based on the score of basic technique respectively. Especially, a image with a good label has high score of good content and technique. On the other hand, an image with bad label has low score of basic technique and vague aim of filming.

2.2 Label of Dataset

Figure 1 and Fig. 2 show the example of sky with good and bad label respectively. Aspects of evaluation are shown as follows.

- Figure 1 has a good label based on clear aim with wide angle by fish-eye lens and clear blue color.
- Figure 2 has a bad label because of vague main theme.

3 Experiment

3.1 Classification

Two-class classification is carried out by Deep Convolutional Neural Networks (DCNN) with AlexNet model.

Fig. 1. Good labeled image

Fig. 2. Bad labeled image

Table 2. Network parameters of DCNN

Layer	Dimension	Filter	Filter-size	Activation function	Dropout
Conv2D	12412432	32	55	Relu	–
Conv2D	12212232	32	33	Relu	–
MaxPooling2D	616132	–	22	–	–
Dropout	616132	–	–	–	0.25
Conv2D	595964	64	33	Relu	–
Conv2D	575764	64	33	Relu	–
MaxPooling2D	282864	–	22	–	–
Dropout	282864	–	–	–	0.25
Flatten	–	–	–	–	–
Dense	512	–	–	Relu	–
Dropout	512	–	–	–	0.25
Dense	2	–	–	Softmax	–

Table 2 and Table 4 show the parameter of the experiment.

Table 4 shows the accuracy which is calculated by eight-fold cross validation (Table 3).

Table 3. Parameter of DCNN

Epoch	112
Batch size	30
Optimization method	SGD
Loss function	Categorical cross entropy

Table 4. The accuracy of the experiment

Test accuracy	0.86

Fig. 3. Correctly classified image

Fig. 4. Incorrectly classified image

3.2 Visualization

In order to reveal features which are contribute to classification by DCNN, Gradient weighted Class Activation Mapping (Grad-CAM) [3,4] is used.

4 Consideration

Good labeled images with partially features of bad images tend to mis classified.

Results of similar images are taken as examples. Figure 3 and Fig. 4 show correctly and incorrectly classified examples respectively. Figure 3 with bad label and Fig. 4 with good label were also classified to bad classes. These images commonly represent the large region of the sky as the main theme and the sun as the sub theme. The structure of photographs is also similar. In order to classify such images, we have to introduce the method considering the relationship between main theme and sub theme.

We compared features focused by DCNN and the first author. Figure 5 and 6 show precisely classified images based on common features between DCNN and the author.

The author evaluated Fig. 5 based on the proportion of cloud and shape of figure, relationship between sky and trees.

The author evaluated Fig. 6 based on the proportion of cloud and the slightly lighted sun.

Fig. 5. Image:num.32

Fig. 6. Image:num.205

Fig. 7. Image::num.46

Fig. 8. Image:num.144

Figure 7 and Fig. 8 show partially common evaluation between the author and DCNN. The author evaluate Fig. 7 based on beautiful light among clouds above the mountain. Figure 8 were evaluated based on the proportion of clouds against the sky and expressionless clouds by the author.

5 Conclusion

In this research, 1) we construct original scenery photograph datasets with detailed scores on the aspects of photographers, 2) the computer experiment is carried out for the dataset using AlexNet, 3) we compare features which are contribute to classify between DCNN and the author.

We indicated that AlexNet can classify scenery images of sky on the aspect of filming aim and technique. Future tasks are described below:

- Annotate layered class labels to be able to consider the sub theme
- Changing setting filter size of DCNN to be able to consider the relationship between the main and sub theme of photograph
- Prepare large-scaled dataset

References

1. Mizushima, A.: Consideration on the viewpoint of the evaluation of the landscape photograph competition incorporating the elements of the tournament competition. Bull. Jpn. Soc. Arts Hist. Photogr. **28**(1), 19–24 (2019)
2. Noguchi, K.: Psychology of Beauty and Kansei: New Horizons of Gestalt Perception. The Japanese Psychonomic Society, pp. 172–176 (2007)
3. Selvaraju, R.R., Cogswell, M., Das, A., Vedantam, R., Parikh, D., Batra, D.: Grad-CAM: visual explanations from deep networks via gradient-based localization. In: The IEEE International Conference on Computer Vision (ICCV), pp. 618–626, October 2017
4. Akagi, T.: Deep learning - video diagnosis and visualization of the judgment factor to begin in the viewpoint of the Wet researcher. Japanese Society of Plant Physiologists (2019)

In-Vehicle Violence Detection in Carpooling: A Brief Survey Towards a General Surveillance System

Francisco S. Marcondes[1(✉)], Dalila Durães[1,2], Filipe Gonçalves[1,3], Joaquim Fonseca[3], José Machado[1], and Paulo Novais[1]

[1] Centro ALGORITMI, University of Minho, Braga, Portugal
francisco.marcondes@algoritmi.uminho.pt, {jmac,pjon}@di.uminho.pt
[2] CIICESI, ESTG, Politécnico do Porto, Felgueiras, Portugal
dad@estg.ipp.pt
[3] Bosch Car Multimedia, Braga, Portugal
{filipe.goncalves,joaquim.fonseca2}@pt.bosch.com

Abstract. Violence is a word that encompasses several meanings ranging from an actual fight to theft and several types of harassment. Therefore, violence detection through surveillance systems can be a quite difficult yet important task. The increasing use of carpooling services and vehicle sharing brought the need to implement a sufficient general surveillance system for monitoring these vehicles for assuring the passengers' safety during the ride. This paper raised the literature for this matter, finding fewer research papers than it was expected for the in-vehicle perspective, noticeably to sexual harassment. Most of the research papers focused on out-vehicle issues such as runs over and vehicle theft. In-vehicle electronic components security and cockpit user experience were perceived as major concern areas. This paper discusses these findings and presents some insights about in-vehicle surveillance.

Keywords: Car cockpit · Surveillance · In-vehicle violence

1 Introduction

Violence detection has being researched at least since the 1990's [15] as surveillance cameras become increasingly popular. The motivation is despite the wide coverage that can be achieved with cameras, given the amount of resulting data, it is not easy for security personnel to watch each camera and to recognize violence in real-time [14]. Several types of research were performed in order to help security operators to select which cameras to look at. The increasing use of carpooling services raised the need for general-purpose cockpit surveillance system. The objective is to perform a review to answer the questions:

1. Which approach presents the highest violence detection accuracy in 2019?
   (a) Is that approach optimum? Is there a known lower bound?

Y. Dong et al. (Eds.): DCAI 2020, AISC 1237, pp. 211–220, 2021.
https://doi.org/10.1007/978-3-030-53036-5_23

   (b) Does that same accuracy can be achieved for a vehicle cockpit?
   (c) Does it require high performing hardware?
2. Which is a suitable sensor set for a general car cockpit surveillance?
3. Which sensor set is best suited for violence detection within a cockpit?

This paper follows a traditional systematic review procedure *cf.* [23], *i.e.*, given a set of research questions a query will be built in order to answer them. The query results will be filtered by reading their titles and abstracts for fitting into the questions. Into the sample, other papers that authors found them relevant but escaped from the query will be included. The sample will be then used for answering the questions or to identify the absence of answers what would demand further researches.

2 State of the Art

Violence Detection Definition. Violence is a human action involved in fighting, beating, stealing, *etc.* Violence detection is a subset of *abnormal or suspicious behavior*, which in turn is a subset of *human action recognition* [38]. *Multi-sensor Surveillance for Violence Detection Overview.* The first generation of automated surveillance was based on detecting anomalous events through computer vision [10] for recognize gestures, actions and interactions [41]. This approach drawbacks are those linked with the cameras, *e.g.* adverse weather, light variation, occlusion, *etc.* For overcoming these drawbacks, on the second generation, different sensor types are used both alone or in sets that may or may not include the video sensor [10]. The cornerstone for the second generation of automated surveillance is to *suit the sensor set with the application environment.*

Implicit yet valuable advantage of multi-sensor surveillance is the possibility of adaptive processing behavior *e.g.* light variation may trigger shifts between active sensor types and surveillance strategy at each moment [28]. But raising the number of sensors set also raises the system complexity (human and computational) [41], therefore it is not a "the more the merrier" situation but a *design* one. Nevertheless, multi-sensor type surveillance provides, *a priori*, a more accurate performance than mono-sensor [10].

Legal Constraints. A car cockpit is a private environment therefore it imposes several constraints that do not exists within public environments. Among others, in-vehicle surveillance must conform to data privacy and traffic laws. For the data privacy side, in short, according to the *protection of personal data* [12], data can be gathered for "legitimate basis laid down by law" (article 8); shared-car surveillance aims to improve the security of person (article 6), this, therefore, at first, can be considered a legitimate concern yet require the user agreement before boarding in the surveilled car.

3 The Sample Setup

This survey took place in February 2020, within the NEXT ACM DL (dlnext.acm.org). The query and the key-word weights are presented in Fig. 1.

a)

b)

key-word	comp.	abs.	rel.	norm.
violence	223	19	7.9	2.4
aggression	213	29	12.0	3.7
theft	165	77	31.8	9.7
rape	238	4	1.7	0.5
fight	142	100	41.3	12.6
surveillance	233	9	3.7	1.1
detection	163	79	32.6	10.0
recognition	207	35	14.5	4.4
car (title)	194	48	19.8	6.0
car (abstract)	211	31	12.8	3.9
vehicle (title)	53	189	78.0	23.8
vehicle (abst.)	69	173	71.5	21.8

Fig. 1. a) Shows the query as submitted for the ACM DL search engine. b) Shows the weight of each key-word in the query. The first column, *complement*, depicts the number of papers retrieved by removing the key-word from the query; the second, *absolute weight*, is the difference between the number of papers retrieved by the query and the complement; the third, *relative weight*, is the percentage representation for the absolute weight in relation to the number of retrieved papers; and the fourth, *normalized weight*, is the percentage representation of relative weight.

The query presented in Fig. 1a retrieved 242 results published between 1973 and 2019. The sample was then reduced for the last five years resulting in 122 papers. The title and abstract for each paper in the sample were then read for removing those that do not aid in answering the questions. Therefore, papers about unmanned aerial vehicles, routing, traffic management, accident prevention, and violent driving were removed resulting in a filtered sample with 37 papers. To the filtered sample 05 papers were included by being considered relevant. The resulting sample is composed of 43 papers from the last five years.

4 Results and Discussion

It is out of scope for this paper but the proportion of paper related to *rape* compared to the other violence related key-words is noticeable. The *theft* key-word in this sample is more related to stealing the whole vehicle than to in-vehicle theft of goods. The *fight* key-word has a broad meaning than of battle including the figurative meaning of opposing to something, this explains its weight. It was not find in this sample papers related to in-vehicle surveillance.

Some other exploratory queries looking only in the title and abstract (*e.g.* `Title:((violence OR aggression OR theft OR rape OR fight)) AND Title: ((surveillance OR detection OR recognition)) AND Title:(''car'' OR vehicle)` retrieved very few papers and also none related to this paper subject. This may be explained by a privacy concern in private vehicles yet for collective transportation and carpooling services this appears to not be an actual concern. Anyway, this finding, even if considered a weak claim, supports the need for this research.

The first approach to the sample revealed some concerns that were not considered in the first place. The security issue is one of them, even considering an in-vehicle general-purpose surveillance system, it may be hacked by the thug before the violent act takes place. Another is to track the emotional state of passengers and react accordingly in order to prevent violence; this is currently used to prevent violent driving but may be also used for cockpit situations. In order to provide a proper overview, also based on title and abstract, the sample was classified into a couple categories and presented in Table 1.

Table 1. Categories description: **a)** *In-Vehicle User Experience* - electronic components UX; **b)** *In-Vehicle Component Network* - cockpit electronic components and interactions; **c)** ~~*In-Vehicle*~~ *Violence Prevention* - big-data on electronic components information, the "in-vehicle" constraint was removed due to lack of papers; **d)** *In-Vehicle Sensor Big-Data* - use of electronic components data in order to improve the user experience; and **e)** *Out-Vehicle Emergency Signal* - communication possibilities to be used when violence is detected. It was found two crosscutting categories: **a)** *Passenger Privacy* - passengers privacy; and **b)** *Economic Resources* - hardware constraints.

Categories	Papers
In-Vehicle User Experience	[2,4,5,13,18,19,24–27,33,42,49,50]
In-Vehicle Component Network	[1,3,6–8,16,20,22,29,32]
~~In-Vehicle~~ Violence Prevention	[10,17,31,38,39,41]
In-Vehicle Sensor Big-Data	[11,35,41,44–46,48]
Out-Vehicle Connection	[9,21,34,37,40]
Passenger Privacy (cross-cut)	[2,16,29,37]
Economic Resources (cross-cut)	[7,9,40]

Findings for Question (1). Violence detection performance varies according to how the word violence is being used [38]. Violence encompasses a wide range of concepts such as fight, crowd brawl, robbery, kidnapping, rape, *etc.*; it may also include verbal aggression, sexual harassment, threats, bullying, *etc.* Therefore, an important milestone for future works in this subject is to properly define the word *violence* within a car cockpit context. It must be also considered specific aggressions by considering people's disabilities like blindness or handicapped [5].

The detection for the first type of violence instances differs from the detection of the second since, roughly, the first is mostly video-based [38,41] whereas the second audio-based [10]; yet existing some fuzzy violence types such as kidnapping and bullying. For instance, a gunshot may be better detected with audio surveillance whereas a theft would be better detected by the video signal. This corroborates with the literature that a multi-sensor type approach provides better performance [41] (at the expense of increasing the overall complexity). However, it was not possible to find any in-vehicle violence data-set [41]. Therefore, a data-set must be built even if by simulation, *i.e.* by synthetic data [24].

Most efforts on violence detection are directed towards the video signal [38], due to these efforts, most of the current video surveillance is performing between 82–98% being considered an optimum result. From this sample was not possible to find a performance rate but it can be suggested that audio surveillance results are not behind the video performance (especially for in-door surveillance) [10]. This same perception cannot be achieved for other sensor types possibly due to the lack of papers in the sample. Nevertheless, it is feasible, for instance, to replace the video signal for a bio-radar with a fair performance [11].

The papers discussing in-vehicle surveillance are scarce in this sample, therefore there is no evidence that this same success rate can be or not achieved for a car cockpit. In addition, the sample also revealed that in-vehicle emotion detection is quite developed [4,33] yet not being used for surveillance but to anger and frustration management [24,27,35,49], especially through music [13,50]. In short, the success rate of cockpit surveillance is an open question in the literature supporting further researches on the theme. Also, presumably, emotion recognition can compose in-vehicle violence detection.

In short, multi-sensor type surveillance provide better performance and must be shaped for each specific environment, also, targeting specific violence types. For video and audio surveillance the sample suggests that they are optimized yet it is not clear if this performance can be reproduced within a car cockpit whereas emotion detection is being widely used within such environment.

Findings for Question (2). Within the raised literature, the cockpit sensors are mostly directed towards improving the cockpit experience than to surveillance, then, it was not possible to find any reference architecture to be used.

Violence detection for the raised sample includes driver recognition [2,31,44] the car theft or unauthorized usage [17] and the detection of roaming trajectories and road recording [21,39]. Most in-vehicle human-computer interfaces are joined with information systems being presented as the driver and auditory displays [18,19,26], whose behavior, as a possibility, can dynamically shift into a violence prevention behavior [30]. Adaptive resets of sensor sets for suiting and cope with several environment constraints are also a cornerstone [43].

Another widespread issue is concerning the VANET (Vehicular *Ad-hoc* Network) security [1,6,22,32] and privacy [16,29]. This willing to avoid, for instance, a remote stoppage of the breaking system yielding to a car "accident" [8,20]. This concern is also valid for in-vehicle surveillance, since an aggressor may hack the surveillance system before performing violence. Therefore some proposed security measures should also be taken [3,7,20]. In short, there was not possible to find any reference architecture for in-vehicle surveillance.

Findings for Question (3). There are several regulations to meet when developing in-vehicle surveillance applications. Two of them to be highlighted are Information Privacy and Traffic Laws. Information privacy requires that all data must be anonymous or do not receive human processing, therefore, the surveillance system is a hard real-time system yet restricted to providing alerts when detecting abnormal behavior [9,25,34,37,40]. Literature showed that current state of the art embedded processors and FPGA are able to process deep-learning and convolution networks [36,47], therefore, on the first sight, the hardware is not restrictive for in-vehicle violence detection.

There are also traffic rules for the cockpit that may. hinder some sensor performance [10], for instance, video surveillance requires light for working but the traffic law forbids driving with the inner-light on. Also, another car headlight may blind some video signal frames. This suggests that video surveillance for a car cockpit may not suits well, especially for night driving.

In another hand, the car cockpit is a highly controlled environment that may favor another set of sensors. For instance, background subtraction for audio is manageable since it is possible to gather information about the music is playing and outdoor noises by controlling if the doors or windows are open [46,48]. In this sense, a heat sensor with a distance sensor may be a better solution for a car cockpit than the video signal (better yet if the seat heater can be controlled by the surveillance system [45]).

In short, the third question could also not be properly answered by this sample, however, it highlighted that it is possible to take advantage of the cockpit being a highly controlled environment for enhancing data collection and sensors performance, working around the mentioned restrictions. Also, since most of the cockpits are similar, it is a generalization-prone environment favoring the conception of a general solution. Finally, current embedded processors and FPGA suits for convolution networks for computer vision, therefore, current hardware suits for in-vehicle violence detection.

5 Conclusion

A key finding for this survey is the lack of studies related to in-Vehicle surveillance, supporting the need for further research in this subject. It must be stated a proper definition for in-vehicle violence, what does it includes and what does it exclude. This is important to direct the research effort and sensor set setup, highlighting that some alternative sensors are promising for replacing video-audio pairs. Further research on adaptive technologies for re-configuring sensor sets is also a cornerstone for cope with environment variations in run-time. Emotion recognition is also promising for forecasting and eventually prevent in-vehicle violence. VANET security is also a concern to be handled for a reliable surveillance system. Finally, given the lack of in-vehicle data-sets, synthetic data generation may provide a starting point for developing this research path.

For future works, due to the specificities related to each environment and violence, it can be suggested the exploration of *surveillance patterns*. A *gunshot pattern*, for instance, would propose audio, heat and particle sensors and also would provide gunshot sounds, heat and particle descriptions allowing the creation of synthetic data for the target environment. Also, a conception of reference architectures incorporated with security measurements is also useful. Finally, both the architecture and patterns should be based on adaptive theory for proper handling of environmental variations.

Acknowledgments. This work is supported by the European Structural and Investment Funds in the FEDER component, through the Operational Competitiveness and Internationalization Programme (COMPETE 2020) [Project n° 039334; Funding Reference: POCI-01-0247-FEDER-039334].

This work has been supported by national funds through FCT – Fundação para a Ciência e Tecnologia through project UIDB/04728/2020.

References

1. Lee, S., Lee, J.H., Koh, B.: Threat analysis for an in-vehicle telematics control unit. Int. J. Internet Technol. Secur. Trans. **8**(4), 653–663 (2018)
2. El Ali, A., Ashby, L., Webb, A.M., Zwitser, R., Cesar, P.: Uncovering perceived identification accuracy of in-vehicle biometric sensing. In: Proceedings of the 11th International Conference on Automotive User Interfaces and Interactive Vehicular Applications: Adjunct Proceedings. ACM, New York (2019)
3. Bertolino, A., Calabro', A., Giandomenico, F., Lami, G., Lonetti, F., Marchetti, E., Martinelli, F., Matteucci, I., Mori, P.: A tour of secure software engineering solutions for connected vehicles. Softw. Qual. J. **26**(4), 1223–1256 (2018)
4. Bosch, E., Oehl, M., Jeon, M., Alvarez, I., Healey, J., Ju, W., Jallais, C.: Emotional garage: a workshop on in-car emotion recognition and regulation. In: Adjunct Proceedings of the 10th International Conference on Automotive User Interfaces and Interactive Vehicular Applications. ACM, New York (2018)
5. Brinkley, J., Posadas, B., Woodward, J., Gilbert, J.E.: Opinions and preferences of blind and low vision consumers regarding self-driving vehicles: results of focus group discussions. In: Proceedings of the 19th International ACM SIGACCESS Conference on Computers and Accessibility. ACM, New York (2017)
6. Cheah, M., Shaikh, S.A., Haas, O., Ruddle, A.: Towards a systematic security evaluation of the automotive bluetooth interface. Veh. Commun. **9**(C), 8–18 (2017)
7. Cheng, X., Lu, J., Cheng, W.: A survey on RFID applications in vehicle networks. In: Proceedings of the 2015 International Conference on Identification, Information, and Knowledge in the Internet of Things (IIKI), USA. IEEE Computer Society (2015)
8. Cho, K.-T., Shin, K.G.: Error handling of in-vehicle networks makes them vulnerable. In: Proceedings of the 2016 ACM SIGSAC Conference on Computer and Communications Security. ACM, New York (2016)
9. Coppola, R., Morisio, M.: Connected car: technologies, issues, future trends. ACM Comput. Surv. **49**(3), 1–36 (2016)
10. Crocco, M., Cristani, M., Trucco, A., Murino, V.: Audio surveillance: a systematic review. ACM Comput. Surv. (CSUR) **48**(4), 1–46 (2016)
11. Du, H., Jin, T., Song, Y., Dai, Y.: DeepActivity: a micro-doppler spectrogram-based net for human behaviour recognition in bio-radar. J. Eng. **2019**(19), 6147–6151 (2019)
12. European Union: Charter of Fundamental Rights of the European Union, vol. 53. European Union, Brussels (2010)
13. FakhrHosseini, M., Jeon, M.: The effects of various music on angry drivers' subjective, behavioral, and physiological states. In: Adjunct Proceedings of the 8th International Conference on Automotive User Interfaces and Interactive Vehicular Applications, AutomotiveUI 2016 Adjunct, pp. 191–196. ACM, New York (2016)

14. Febin, I.P., Jayasree, K., Joy, P.T.: Violence detection in videos for an intelligent surveillance system using MoBSIFT and movement filtering algorithm. Pattern Anal. Appl. 1–13 (2019)
15. Gracia, I.S., Suarez, O.D., Garcia, G.B., Kim, T.-K.: Fast fight detection. PLoS ONE **10**(4), e0120448 (2015)
16. Gupta, M., Sandhu, R.: Authorization framework for secure cloud assisted connected cars and vehicular internet of things. In: Proceedings of the 23nd ACM on Symposium on Access Control Models and Technologies. ACM, New York (2018)
17. Guravaiah, K., Thivyavignesh, R.G., Leela Velusamy, R.: Vehicle monitoring using internet of things. In: Proceedings of the 1st International Conference on Internet of Things and Machine Learning, IML 2017. ACM, New York (2017)
18. Haeuslschmid, R., Pfleging, B., Alt, F.: A design space to support the development of windshield applications for the car. In: Proceedings of the 2016 CHI Conference on Human Factors in Computing Systems. ACM, New York (2016)
19. Heijboer, S., Schumann, J., Tempelman, E., Groen, P.: Physical fights back: introducing a model for bridging analog digital interactions. In: Proceedings of the 11th International Conference on Automotive User Interfaces and Interactive Vehicular Applications: Adjunct Proceedings. ACM, New York (2019)
20. Jellid, K., Mazri, T.: Security study on three modes of connection for a connected car. In: Proceedings of the 3rd International Conference on Smart City Applications, SCA 2018. ACM, New York (2018)
21. Kadu, S., Cheggoju, N., Satpute, V.R.: Noise-resilient compressed domain video watermarking system for in-car camera security. Multimedia Syst. **24**(5), 583–595 (2018)
22. Khodari, M., Rawat, A., Asplund, M., Gurtov, A.: Decentralized firmware attestation for in-vehicle networks. In: Proceedings of the 5th on Cyber-Physical System Security Workshop. ACM, New York (2019)
23. Kitchenham, B., Brereton, P.: A systematic review of systematic review process research in software engineering. Inf. Softw. Technol. **55**(12), 2049–2075 (2013)
24. Krome, S., Goedicke, D., Matarazzo, T.J., Zhu, Z., Zhang, Z., Zamfirescu-Pereira, J.D., Ju, W.: How people experience autonomous intersections: taking a first-person perspective. In: Proceedings of the 11th International Conference on Automotive User Interfaces and Interactive Vehicular Applications. ACM, New York (2019)
25. Kun, A.L., Wachtel, J., Thomas Miller, W., Son, P., Lavallière, M.: User interfaces for first responder vehicles: views from practitioners, industry, and academia. In: Proceedings of the 7th International Conference on Automotive User Interfaces and Interactive Vehicular Applications. ACM, New York (2015)
26. Lamm, L., Wolff, C.: Exploratory analysis of the research literature on evaluation of in-vehicle systems. In: Proceedings of the 11th International Conference on Automotive User Interfaces and Interactive Vehicular Applications. ACM, New York (2019)
27. Löcken, A., Ihme, K., Unni, A.: Towards designing affect-aware systems for mitigating the effects of in-vehicle frustration. In: Proceedings of the 9th International Conference on Automotive User Interfaces and Interactive Vehicular Applications Adjunct. ACM, New York (2017)
28. Maheshwari, S., Heda, S.: A review on crowd behavior analysis methods for video surveillance. In: Proceedings of the Second International Conference on Information and Communication Technology for Competitive Strategies, p. 52. ACM (2016)

29. Malina, L., Vives-Guasch, A., Castellà-Roca, J., Viejo, A., Hajny, J.: Efficient group signatures for privacy-preserving vehicular networks. Telecommun. Syst. **58**(4), 293–311 (2015)
30. Marcondes, F.S., Almeida, J.J., Novais, P.: Chatbot theory. In: International Conference on Intelligent Data Engineering and Automated Learning, pp. 374–384. Springer (2018)
31. Markwood, I.D., Liu, Y.: Vehicle self-surveillance: sensor-enabled automatic driver recognition. In: Proceedings of the 11th ACM on Asia Conference on Computer and Communications Security. ACM, New York (2016)
32. Mazloom, S., Rezaeirad, M., Hunter, A., McCoy, D.: A security analysis of an in vehicle infotainment and app platform. In: Proceedings of the 10th USENIX Conference on Offensive Technologies, USA. USENIX Association (2016)
33. Moore, G.: Emotional drive wearing your heart on your car. In: Proceedings of the 31st British Computer Society Human Computer Interaction Conference, HCI 2017, Swindon, GBR. BCS Learning & Development Ltd. (2017)
34. Nugra, H., Abad, A., Fuertes, W., Galárraga, F., Aules, H., Villacís, C., Toulkeridis, T.: A low-cost IoT application for the urban traffic of vehicles, based on wireless sensors using GSM technology. In: Proceedings of the 20th International Symposium on Distributed Simulation and Real-Time Applications. IEEE Press (2016)
35. Paredes, P.E., Ordonez, F., Ju, W., Landay, J.A.: Fast & furious: detecting stress with a car steering wheel. In: Proceedings of the 2018 CHI Conference on Human Factors in Computing Systems, CHI 2018. ACM, New York (2018)
36. Qiu, J., Wang, J., Yao, S., Guo, K., Li, B., Zhou, E., Yu, J., Tang, T., Xu, N., Song, S., et al.: Going deeper with embedded FPGA platform for convolutional neural network. In: Proceedings of the 2016 ACM/SIGDA International Symposium on Field-Programmable Gate Arrays, pp. 26–35 (2016)
37. Ramakrishnan, B., Selvi, M., Bhagavath Nishanth, R., Milton Joe, M.: An emergency message broadcasting technique using transmission power based clustering algorithm for vehicular ad hoc network. Wirel. Pers. Commun. **94**(4), 3197–3216 (2017)
38. Ramzan, M., Abid, A., Khan, H.U., Awan, S.M., Ismail, A., Ahmed, M., Ilyas, M., Mahmood, A.: A review on state-of-the-art violence detection techniques. IEEE Access **7**, 107560–107575 (2019)
39. Shen, M., Liu, D.-R., Shann, S.-H.: Outlier detection from vehicle trajectories to discover roaming events. Inf. Sci. **294**(C), 242–254 (2015)
40. Siddiqui, S.A., Mahmood, A.: Towards fog-based next generation internet of vehicles architecture. In: Proceedings of the 1st International Workshop on Communication and Computing in Connected Vehicles and Platooning. ACM, New York (2018)
41. Singh, T., Vishwakarma, D.K.: Video benchmarks of human action datasets: a review. Artif. Intell. Rev. **52**(2), 1107–1154 (2019)
42. Solís-Marcos, I., Kircher, K.: Event-related potentials as indices of mental workload while using an in-vehicle information system. Cogn. Technol. Work **21**(1), 55–67 (2019)
43. Stange, R.L., Cereda, P.R.M., Neto, J.J.: Survival of the mutable: architecture of adaptive reactive agents. In: XXIII Congreso Argentino de Ciencias de la Computación, La Plata, 2017 (2017)
44. Tamoto, A., Itou, K.: Voice authentication by text dependent single utterance for in-car environment. In: Proceedings of the Tenth International Symposium on Information and Communication Technology. ACM, New York (2019)

45. Tennent, H., Moore, D., Ju, W.: Character actor: design and evaluation of expressive robot car seat motion. In: Proceedings of the ACM on Interactive, Mobile, Wearable and Ubiquitous Technologies, vol. 1, no. 4, January 2018
46. Tian, D., Zhu, Y., Zhou, J., Duan, X., Wang, Y., Song, J., Rong, H., Guo, P.: A novel data quality assessment framework for vehicular network testbeds. In: Proceedings of the 12th EAI International Conference on Testbeds and Research Infrastructures for the Development of Networks & Communities, Brussels, BEL. ICST (Institute for Computer Sciences, Social-Informatics and Telecommunications Engineering) (2018)
47. Verhelst, M., Moons, B.: Embedded deep neural network processing: algorithmic and processor techniques bring deep learning to iot and edge devices. IEEE Solid State Circuits Mag. **9**(4), 55–65 (2017)
48. Won, M., Alsaadan, H., Eun, Y.: Adaptive audio classification for smartphone in noisy car environment. In: Proceedings of the 25th ACM International Conference on Multimedia. ACM, New York (2017)
49. Wurhofer, D., Krischkowsky, A., Obrist, M., Karapanos, E., Niforatos, E., Tscheligi, M.: Everyday commuting: prediction, actual experience and recall of anger and frustration in the car. In: Proceedings of the 7th International Conference on Automotive User Interfaces and Interactive Vehicular Applications. ACM, New York (2015)
50. Zhu, Y., Wang, Y., Li, G., Guo, X.: Recognizing and releasing drivers' negative emotions by using music: evidence from driver anger. In: Adjunct Proceedings of the 8th International Conference on Automotive User Interfaces and Interactive Vehicular Applications. ACM, New York (2016)

Multiagent Systems and Role-Playing Games Applied to Natural Resources Management

Vinícius Borges Martins(✉) and Diana Francisca Adamatti

Programa de Pós-graduação em Engenharia da Computação,
Universidade Federal do Rio Grande, Rio Grande, RS, Brazil
vimiciusbormar27@gmail.com, dianaada@gmail.com

Abstract. The growing use of chemicals in plantations and high use of car/trucks are being a problem to be fought all over the globe. In this paper, we present the theoretical basis and the creation of a game that is going to be used with the Committee of the Mirim Lagoon and of the São Gonçalo River (Brazil), where each member are going to play the role of others members to better understand their actions and to make better decisions to the regional resources. In this way, the game was created physically, where, after every stage, the actions were transferred to be calculated by a JAVA software (methodologically defined each role as an agent) on a computer, returning the results to the players in printed papers. Further, a graphical interface will be implemented in the calculation program and, starting from that, a computational version of the game will be created. Besides that, random roles are going to be played with intelligent agents.

Keywords: Multiagent Systems · Role-Playing Games · Natural Resources Management

1 Introduction

The growth of the pollution in the world, due to the increase of vehicles, garbage, and chemicals in the agriculture, bring the expectation of water impurity, trash islands, air pollution, etc., showing the need of immediate intervation with projects that fight it [12].

Natural Resources Management (NRM) is a field which aims to solve the problem presented, that is, it seeks the best quality of future's resources based on the quality of today's population, managing fields (nature reserves, private and public properties), water (watersheds, oceans), plants (forests, city afforestation) and animals (wild and domestic life) [16].

However, the NRM can bring three challenges [16]: (1) optimization and control, (2) management and communication and (3) data analysis. For this reason, there is the need for tools to deal with such challenges.

Y. Dong et al. (Eds.): DCAI 2020, AISC 1237, pp. 221–230, 2021.
https://doi.org/10.1007/978-3-030-53036-5_24

Artificial Intelligence (AI), more specifically Multiagent Systems (MAS), aims at using intelligent objects (called *agents*) that, based on the known information about the environment, decide the best plan to reach a personal goal and, consequently, the system goal, independently and heterogeneously [23]. The MAS, when used to simulations, are called Multiagent Based Simulations (MABS). These try to simulate an environment where the agents interact with each other trying to reach or to predict the solution of the problem, typically in an interdisciplinary way [17,19,20].

Such interdisciplinarity and tools available in the MABS make them a suitable instrument to design and simulate problems in NRN scope.

Another tool that has been used in this field is the Role-Playing Game (RPG) [27]. In RPG, the player, at the beginning of the game, receives a role (character) that is performed until the end. This kind of game is usually divided in turns, which allows all the players make their actions and all these actions influence the environment before the next turn.

A RPG can be played in printed paper (with boards/cards), orally (with storyteller and oral choices) and computationally. There are projects [2,5,11,20, 21] that use RPG to solve problems for workgroups, making the players interpret the coworkers roles in the game. Doing so, the players start to consider the coworker's needs before making any choices and understand the reason for the others' choices.

For this reason, in the NRM field, the RPG can be used as a tool to help the decision making of the entities designated for this type of decision, such as the Committees.

This paper presents the theoretical basis assembled to build a RPG, which is also presented. This RPG, called Gorim, is going to be built computationally to be used with the Committee of the Mirim Lagoon and of the São Gonçalo River (Brazil). The main goal of this paper is the explanation of areas in this work as well as the definition of our game, presenting the conceptual model and the first results of it. We expect the game helps the Committee members in their decision making since they are playing in each other's roles, showing the possibilities and needs of the others, so, when they take any decision, they will think in the solution that brings the greater number of advantages to everyone, and, specially, to the environment. Besides that, intelligent agents will be inserted to play with the players as Non-Player Characters (NPC).

2 Theoretical Basis

2.1 Multiagent Systems

In conventional programming, *objects* are used in software. But, in the agent-oriented programming, *agents* are used. The difference between the two structures is that, while an object has all the functionalities predefined, that is, has exactly the method to be called for each input, the agents have a set of options to be decided by themselves in execution time, trying to reach the objective set by the user [25,30, p. 26]. Additionally, an autonomous agent has to be capable of interacting over time with the environment it was inserted to import relevant information for its decision making, reaching its objectives [14].

Multiagent Systems are systems where many agents co-work independently to hit their goals, and reaching main goals of the system. The agents, inside the MAS, can be homogeneous (that is, learn in the same way) or heterogeneous (in different ways), however, in both approaches, they work asynchronously from each other's threads [23]. Even working individually, agents must be capable to act organized, due to the most issues solved by them are solved in a distributed way. Therefore, it must be implemented characteristics as cooperation, coordination, competition, negotiation and communication [4,8,10,15].

Simulations, by showing details with great accuracy, are useful tools for decision making [15]. There is a field into the MAS that it is merged to simulations, forming the Multi-Agent Based Simulations (MABS), which has as its objective the study of the consequences of different parameters in an environment using many fields in the process (such as psychology and social biology, computer science, sociology and economy) [17,19,20].

Summarizing, the use of MAS can be profitable in many ways, such as the speed in the problem solving (due to its distributed execution in the agents activities); the flexibility and scalability connecting many systems; and, the presence of a greater scope of informations that are returned of the system (since all the resources are in the same environment) [22].

2.2 Role-Playing Games

A gender of game is called Role-Playing Game (RPG), which is localized between the game and the theatre fields, because it joins the players' interaction and their acting of the available characters, making them act in an environment with rules and behaviour defined previously to complete the given story [1,26]. This way, the most of RPG do not have winners or losers, being a collaborative way of game, revealing aspects of social relations and allowing the straight observation of the interactions between the players [1,7].

The RPG is being constantly used in training, due to its approach of simulation of situations without effects in the real world, but the gained learning, making it a ludic tool, facilitating the learning of the subject [27].

RPG and the MAS are very similar [6]: the agents can be understood as the roles, the environment as the board, the steps of the system as the game rounds, etc, making them suitable tools.

2.3 Natural Resources Management

Water is the most vital natural resource for the ecosystem and its populations, due to the need of this mineral to life and health of plants and animals, besides the importance of water in social and economical human matters [28]. The Natural Resources Management (NRM) field is the area where is studied the ways of manage water, animals, plants and fields, basing in the quality of the environment to the next generations [18]. The NRM has three computational challenges

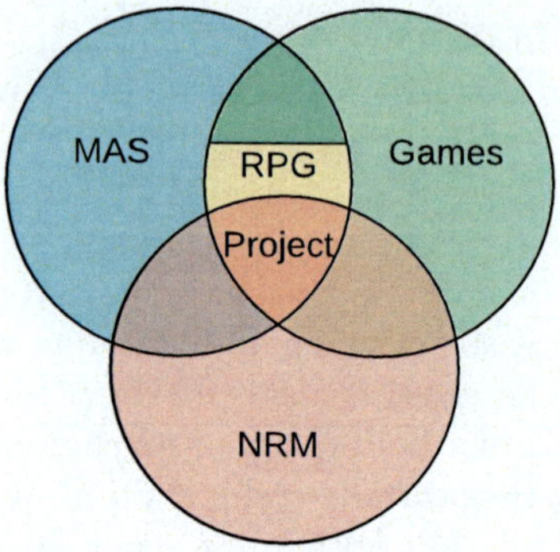

Fig. 1. Venn diagram illustrating where this project is inserted.

[16]: (1) otimization and control, (2) management and communication and (3) data analysis.

One of the areas of NRM is the Hydric Resources Management (HRM), which discuss the best usage and distribution of water, envolving organizations and different social groups. Besides, an HRM is responsible for the social and environmental risks situations, such as floods and droughts, such situations that need the preventive actions of the authorities [29].

Watersheds are geographical fields where, according to their relief, the rain-water or from sources drain until the bottom of the valley, leaving by a single point (usually a river) or, if there is a crest along the way, the water is divided between two different watersheds [3].

The state of Rio Grande do Sul (Brazil) has the Mirim Lagoon and of the São Gonçalo River Watershed, which has as its managers the local Management Committee. This watershed is the focus of our work.

2.4 Integration of the Areas

The interdisciplinarity of MAS elects them as a tool for the modelling of problems related to natural management if considering the computational challenges of NRM. Besides that, MAS and RPG are similar and from the formal view, RPG is a kind of MAS due to its composition by entities immersed into an environment pursuing their specific objectives [6, 21]. For those particularities, the project was inserted in the intersection of the three areas, as shown in Fig. 1.

2.5 Methodologies

There are methodologies created for studies using the three areas (MAS, RPG and NRM): ComMod and GMABS.

Companion Modelling (ComMod). This methodology was proposed by the members of CIRAD[1] (*Centre de coopération Internationale en Recherche Agronomique pour le Développement*) [9]. It was created to standardize a methodology for the development of systems/simulations that are going to be used to help managers in their decision making. It is divided into three stages:

- *Building an Artificial World.* This stage relates one or more researchers and must be used to collect information of the system proposed, that is, identify different actors and perceptions related and use MAS for the modelling.
- *Validation of the Cognitive Model.* Stage for testing the model proposed. The test must be realized to analyse the representations and intentions of the agents, using an RPG.
- *Simulations.* This stage shows how the dynamics of the system arise from the interactions of the actors with different weights and representations.

Games and Multi-Agent-Based Simulation (GMABS). This methodology was proposed on [6] and it was named by ADAMATTI et al. as GMABS (*Games and Multi-Agent-Based Simulation*) in [1], because the original authors did not name the approach. In this methodology, the main aim is the dynamic description of the integration techniques between RPG and MABS. This approach is divided into five steps:

1. Rules, available actions and game environment are presented to the players. Also in this step, the characters are sorted for them.
2. The game begins. The players have time to decide their actions and negotiate.
3. The information from the previous step is put in the MAS system.
4. The environment changes with the data computed, creating a new scenario.
5. The fresh environment is informed to the players, who take new actions, and repeat it until the game's finished.

This project uses a merge of the two given approaches. For the first version of the game (board game, current project state), that the computational module for calculations is used, it was implemented the GMABS methodology. The project in general, for a better relationship with the stakeholders, uses the ComMod approach.

2.6 Related Projects

In [13], a Systematic Literature Review (SLR) was presented, using the methodology proposed in [24], and it tries to join all the researches in the main databases considering three main fields (RPG, MAS and NRM). They found 10 papers, which were divided in 4 groups:

- *RPG -> MAS.* Group characterized by the papers which the RPG is played by the stakeholders in cards or paper and, in the end of the game, the remain information is turned into data in a MABS.

[1] www.cirad.fr.

- *MAS -> RPG*. Characterized by projects which the researchers build a SMA and, then, an RPG with the same metrics used in the MAS to be validated by the stakeholders.
- *RPG + MAS*. The systems presented in the papers of this group had in common the RPG in paper mode with a computational calculation module, due to the complexity of the calculation done, that is, the RPG and the MAS work together.
- *RPG ++ MAS*. This last group shows projects where both RPG and MAS are computational and work together.

The papers of the last two groups were chosen as related projects due to the current state of the project (RPG + MAS) and its final objective (RPG ++ MAS). Some of the papers found in this SLR are presented below. In [11], is developed an environment that facilitates the communication and negotiation between different stakeholders about their interests, consequences of strategies and identification of intervention areas. The authors opted for the use of the ComMod approach.

The project proposed in [21] creates a new approach for environmental modelling. To test their method, the authors used ComMod for the creation of a MAS and an RPG aiming the decision making of a village facing the possible incidents in plantations.

In [2] it is described a project where the authors use a tool made by them for the players to determine the quality and quantity of water in a capture system present in an urban perimeter. There were made two prototypes, where, in the first one, the participants played with other participants, and, in the second one, the participants could, randomly, play with intelligent agents. The approach used in this game was GMABS.

3 The Proposed RPG

The game, called Gorim, consists of five different character classes (`Businessman`, `Farmer`, `Environmental Police`, Mayor and City Counsellor) and two cities (Atlantis and Cidadela). Each city has two businessmen, three farmers, an environmental police, a mayor and a city councillor. The Businessman class is divided into four types: Seeds, Machines, Fertilizers and Pesticides, where the two first types of businessmen are from Atlantis and, the other two, from Cidadela. The characters from this class are responsible for selling their products for farmers, which do their own choices for combinations of products to plant in their lands since each combination makes a different mountant of pollution and productivity. The members of the Environmental Police class must observe the pollution of each previous player to apply a fine if needed. Besides that, they are responsible for the Green Seal license to the farmers, giving them a tax discount. Mayors have to apply the money from taxes and fines in environmental solutions, and, also, increase or decrease the taxes. The City Counsellors, though, must give ideas to the mayors about what to do in their turn.

Each type of **businessman** has different products to offer to the farmers. The Seed businessman *sells* seeds of Vegetable, Rice and Soy. The Fertilizer one *sells* Common, Premium and Super Premium fertilizers, the same as the Pesticides one. Although, the Machine businessman can *rent* three packages of machines: (1) with just a Seed Drill; (2) with a Seed Drill and a Harvester pack; and (3) with a Seed Drill, a Harvester and a Drone pack. This businessman can, also, rent a Sprayer, which does not increase productivity, but decrease the farmer's pollution in the half (50%).

The **farmers** have six parcels of field, where they can plant a seed with fertilizer and/or pesticide or machine package, that is, it is not possible to put a pesticide and a machine package in the same parcel. Although, it is possible to put a Sprayer with a pesticide or any machine package. It is allowed to leave the field empty too, that is, do not plant anything. Each product has a price and each combination of them generates different productivity, which, depending on the mountant, has a different mountant of tax to pay.

Each turn of the game is divided into two stages, where in the first one play the **businessmen** and the **farmers** and, in the second, play the **environmental police**, the **mayors** and the **city counsellors**. In the current stage of the project, each game is played for 10 players, that is, the electable roles (environmental police, mayor and counsellor) are played by the same players from the first stage, which are elected every two turns by the players of the same city (just the characters from Atlantis can vote in their mayor, for example).

Figure 2a shows students and teachers playing in the fourth simulation-test of the RPG, and Fig. 2b presents the pins were made for players, identifying.

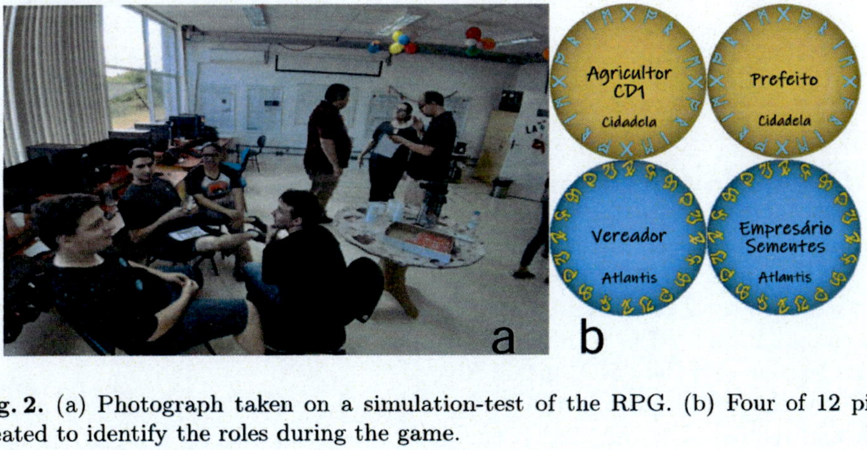

Fig. 2. (a) Photograph taken on a simulation-test of the RPG. (b) Four of 12 pins created to identify the roles during the game.

3.1 The Calculation Engine

Even starting the game with rounded "Moneys" (game currency) for each player, after the tax charge and buying/selling, it has shown impracticable the hand calculation of each player, of the global and individual pollutions, which shown the need to develop a computational calculation module to do it. This calculation engine was programmed in JAVA language using agent-oriented programming. The engine structure is explicit in the class diagram of Fig. 3. Besides the classes showed in the diagram, there is the World class, that receive the inputs from the interface and calls the methods of each class. In the end of each stage, the engine has as output the summarized files of the players, which shows what was made in the last stage. Moreover, during the whole game, log files are created to have stored all the game actions.

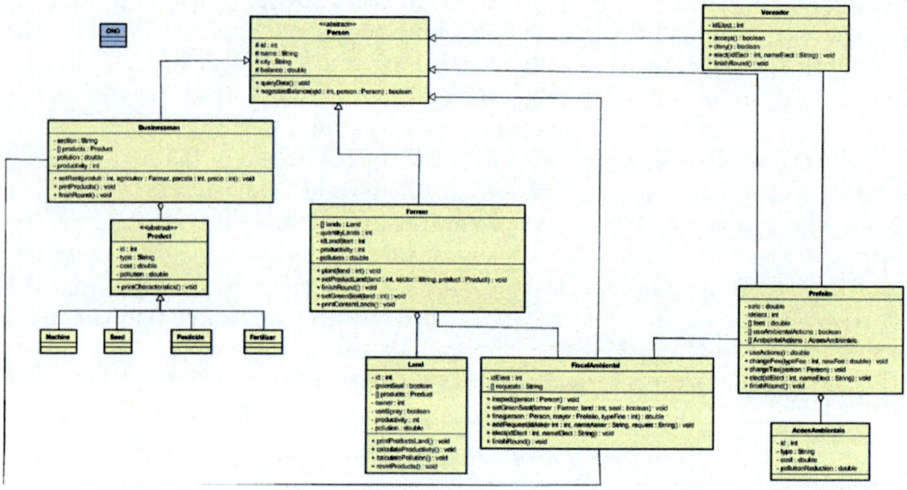

Fig. 3. Class Diagram which shows the calculation engine structure.

4 Conclusions and Further Works

This paper showed the theoretical basis, the related projects and the current state of the project that involves the fields of Multiagent Systems, Role-Playing Games and Natural Resources Management. This project aims the development of a computational RPG to help the decision making of the Committee of the Mirim Lagoon and the São Gonçalo River.

Currently, the project is in development, where the board game is being built and refined. The calculation engine is already developed in JAVA, which helps the next implementation of the game: the graphical interface. After this, the whole software will be put onto the web, to be played by people that are not in the same place. Concomitantly, it will be performed a data mining with the data logs of the simulations, tracing the strategies of players to build non-players characters (NPCs) agents. Those will play with real players when it does not have enough players to start a match.

Acknowledgements. We would like to thanks Coordenação de Aperfeiçoamento de Pessoal de Nível Superior (CAPES) and Agência Nacional das Águas (ANA) for the financial support of the project number 16/2017.

References

1. Adamatti, D.F., Sichman, J.S., Coelho, H.: Utilização de rpg e mabs no desenvolvimento de sistemas de apoio a decisão em grupos. In: Anais do IV Simpósio Brasileiro de Sistemas Colaborativos SBSC 2007, p. 15 (2007)
2. Adamatti, D.F., Sichman, J.S., Coelho, H.: An analysis of the insertion of virtual players in GMABS methodology using the Vip-JogoMan prototype. J. Artif. Soc. Soc. Simul. **12**(3), 7 (2009). http://jasss.soc.surrey.ac.uk/12/3/7.html
3. Ahmad, S., Simonovic, S.P.: An artificial neural network model for generating hydrograph from hydro-meteorological parameters. J. Hydrol. **315**(1–4), 236–251 (2005)
4. Alvares, L.O., Sichman, J.S.: Introduçao aos sistemas multiagentes. In: XVII Congresso da SBC-Anais JAI 1997 (1997)
5. Barnaud, C., Promburom, T., Trébuil, G., Bousquet, F.: An evolving simulation/gaming process to facilitate adaptive watershed management in northern mountainous thailand. Simul. Gaming **38**(3), 398–420 (2007)
6. Barreteau, O., Bousquet, F., Attonaty, J.M.: Role-playing games for opening the black box of multi-agent systems: method and lessons of its application to Senegal River Valley irrigated systems. J. Artif. Soc. Soc. Simul. **4** (2001)
7. Barreteau, O., Le Page, C., D'aquino, P.: Role-playing games, models and negotiation processes. J. Artif. Soc. Soc. Simul. **6**(2) (2003)
8. Bordini, R.H., Vieira, R., Moreira, A.F.: Fundamentos de sistemas multiagentes. In: Anais do XXI Congresso da Sociedade Brasileira de Computação (SBC 2001), vol. 2, pp. 3–41 (2001)
9. Bousquet, F., Barreteau, O., Le Page, C., Mullon, C., Weber, J.: An environmental modelling approach: the use of multi-agent simulations. Adv. Environ. Ecol. Model. **113**(122) (1999)
10. Bousquet, F., Le Page, C.: Multi-agent simulations and ecosystem management: a review. Ecol. Model. **176**(3–4), 313–332 (2004)
11. Campo, P.C., Mendoza, G.A., Guizol, P., Villanueva, T.R., Bousquet, F.: Exploring management strategies for community-based forests using multi-agent systems: a case study in Palawan, Philippines. J. Environ. Manag. **90**(11), 3607–3615 (2009). http://www.sciencedirect.com/science/article/pii/S0301479709002321
12. European Environment Agency: Increasing environmental pollution (gmt 10), May 2018. https://www.eea.europa.eu/soer-2015/global/pollution. Disponible at: https://www.eea.europa.eu/soer-2015/global/pollution. Accessed 15 Nov 2019
13. Farias, G., Leitzke, B., Born, M., Aguiar, M., Adamatti, D.: Systematic review of natural resource management using multiagent systems and role-playing games. In: 18th Mexican International Conference on Artificial Intelligence - MICAI. Springer, Xalapa (2019)
14. Franklin, S., Graesser, A.: Is it an agent, or just a program?: A taxonomy for autonomous agents. In: International Workshop on Agent Theories, Architectures, and Languages, pp. 21–35. Springer (1996)
15. Frozza, R.: SIMULA: Ambiente para desenvolvimento de sistemas multiagentes reativos. Master's thesis, Universidade Federal do Rio Grande do Sul (UFRGS) (1997)

16. Fuller, M.M., Wang, D., Gross, L.J., Berry, M.W.: Computational science for natural resource management. Comput. Sci. Eng. **9**(4), 40 (2007)
17. Gilbert, N., Troitzsch, K.: Simulation for the Social Scientist. McGraw-Hill Education, New York (2005)
18. Holzman, B.: Natural resource management (2009). http://online.sfsu.edu/bholzman/courses/GEOG%20657/. Accessed 30 2019
19. Le Page, C., Bobo Kadiri, S., Towa, K., William, O., Ngahane Bobo, F., Waltert, M.: Interactive simulations with a stylized scale model to codesign with villagers an agent-based model of bushmeat hunting in the periphery of Korup National Park (Cameroon). J. Artif. Soc. Soc. Simul. **18**(1) (2015)
20. Le Page, C., Dray, A., Perez, P., Garcia, C.: Exploring how knowledge and communication influence natural resources management with rehab. Simul. Gaming **47**(2), 257–284 (2016). https://doi.org/10.1177/1046878116632900
21. Le Page, C., Perrotton, A.: KILT: a modelling approach based on participatory agent-based simulation of stylized socio-ecosystems to stimulate social learning with local stakeholders. In: Sukthankar, G., Rodriguez-Aguilar, J.A. (eds.) Autonomous Agents and Multiagent Systems, pp. 31–44. Springer, Cham (2017)
22. Leitzke, B., Farias, G., Melo, M., Gonçalves, M., Born, M., Rodrigues, P., Martins, V., Barbosa, R., Aguiar, M., Adamatti, D.: Sistema multiagente para gestão de recursos hídricos: Modelagem da bacia do são gonçalo e da lagoa mirim. In: Anais do X Workshop de Computação Aplicada a Gestão do Meio Ambiente e Recursos Naturais. Belém, Pará, Brasil (2019)
23. Lesser, V.R.: Cooperative multiagent systems: a personal view of the state of the art. IEEE Trans. Knowl. Data Eng. **11**(1), 133–142 (1999)
24. Mariano, D.C.B., Leite, C., Santos, L.H.S., Rocha, R.E.O., de Melo-Minardi, R.C.: A guide to performing systematic literature reviews in bioinformatics. arXiv preprint arXiv:1707.05813 (2017)
25. Moreno, A.: Medical applications of multi-agent systems. Computer Science and Mathematics Department, University of Rovira, Spain (2003)
26. Pereira, C.E.K.: Construção de personagem & aquisição de linguagem: O desafio do rpg no ines, vol. 10 (jul/dez) Rio de Janeiro INES, 2004 Semestral ISSN 1518-2509 1–Forum–Instituto Nacional de Educação de Surdos, p. 7 (2003)
27. Perrotton, A., Garine-Wichatitsky, D., Fox, H.V., Le Page, C.: My cattle and your park: codesigning a role-playing game with rural communities to promote multistakeholder dialogue at the edge of protected areas. Ecol. Soc. **22**(1) (2017)
28. Ponte, B., De la Fuente, D., Parreño, J., Pino, R.: Intelligent decision support system for real-time water demand management. Int. J. Comput. Intell. Syst. **9**(1), 168–183 (2016)
29. Tundisi, J.G.: Novas perspectivas para a gestão de recursos hídricos. Revista USP **1**(70), 24–35 (2006)
30. Wooldridge, M.: Introduction to MultiAgent Systems. Wiley, Hoboken (2002)

A Systematic Review to Multiagent Systems and Regulatory Networks

Nilzair Barreto Agostinho(✉), Adriano Velasque Wherhli(✉), and Diana Francisca Adamatti(✉)

Programa de Pós-Graduação em Modelagem Computacional, Universidade Federal do Rio Grande (PPGMC/FURG), Av. Itália, km 8, Bairro Carreiros, Rio Grande, RS, Brazil
nilzairmb@gmail.com, werhli@gmail.com, dianaada@gmail.com
http://www.c3.furg.br

Abstract. Today, it is natural that great efforts are directed towards the development of tools to improve our knowledge about molecular interactions. The representation of biological systems as Genetic Regulatory Networks (GRN) that form a map of the interactions between the molecules in an organism is a way of representing such biological complexity. In the past few years, for simulation and inference purposes, many different mathematical and algorithmic models have been adopted to represent the GRN. Among these methods, Multiagent Systems (MAS) are somewhat neglected. Thus, in this paper was performed a Systematic Literature Review (SLR) to clarify the use of MAS in the representation of GRN. The results show that there are very few studies in which the MAS are applied in the task of modeling the GRN. Therefore, given the interesting properties of MAS, it is expected that it can be further investigated in the task of GRN modelling.

Keywords: Multiagent systems · Genetic regulatory network · Simulation · Systematic literature review

1 Introduction

The considerable advances in measurement and storage of *omics* data and the potential applications in medicine, agriculture and industry, have vastly increased the interest in the area of Biological Systems. A great challenge, being addressed in face of this deluge of data, is the development of mathematical, statistical and computational tools that can shed some light in the intricate mechanisms of molecular interactions.

It is now common sense that the high complexity of living organisms is closely linked to the organization of its components in a cascade of interconnected events, as a network. Because of this complexity, researchers have been devoting great efforts to the investigation of biological networks, as for example, the GRN. GRN are a representation of molecules (nodes) and their interactions (edges) as a graph [5,9].

Y. Dong et al. (Eds.): DCAI 2020, AISC 1237, pp. 231–240, 2021.
https://doi.org/10.1007/978-3-030-53036-5_25

In the representation of a GRN, genes control the expression of other genes, nonetheless, not all genes interact with other genes. Normally, a gene controls the expression of only a subset of other genes present in a network. Likewise, its expression is controlled not for all genes but for a subset of genes [9].

Some methods that have been used to perform the inference and representation of GRN are: Differential equations (DE) [15]; Bayesian Network [2]; Boolean Network (NB) [13]; Boolean Probabilistic Networks (BPN) [14] and Bayesian networks using Markov Chain Monte Carlo and Graphical Gaussian Models [1].

Systematic Literature Review (SLR) is a type of research focused on a well-defined subject, which aims to identify, select, evaluate and synthesize evidences of a specific research topic [3].

Reviews that can be considered as SLR have started to be published around 1950s. However, the methodological development of this type of research, e.g. in the health area, was consolidated only in the 1980s. A good example is the book "Effective care during pregnancy and childbirth" and the Cochrane collaboration. The SLR should be in-depth and not biased in their preparation. Also, the chosen criteria must be clear to permit that other researchers repeat the process [3].

There is not a standardization for SLR in the area of Bioinformatics. In [10], the authors have proposed a standardization of steps and procedures for the development of this type of research. The SLR purpose is to build an overview of a specific issue and provide a summary of the literature, guaranteeing standardization and precision. This procedure it can help to attain an unbiased overview of the studies that have been published in several sources (conferences and journals) as well as robustness and security. The steps for performing a SLR following a standardization are described in [10] and are presented in the Fig. 1.

In the preparation step was defined the main subject of study and the keywords. Because different databases can present different search methods it is recommended the use of search strategies. After the preparation step, the development of the review follows the four sub-steps: title evaluation, abstract evaluation, diagonal reading and complete paper reading.

In the process of systematic review, it is common to find a large amount of papers that comply with the keywords. Therefore, often turned some steps to redefine the keywords or their variants and how they are inserted into the searches, whereas the process is a cycle and, in some cases, it is necessary to be re-calibrated.

The filters of abstract reading step will "feed" the next step, the diagonal reading, and consequently, it will feed the next step, the complete paper reading. In the last steps, it is necessary the participation of, at least, one researcher from each involved area.

The main goal of this SLR is to obtain a good theoretical background regarding the intersection of areas MAS and GRN, i.e., obtain a theoretical background for development of a Multiagent Simulator to GRN.

The paper is structured as follows: in Sect. 2 the Methodology is presented. Section 3 presents selected papers that have been found after of SLR, as well as the difference between these papers and our approach. Section 4 Concludes the Paper.

2 Methodology

All the steps taken to carry out the SLR about MAS and GRN are shown in Fig. 1. Firstly, the main question should be defined and after the keywords and databases protocol are outlined. In this protocol, the goal, the inclusion and exclusion criteria and the specific questions have to be clearly defined. In our study, the protocol defined is shown in Table 1 and the keywords are presented in Table 2.

In the first moment, was used the keywords "network" AND "multiagents" in Google Academic, which resulted in 19.700 papers. These resulting papers are from any area of computer science or bioinformatics and have either in the title or in the keywords these two terms. Considering the high number of papers and low quality of the results, it was decided to refine the keywords and the search databases.

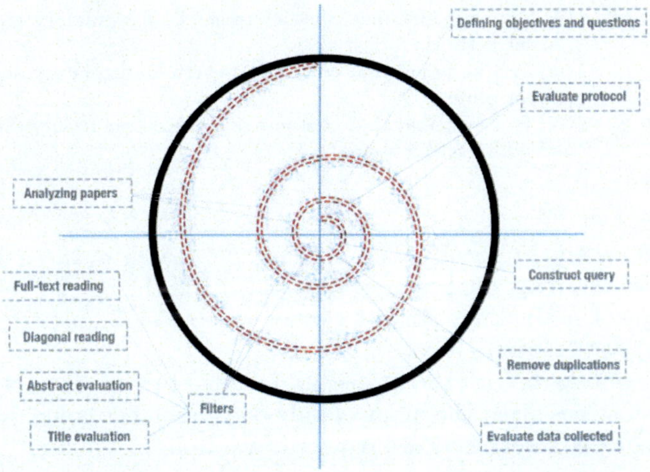

Fig. 1. Spiral review template adapted of [10]. Two steps of the final step were removed for this paper.

The databases that we have searched were: Pubmed[1], PMC(see Footnote 1), Science Direct[2], Scopus[3], ACM[4] and Scielo[5]. The keywords that have been defined are: "regulatory network" AND "multiagent system". However, these keywords can have variants (in writing) or the search could be different in each database (the use of relational operator and regular expression).

[1] https://www.ncbi.nlm.nih.gov/pubmed/.
[2] https://www.sciencedirect.com/.
[3] https://www.scopus.com/.
[4] https://dl.acm.org/.
[5] http://www.scielo.org/php/index.php.

Table 1. Defined protocol for the systematic review.

Main question	Is there a multiagent system for regulatory networks?
Goal	The goal is to find researches that involve multiagent systems and regulatory networks
Inclusion criteria	The inclusion criteria are: • Studies that mention multiagent systems and (regulatory networks or genetic networks) • Studies that contain the key words: regulatory networks, multiagents, multiagent systems
Exclusion criteria	The exclusion criteria are: • The study is not in English • The papers does not contain any studies on (regulatory networks or genetic networks) and (multiagents or multiagent systems)
Specific questions	Specific questions (i) Is it an paper that contains research on regulatory networks and multiagent systems? (ii) Is it an paper that contains research on regulatory and multiagent networks? (iii) Is it an paper that contains research on regulatory and multiagent routes? (iv) Is it an paper that contains research on regulatory avenues and multiagent systems?

Having defined the protocol, was performed the search in the databases to obtain the papers. Firstly, duplications are verified and the papers are prepared for the subsequent evaluations steps. In order to obtain the papers, the tools Mendeley[6] and JabRef[7] were used.

The filters step starts with the title filter, followed by the abstract filter. Up to the step of abstract filter, the input of only one researcher is required. In the following steps, three researchers are required, two from bioinformatics and one from multiagent systems.

The next step is the diagonal reading, where three researchers must read the introduction, figures and tables titles and conclusions. In the diagonal reading, the researchers can improve the knowledge about the area. Therefore, in this step, researchers must check if the specific questions are in agreement with the goal defined in the protocol. In this step it is expected that the amount of studies to be evaluated in next step is reduced drastically.

[6] https://www.mendeley.com/newsfeed.
[7] http://www.jabref.org/.

Table 2. Databases and all variants of the keywords

Database	Keywords
PMC	(regulatory network AND multiagents systems)
PMC	regulatory network AND multiagents
PMC	regulatory network AND multiagents system
PMC	regulatory network AND multiagent system
PMC	regulatory network AND multiagent systems
PMC without quotation marks	(regulatory pathways AND multiagent)
PMC without quotation marks	(regulatory pathways AND multiagents)
PMC without quotation marks	(regulatory Pathways AND multiagents system)
PMC without quotation marks	(regulatory pathways AND multiagent system)
PMC without quotation marks	(regulatory pathways AND multiagent systems)
PMC	regulatory network AND multiagent
Pubmed	(regulatory network) AND multiagent
Pubmed	(regulatory network) AND multiagents system
Pubmed	(regulatory network) AND multiagent systems
Pubmed	(regulatory pathways) AND multiagent system
Science	(regulatory network) AND multiagent
Science	(regulatory network) AND multiagent
Science	(regulatory network) AND multiagents system
Science	(regulatory network) AND multiagents systems
Science	(regulatory network) AND multiagent system
Science	(regulatory network) AND multiagent systems
Scopus	regulatory network AND “multiagent system”
Scopus without quotation marks	(regulatory pathways AND multiagent)

The following step is the complete reading, that will define the studies that are included in the systematic review. In [10] the complete SLR process is defined and presented by the authors. This process is presented in the present work in Fig. 2.

In our process of SLR, was used a variant of the scheme presented in the Fig. 2. According to the original process, each researcher approves (labeled with 1) or disapproves (labeled with 0) each paper[8]. Only if all researchers approve one paper it goes to the next step of evaluation, otherwise it is discarded. The Table 3 presents the number of papers in each step of systematic review. And the Tables 4 and 5 show the seven final papers of this review.

[8] The “?” label means that the research does not have confidence to approve or disapprove the paper.

Table 3. Number of paper in each step

Step	Number of papers
Title filter	393
Abstract filter	67
Diagonal reading filter	21
Complete reading Filter	11
Result	7

3 Results Analysis

In this section, shows up the seven papers that are the final analysis of our systematic review.

Tables 4 and 5 present an overview of the selected papers, showing their goals, techniques and results. By comparing all the papers found in this study, it is possible to note that they present distinct goals. In [8], the main idea is to rebuild GRNs from time-series expression profiles based on fuzzy cognitive maps (FCMs). In [7], they describe the simulation of signal transduction (ST) networks using the DECAF MAS architecture.

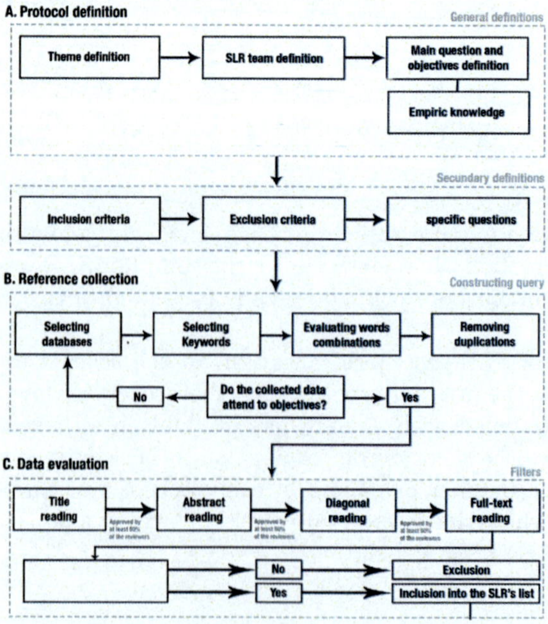

Fig. 2. SLR Model adapted of [10]

Table 4. Comparison between papers

Aut.	Goal	Techniques	Results
[8]	Reconstruct GRNs from time-series expression profiles based on fuzzy cognitive maps (FCMs)	The algorithm is labeled as dMAGAFCM -GRN, where agents and their behaviors are designed with intrinsic properties of GRN reconstruction problems	The experimental results show that dMAGAFCM -GRN is able to learn FCMs until 200 nodes in an effective way. For that, 40.000 weights are optimized
[7]	Describe the simulation of signal transduction (ST) networks using the DECAF MAS architecture	Agents maintain internal state representations of reactant concentration, there is vital to model ST domain. For each agent, there is a rule file with inter-agent communication	The epidermal growth factor (EGF) pathway is modelled, showing the MAS technology viability to simulate biological networks
[6]	Introduce a generic simulation framework for BioPAX	It uses the discrete event and multiagent systems. It is capable of automatically generating a hierarchical multiagent system to a specific BioPAX model	All simulations were consistent with original models
[4]	Propose a novel algorithm for gene regulatory network inference	Uses multiagent systems to train a recurrent neural network (RNN)	They compare their algorithm with a similar algorithm that uses RNN and Particle swarm optimization (PSO) to E. coli SOS dataset

In the work of [6], the goal is to develop a generic software framework for simulating BioPAX models. To fill in this gap, they have introduced a generic simulation framework for BioPAX. In the work [4], a novel algorithm for gene regulatory network inference is proposed.

The work of [12] presents an approach where emergence modeling is extended to regulatory networks and demonstrates its application to the discovery of neuroprotective pathways. The hypothesis is that neuroprotective biological networks emerge from self organizational regulatory properties of genes, akin to how complex systems arise in nature from swarming behaviors – e.g. food source selection and cooperative transport in ant colonies, hive construction in termites, formation of slime mold colonies.

The work of [11] proposes a multiagent based approach for simulating large random Boolean Networks, which promises to give an improvement of the performance (using multiprocessors systems). However, this performance always requests synchronization for agents and depends of the communication between agents. They believe that it is an appropriate approach for simulating a large

Table 5. Comparison between papers

Aut.	Goal	Techniques	Results
[12]	Present an approach where emergence modeling is extended to regulatory networks and demonstrate its application to the discovery of neuroprotective pathways	The focus on modeling emergence is of particular interest as it supports the discovery of new network pathways that emerge iteratively from self-organizational properties of gene and gene clusters, as opposed to other dynamic modeling approaches such as dynamic Bayesian nets and system dynamics, where network structure remains unchanged through the simulation process	An initial evaluation of the approach indicates that emergence modeling provides novel insights for the analysis of regulatory networks which can advance the discovery of acute treatments for stroke and other diseases
[11]	Propose a multiagent based approach to simulate large random Boolean networks	Each node of the Random Boolean Network is modelled as an agent. Agents change step by step, basing to the state of all their neighborhoods	A simple multiagent based approach for implementing large-scale simulation of random Boolean network was proposed. To evaluate, the theoretic analysis of dependence between the performance and the ratio of computation to communication were presented
[16]	Present a new method to reconstruct the network, using multiagent systems to fuse the gene expression data and transcription factor (TF) binding data	Basing in the fusion, an initial network is generated. To obtain the final network, Dynamic Bayesian Network (DBN) method is used to learn	The experiments were with 25 genes and the results were compared with XX algorithms. Using MAS and DBN, the results had a better performance

random Boolean network. And finally, [16] presents a new method to reconstruct the GRN. In this method, they use MAS to fuse the gene expression data and TF binding data and generate an initial network.

The systematic review process is highly important because it provides the means for a full and impartial bibliographic review, producing a result of great scientific value. In this review process, the papers presented in Tables 4 and 5 were selected because they fulfilled all the requirements of the protocol defined in the Table 1 and attended all the steps described in Fig. 1.

4 Conclusions

The main goal of this work is to perform a SLR of the use of MAS in the modelling of GRN. Thus, discovering studies in these areas and what are their advantages and drawbacks of joint use. The result is shown in Table 3.

Among the seven final papers, only two [11,16] employed MAS to simulate some type of network, however, they have not used it for the final modeling. The other studies [4,6–8,12], have applied MAS as a support tool. The first two cited use MAS to simulate networks, however [11] does not present details about how it was developed and the MAS results are not accurate. The paper [16] used data and parameters for generating the initial network from other studies.

The final result obtained was seven papers that are related to both areas and thus, the objective of SLR has been achieved. The present study presents results of the bibliographic research by applying a systematic method that guarantees greater accuracy, certainly important for future works these areas.

Acknowledgments. We would like to thank CAPES (Coordination for the Improvement of Higher Education Personnel) for the financial support to Doctorate Scholarship.

References

1. Agostinho, N.B., Machado, K.S., Werhli, A.V.: Inference of regulatory networks with a convergence improved MCMC sampler. BMC Bioinform. **16**(1), 306 (2015). https://doi.org/10.1186/s12859-015-0734-6
2. Friedman, N., Linial, M., Nachman, I., Peter, D.: Using Bayesian networks to analyze expression data. J. Comput. Biol. **7**(3–4), 601–620 (2000)
3. Galvao, T., Pereira, M.: Etapas de busca e selecao de artigos em revisoes sistematicas da literatura. Epidemiologia e Servico de Saude **23**, 369–371 (2014). https://doi.org/10.5123/S1679-49742014000200019
4. Ghazikhani, A., Akbarzadeh, T., Monsefi, R.: Genetic regulatory network inference using recurrent neural networks trained by a multi agent system. In: International eConference on Computer and Knowledge Engineering (ICCKE) (2011). https://doi.org/10.1109/ICCKE.2011.6413332
5. Gibas, C., Jambeck, P.: Developing bioinformatics computer skills. Yale J. Biol. Med. **75**(2), 117 (2002)
6. Haydarlou, R., Jacobsen, A., Bonzanni, N., Feenstra, K.A., Abeln, S., Heringa, J.: BioASF: a framework for automatically generating executable pathway models specified in BioPAX. Bioinformatics **32**, i60–i69 (2016). https://doi.org/10.1093/bioinformatics/btw250
7. Khan, S., Makkena, R., McGeary, F., Decker, K., Gillis, W., Schmidt, C.: A multi-agent system for the quantitative simulation of biological networks. In: AAMAS (2003). https://doi.org/10.1145/860575.860637
8. Liu, J., Chi, Y.Z.C., Zhu, C.: A dynamic multiagent genetic algorithm for gene regulatory network reconstruction based on fuzzy cognitive maps. IEEE Trans. Fuzzy Syst. **24**, 419–431 (2016). https://doi.org/10.1109/TFUZZ.2015.2459756
9. Lopes, F.M.: Redes complexas de expressao genica: sintese, identificacao, analise e aplicacoes. Master's thesis, Universidade de Sao Paulo (2011)

10. Mariano, D., Leite, C., Santos, L., Rocha, R., Melo-Minardi, R.: A guide to performing systematic literature reviews in bioinformatics. Technical report, RT.DCC.002/2017, Universidade Federal de Minas Gerais (2017)
11. Pham, D.: Multi-agent based simulation of large random boolean network. MSC-LES (2008)
12. Sanfilippo, A., Haack, J., McDermott, J., Stevens, S., Stenzel-Poore, M.: Modeling emergence in neuroprotective regulatory networks. In: International Conference on Complex Sciences, pp. 291–302 (2012). https://doi.org/10.1007/978-3-319-03473-7_26
13. Shawn Martin, Z.Z., Martino, A., Faulon, J.: Boolean dynamics of genetic regulatory networks inferred from microarray time series data. Bioinformatics **23**(7), 866–874 (2005). https://doi.org/10.1093/bioinformatics/btm021
14. Shmulevich, I., Dougherty, E., Kim, S., Zhang, W.: Probabilistic boolean networks: a rule-based uncertainty model for gene regulatory networks. Bioinformatics **18**(2), 261–274 (2002). https://doi.org/10.1093/bioinformatics/18.2.261
15. Werhli, A.V.: Reconstruction of gene regulatory networks from postgenomic data. Ph.D. thesis, School of Informatics University of Edinburgh (2007)
16. Yang, T., Sun, Y.: The reconstruction of gene regulatory network based on multi-agent system by fusing multiple data sources. In: IEEE International Conference on Computer Science and Automation Engineering, vol. 4 (2011). https://doi.org/10.1109/CSAE.2011.5952438

The Reversibility of Cellular Automata on Trees with Loops

A. Martín del Rey[1](✉), E. Frutos Bernal[2], D. Hernández Serrano[3], and R. Casado Vara[4]

[1] Department of Applied Mathematics, Institute of Fundamental Physics and Mathematics, University of Salamanca, 37008 Salamanca, Spain
delrey@usal.es

[2] Department of Statistics, University of Salamanca, 37007 Salamanca, Spain
efb@usal.es

[3] Department of Mathematics, Institute of Fundamental Physics and Mathematics, University of Salamanca, 37008 Salamanca, Spain
dani@usal.es

[4] BISITE Research Group, University of Salamanca, 37007 Salamanca, Spain
rober@usal.es

Abstract. In this work the notion of linear cellular automata on trees with loops is introduced and the reversibility problem in some particular cases is tackled. The explicit expressions of the inverse cellular automata are computed.

Keywords: Cellular automata on graphs · Reversibility · Trees with loops · Evolutionary computation

1 Introduction

A cellular automaton can be considered as a finite state machine formed by a finite number of identical memory units (called cells) which are endowed with a state at every step of time. The state of each cell is updated according to a local transition rule whose variables are the states of its neighbor cells. Cellular automata are simple models of computation capable to simulate complex behaviors; consequently, several applications to all fields of Science and Technology can be found in the scientific literature [6,9].

Of special interest are those cellular automata whose state set is $\mathcal{F}_2 = \{0, 1\}$ (boolean cellular automata). The vector whose coordinates stand for the states of the ordered set of cells at time t is called configuration of the cellular automata at t: $C^t \in \mathbb{F}_2^n$, where n is the total number of cells. When the dynamics of the cellular automaton is governed by means of deterministic transition rules, every configuration will have an unique successor in time. However, every configuration may have several distinct predecessors and, consequently, the time evolution mapping of the cellular automaton is not invertible in general. When there is only one predecessor, the cellular automaton is called reversible; that is, the global

Y. Dong et al. (Eds.): DCAI 2020, AISC 1237, pp. 241–250, 2021.
https://doi.org/10.1007/978-3-030-53036-5_26

transition function $F\colon \mathbb{F}_2^n \to \mathbb{F}_2^n$, that yields the configuration at the next-step during the evolution of the cellular automata, is injective [8]. The reversibility is an interesting and desirable property in some applications of cellular automata in Cryptography [13] and Computer Science [7].

The reversibility problem for cellular automata consists of both determining when a certain cellular automaton is reversible, and computing the inverse cellular automaton -if it is possible-. This is not a new problem [5] and it has been tackled for different classes of cellular automata: elementary cellular automata [10], linear cellular automata (see, for example, [3]), two-dimensional cellular automata with hexagonal cellular space [1], memory cellular automata [11], multidimensional finite cellular automata [2,4], etc.

The main goal of this work is to study the reversibility problem of a particular and interesting type of cellular automata on graphs: the linear cellular automata defined on full trees with loops. Specifically, we will show that some cellular automata of this type are reversible, and the corresponding inverse cellular automata will be explicitly computed.

The rest of the paper is organized as follows: the basic definitions and results concerning linear cellular automata on trees with loops are introduced in Sect. 2; in Sect. 3 the reversibility problem for this type of cellular automata is tackled. Finally, the conclusions and further work are shown in Sect. 4.

2 Linear Cellular Automata on Trees with Loops

2.1 General Considerations

Let $G = (V, E)$ be an undirected multigraph such that $V = \{v_1, v_2, \ldots, v_n\}$. A *boolean cellular automaton on* G (CA for short) is a 4-uplet $\mathcal{A}_G = (\mathcal{C}, \mathbb{F}_2, \mathcal{N}, \mathcal{F})$ where:

(1) The set $\mathcal{C} = \{c_1, c_2, \ldots, c_n\}$ is the cellular space of the CA such that the i-th cell c_i stands for the node v_i of G, where $1 \le i \le n = |V|$. For the sake of simplicity will be denote the i-th cell by i in place of c_i.
(2) The Galois field $\mathbb{F}_2 = \{0, 1\}$ is the finite set of states that can be assumed by each cell/node at each step of time. In this sense, the state of the i-th cell at time step t is denoted by $s_i^t \in \mathbb{F}_2$, with $1 \le i \le n$. Moreover, $C^t = (s_1^t, s_2^t, \ldots, s_n^t) \in \mathbb{F}_2^n$ is called configuration of the CA at step of time t.
(3) $\mathcal{N}$ denotes the function which assigns to each node its neighborhood (the adjacent nodes). Then:

$$\mathcal{N}\colon \mathcal{C} \to 2^{\mathcal{C}} \tag{1}$$

$$i \mapsto \mathcal{N}_i = \{i, i_1, i_2, \ldots, i_{m_i}\} \tag{2}$$

Note that as $i \in \mathcal{N}_i$ for every node i, then there is a loop on every node. Moreover, $(i, j) \in E$ iff $i \in \mathcal{N}_j$ or $j \in \mathcal{N}_i$.

(4) $\mathcal{F} = \{f_1, f_2, \dots, f_n\}$ is the transition functions family that governs the dynamic of the cellular automaton. The state of the node i at a particular time step $t+1$ is computed by means of the boolean function f_i whose variables are the states of the neighbor nodes at the previous step of time t:

$$s_i^{t+1} = f_i\left(s_i^t, s_{i_1}^t, s_{i_2}^t, \dots, s_{i_{m_i}}^t\right) \in \mathbb{F}_2. \quad (3)$$

Note that these local transition functions define a global transition function F:

$$\begin{aligned} F\colon \mathbb{F}_2^n &\to \mathbb{F}_2^n \\ C^t &\mapsto C^{t+1} = F\left(C^t\right) \end{aligned} \quad (4)$$

The CA is said to be *linear* if f_i is a linear function for every i, that is:

$$s_i^{t+1} = a_{ii} s_i^t \oplus \bigoplus_{k=1}^{m_i} a_{ii_k} s_{i_k}^t, \quad a_{ii}, a_{ii_k} \in \mathbb{F}_2. \quad (5)$$

In this case, the global dynamics of the CA can be interpreted in terms of matrix algebra as follows: $C^{t+1} = F\left(C^t\right) = A \cdot C^t$, where $A = (a_{ij})_{1\le i,j\le n}$ is the local transition matrix.

Note that the dynamics of the linear cellular automata is biunivocally determined by the multigraph G (where each edge stands for a XOR summation of the states of the corresponding adjacent nodes). This does not happen in the case of non-linear cellular automata since the same multigraph yields to several families of transition functions.

A cellular automaton is *reversible* when F is bijective, that is, the evolution backwards is possible [12]; in this case $F^{-1}\colon \mathbb{F}_2^n \to \mathbb{F}_2^n$ is the global transition function of the inverse cellular automaton. Note that a reversible CA yields an invertible global behavior from a set of local transition functions which are not reversible. The reversibility of a linear CA depends on the nature of its local transition matrix: the linear cellular automaton is reversible iff its local transition matrix A is non-singular and, consequently, A^{-1} is the local transition matrix of the inverse CA.

2.2 Linear CA on Full Trees with Loops

This work deals with linear cellular automata on full trees with loops. A full binary tree is a rooted tree in which each internal vertex has exactly two children. Note that if the full binary tree has k internal vertices then it has $n = 2k+1$ vertices, $n-1$ edges and $\frac{n+1}{2}$ leaves. If there is a loop in each vertex, the notion of full binary tree with loops is derived.

Let $\mathcal{T}_k$ be the set of full binary trees with loops with k internal vertices. Set $T_k \in \mathcal{T}_k$ the tree with loops such that for every node i the following holds:

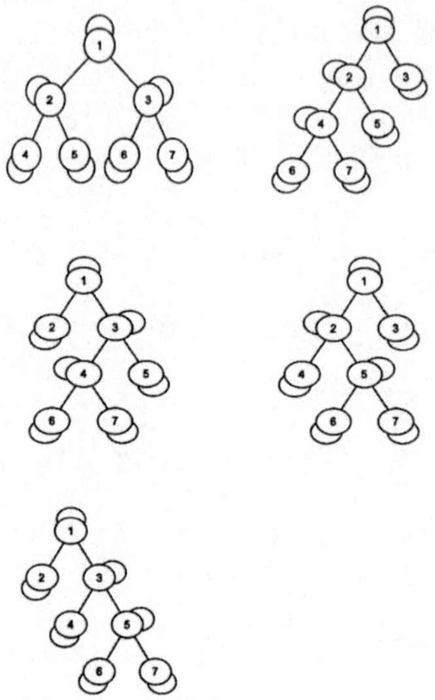

Fig. 1. Family $\mathcal{T}_3$ of full binary trees with loops.

- $i = 1$ is the root.
- If i is even or $i = 2k + 1$ then i is a leaf.
- If i is odd with $i \neq 2k + 1$, then i is an internal node.

This particular tree is called the *characteristic representative* of $\mathcal{T}_k$. In Fig. 1 the family of full tree with loops $\mathcal{T}_3$ is shown (the full binary tree of the last row is the characteristic representative T_3).

A linear cellular automaton on a full tree with loops $T \in \mathcal{T}_k$ is $\mathcal{A}_T = (\mathcal{C}, \mathbb{F}_2, \mathcal{N}, \mathcal{F})$ where $\mathcal{F} = \{f_1, f_2, \ldots, f_n\}$ is a family of boolean linear functions. In particular, when $\mathcal{A}_{T_k}$ is considered, the following hold:

(1) The root node is $1 \in \mathcal{C}$.
(2) The neighborhood function is defined as follows:

$$\begin{aligned} \mathcal{N} : \mathcal{C} &\to 2^{\mathcal{C}} \\ 1 &\mapsto \mathcal{N}_1 = \{1, 2, 3\} \\ i &\mapsto \mathcal{N}_i = \{p_i, i, i^+, i^-\} \\ l &\mapsto \mathcal{N}_l = \{p_l, l\} \end{aligned}$$

where $i \in \mathcal{C}$ is an internal vertex ($i \neq 1$), $l \in \mathcal{C}$ is a leaf, $p_i \in \mathcal{C}$ is the parent of the node i, and $i^+, i^- \in \mathcal{C}$ are the right and left children, respectively, of i.
(3) The local transition functions are as follows:

$$\begin{aligned} s_1^{t+1} &= s_1^t \oplus s_2^t \oplus s_3^t, \\ s_i^{t+1} &= s_{p_i}^t \oplus s_i^t \oplus s_{i^-}^t \oplus s_{i^+}^t, \quad i \neq 1 \text{ internal vertex}, \\ s_l^{t+1} &= s_{p_l}^t \oplus s_l^t, \quad l \text{ leaf}. \end{aligned}$$

Note that the transition matrix of $\mathcal{A}_{T_k}$ is $A = (a_{ij})_{1\le i,j\le 2k+1}$ where:

$$a_{ii} = 1, \text{with } 1 \le i \le 2k+1 \tag{6}$$

$$a_{i,i+1} = \begin{cases} 1, \text{ if } i \text{ is odd} \\ 0, \text{ if } i \text{ is even} \end{cases} \text{with } 1 \le i \le 2k \tag{7}$$

$$a_{i,i+2} = \begin{cases} 1, \text{ if } i \text{ is odd} \\ 0, \text{ if } i \text{ is even} \end{cases} \text{with } 1 \le i \le 2k-1 \tag{8}$$

$$a_{ij} = 0, \text{ if } i+2 \le j \le 2k+1 \tag{9}$$

$$a_{ij} = a_{ji}, \text{ with } 1 \le i,j \le 2k+1 \tag{10}$$

3 Solving the Reversibility Problem

Theorem 1. *The linear cellular automaton $\mathcal{A}_{T_k}$ is reversible for every k, and its inverse cellular automaton is defined by the symmetric transition matrix $B = (b_{ij})_{1\le i,j\le 2k+1}$, where:*

$$b_{ii} = \begin{cases} 1, \text{ if } i = 4l+1 \text{ or } i = 4l+4 \text{ with } l \ge 0, l \in \mathbb{N} \\ 0, \text{ if } i = 4l+2 \text{ or } i = 4l+3 \text{ with } l \ge 0, l \in \mathbb{N} \end{cases} \tag{11}$$

$$b_{ij} = \begin{cases} 1, \text{ if } i = 4l+1 \text{ or } i = 4l+2 \text{ with } l \ge 0, l \in \mathbb{N} \\ 0, \text{ if } i = 4l+3 \text{ or } i = 4l+4 \text{ with } l \ge 0, l \in \mathbb{N} \end{cases} \tag{12}$$

$j \ge i+1.$

Proof. To proof the above statement, it is enough to show that the boolean matrix $B = (b_{ij})_{1\le i,j\le 2k+1}$ defined by (11)–(12) satisfies the following: $A \cdot B = B \cdot A = Id$.

Let us suppose that $A \cdot B = C$ where $C = (c_{ij})_{1\le i,j\le 2k+1}$, then we have to prove that: (1) $c_{ii} = 1$ for $1 \le i \le 2k+1$, and (2) $c_{ij} = 0$ for every $i \ne j$.

(1) For the sake of clarity, we will distinguish seven cases:
- Computation of c_{11}: as $a_{14} = a_{15} = \ldots = a_{1,2k+1} = 0$, then

$$\begin{aligned} c_{11} &= \sum_{h=1}^{2k+1} a_{1h} b_{h1} \\ &= a_{11}b_{11} + a_{12}b_{21} + a_{13}b_{31} + a_{14}b_{41} + \ldots + a_{1,2k+1}b_{2k+1,1} \\ &= 1\cdot 1 + 1\cdot 1 + 1\cdot 1 + 0 + \ldots + 0 = 3 \\ &\equiv 1 (\text{mod } 2). \end{aligned} \tag{13}$$

- Computation of c_{22}: as $a_{23} = a_{24} = \ldots = a_{2,2k+1} = 0$, then

$$\begin{aligned} c_{22} &= \sum_{h=1}^{2k+1} a_{2h} b_{h2} \\ &= a_{21}b_{12} + a_{22}b_{22} + a_{23}b_{32} + \ldots + a_{2,2k+1}b_{2k+1,2} \\ &= 1\cdot 1 + 1\cdot 0 + 0 + \ldots + 0 = 1. \end{aligned} \tag{14}$$

– Computation of c_{33}: as $a_{36} = a_{37} = \ldots = a_{3,2k+1} = 0$, then

$$\begin{aligned} c_{33} &= \sum_{h=1}^{2k+1} a_{3h} b_{h3} \\ &= a_{31}b_{13} + a_{32}b_{23} + a_{33}b_{33} + a_{34}b_{43} + a_{35}b_{53} \\ &\quad + a_{36}b_{63} + \ldots + a_{3,2k+1}b_{2k+1,3} \\ &= 1 \cdot 1 + 0 \cdot 1 + 1 \cdot 0 + 1 \cdot 0 + 1 \cdot 0 + 0 + \ldots + 0 = 1. \end{aligned} \tag{15}$$

– Computation of c_{ii} with $3 \leq i \leq 2k-1$. In this case we can also consider four subcases depending on the value of the subindex i.
  - If $i = 4l + 1$ with $l \geq 0$, then:

$$\begin{aligned} c_{ii} = c_{4l+1,4l+1} &= \sum_{h=1}^{2k+1} a_{4l+1,h} b_{h,4l+1} = \sum_{h=4l-1}^{4l+3} a_{4l+1,h} b_{h,4l+1} \\ &= \sum_{h=4l-1}^{4l+3} b_{h,4l+1}, \end{aligned} \tag{16}$$

    since $i = 4l+1$ is odd and, consequently, $a_{4l+1,h} = 1$ for every value of h. Moreover, as $b_{4l-1,4l+1} = b_{4l,4l+1} = 0$ and $b_{4l+1,4l+1} = b_{4l+2,4l+1} = b_{4l+3,4l+1} = 1$ then $c_{4l+1,4l+1} = 3 \equiv 1 (\mathrm{mod}\, 2)$.
  - If $i = 4l + 2$ with $l \geq 0$, then:

$$\begin{aligned} c_{ii} = c_{4l+2,4l+2} &= \sum_{h=1}^{2k+1} a_{4l+2,h} b_{h,4l+2} = \sum_{h=4l}^{4l+4} a_{4l+2,h} b_{h,4l+2} \\ &= b_{4l+1,4l+2} + b_{4l+2,4l+2} = 1 + 0 = 1, \end{aligned} \tag{17}$$

    since $a_{4l+2,4l} = a_{4l+2,4l+3} = a_{4l+2,4l+4} = 0$ and $a_{4l+2,4l+1} = a_{4l+2,4l+2} = 1$.
  - If $i = 4l + 3$ with $l \geq 0$ then:

$$\begin{aligned} c_{ii} = c_{4l+3,4l+3} &= \sum_{h=1}^{2k+1} a_{4l+3,h} b_{h,4l+3} = \sum_{h=4l+1}^{4l+5} a_{4l+3,h} b_{h,4l+3} \\ &= a_{4l+3,4l+1} + a_{4l+3,4l+2} = 1 + 0 = 1, \end{aligned} \tag{18}$$

    since $b_{4l+1,4l+3} = b_{4l+2,4l+3} = 1$ and $b_{4l+3,4l+3} = b_{4l+4,4l+3} = b_{4l+5,4l+3} = 0$.
  - If $i = 4l + 4$ with $l \geq 0$, then:

$$\begin{aligned} c_{ii} = c_{4l+4,4l+4} &= \sum_{h=1}^{2k+1} a_{4l+4,h} b_{h,4l+4} = \sum_{h=4l+2}^{4l+6} a_{4l+4,h} b_{h,4l+4} \\ &= b_{4l+3,4l+4} + b_{4l+4,4l+4} = 0 + 1 = 1, \end{aligned} \tag{19}$$

    since $a_{4l+4,4l+2} = a_{4l+4,4l+5} = a_{4l+4,4l+6} = 0$ and $a_{4l+4,4l+3} = a_{4l+4,4l+4} = 1$.

– Computation of $c_{2k-1,2k-1}$:

$$\begin{aligned} c_{2k-1,2k-1} &= \sum_{h=1}^{2k+1} a_{2k-1,h} b_{h,2k-1} \\ &= a_{2k-1,1} b_{1,2k-1} + \ldots + a_{2k-1,2k-4} b_{2k-4,2k-1} \\ &\quad + a_{2k-1,2k-3} b_{2k-3,2k-1} + a_{2k-1,2k-2} b_{2k-2,2k-1} \\ &\quad + a_{2k-1,2k-1} b_{2k-1,2k-1} + a_{2k-1,2k} b_{2k,2k-1} \\ &\quad + a_{2k-1,2k+1} b_{2k+1,2k-1} \\ &= b_{2k-3,2k-1} + b_{2k-1,2k-1} + b_{2k,2k-1} + b_{2k+1,2k-1}. \end{aligned} \tag{20}$$

As

$$b_{2k-3,2k-1} = \begin{cases} 1, \text{ if } 2k-3 = 4l+1 \iff k \text{ is even} \\ 0, \text{ if } 2k-3 = 4l+3 \iff k \text{ is odd} \end{cases} \tag{21}$$

$$b_{2k-2,2k-1} = \begin{cases} 1, \text{ if } 2k-2 = 4l+2 \iff k \text{ is even} \\ 0, \text{ if } 2k-2 = 4l+4 \iff k \text{ is odd} \end{cases} \tag{22}$$

$$b_{2k,2k-1} = b_{2k-1,2k} = \begin{cases} 1, \text{ if } k \text{ is even} \\ 0, \text{ if } k \text{ is odd} \end{cases} \tag{23}$$

$$b_{2k+1,2k-1} = b_{2k-1,2k+1} = \begin{cases} 1, \text{ if } k \text{ is even} \\ 0, \text{ if } k \text{ is odd} \end{cases} \tag{24}$$

then

$$c_{2k,2k} = \begin{cases} 1+0 = 1, \text{ if } k \text{ is odd} \\ 0+1 = 1, \text{ if } k \text{ is even} \end{cases} \tag{25}$$

– Computation of $c_{2k,2k}$: since $a_{2k,1} = \ldots = a_{2k,2k-3} = 0$, $a_{2k,2k-2} = a_{2k,2k+1} = 0$, and $a_{2k,2k-1} = a_{2k,2k} = 1$, then

$$\begin{aligned} c_{2k,2k} &= \sum_{h=1}^{2k+1} a_{2k,h} b_{h,2k} \\ &= a_{2k,1} b_{1,2k} + \ldots + a_{2k,2k-3} b_{2k-3,2k} \\ &\quad + a_{2k,2k-2} b_{2k-2,2k} + a_{2k,2k-1} b_{2k-1,2k} \\ &\quad + a_{2k,2k} b_{2k,2k} + a_{2k,2k+1} b_{2k+1,2k} \\ &= b_{2k-1,2k} + b_{2k,2k}. \end{aligned} \tag{26}$$

As

$$b_{2k-1,2k} = \begin{cases} 1, \text{ if } 2k-1 = 4l+1 \iff k \text{ is odd} \\ 0, \text{ if } 2k-1 = 4l+3 \iff k \text{ is even} \end{cases} \tag{27}$$

$$b_{2k,2k} = \begin{cases} 1, \text{ if } 2k = 4l+4 \iff k \text{ is even} \\ 0, \text{ if } 2k = 4l+2 \iff k \text{ is odd} \end{cases} \tag{28}$$

then

$$c_{2k,2k} = \begin{cases} 1+0 = 1, \text{ if } k \text{ is odd} \\ 0+1 = 1, \text{ if } k \text{ is even} \end{cases} \tag{29}$$

- Computation of $c_{2k+1,2k+1}$: since $a_{2k+1,1} = \ldots = a_{2k+1,2k-2} = 0, a_{2k+1,2k} = 0$ and $a_{2k+1,2k-1} = a_{2k+1,2k+1} = 1$, then

$$
\begin{aligned}
c_{2k+1,2k+1} &= \sum_{h=1}^{2k+1} a_{2k+1,h} b_{h,2k+1} \\
&= a_{2k+1,1} b_{1,2k+1} + \ldots + a_{2k+1,2k-2} b_{2k-2,2k+1} \\
&\quad + a_{2k+1,2k-1} b_{2k-1,2k+1} + a_{2k+1,2k} b_{2k,2k+1} \\
&\quad + a_{2k+1,2k+1} b_{2k+1,2k+1} \\
&= b_{2k-1,2k+1} + b_{2k+1,2k+1}.
\end{aligned} \tag{30}
$$

As

$$
b_{2k-1,2k+1} = \begin{cases} 1, \text{ if } 2k-1 = 4l+1 \iff k \text{ is odd} \\ 0, \text{ if } 2k-1 = 4l+3 \iff k \text{ is even} \end{cases} \tag{31}
$$

$$
b_{2k+1,2k+1} = \begin{cases} 1, \text{ if } 2k+1 = 4l+1 \iff k \text{ is even} \\ 0, \text{ if } 2k+1 = 4l+3 \iff k \text{ is odd} \end{cases} \tag{32}
$$

then

$$
c_{2k+1,2k+1} = \begin{cases} 1+0 = 1, \text{ if } k \text{ is odd} \\ 0+1 = 1, \text{ if } k \text{ is even} \end{cases} \tag{33}
$$

(2) Now, we can distinguish six cases:
- Computation of c_{1j} with $j > 1$: as $a_{1h} = 0$ for $4 \leq h \leq 2k+1$ and $a_{11} = a_{12} = a_{13} = 1$ then

$$
c_{1j} = \sum_{h=1}^{2k+1} a_{1h} b_{hj} = b_{1j} + b_{2j} + b_{3j} = 0. \tag{34}
$$

- Computation of c_{2j} with $j > 2$: as $a_{2h} = 0$ for $h \geq 3$ and $a_{21} = a_{22} = 1$ then

$$
c_{2j} = \sum_{h=1}^{2k+1} a_{2h} b_{hj} = b_{1j} + b_{2j} = 1 + 1 = 2 \equiv 0 (\text{mod } 2). \tag{35}
$$

- Computation of c_{3j} with $j > 3$: as $a_{3h} = 0$ for $h \geq 6$, $a_{31} = a_{33} = a_{34} = a_{35} = 1$, and $a_{32} = 0$ then

$$
\begin{aligned}
c_{3j} &= \sum_{h=1}^{2k+1} a_{3h} b_{hj} = b_{1j} + b_{3j} + b_{4j} + b_{5j} \\
&= 1 + 1 + 0 + 0 = 2 \equiv 0 (\text{mod } 2).
\end{aligned} \tag{36}
$$

- Computation of c_{ij} with $4 < i < 2k$ and $j > i$: as $a_{i,j} = 0$ for $j < i-2$ and $j > i+2$ then

$$
\begin{aligned}
c_{ij} &= a_{i,i-2} b_{i-2,j} + a_{i,i-1} b_{i-1,j} + b_{ij} \\
&\quad + a_{i,i+1} b_{i+1,j} + a_{i,i+2} b_{i+2,j}.
\end{aligned} \tag{37}
$$

If i is even then $c_{ij} = b_{i-1,j} + b_{ij}$ since $a_{i,i-2} = a_{i,i+1} = a_{i,i+2} = 0$ and $a_{i,i-1} = 1$. Consequently:

$$\begin{aligned} c_{ij} &= b_{i-2,j} + b_{ij} + b_{i+1,j} + b_{i+2,j} \\ &= \begin{cases} 0+1+1+0 = 2 \equiv 0 (\bmod 2), \text{ if } i = 4l+1 \\ 1+0+0+1 = 2 \equiv 0 (\bmod 2), \text{ if } i = 4l+3 \end{cases} \end{aligned} \tag{38}$$

On the other hand, if i is odd then $c_{ij} = b_{i-2,j} + b_{ij} + b_{i+1,j} + b_{i+2,j}$ since $a_{i-1,i} = 0$ and $a_{i,i-2} = a_{i,i+1} = a_{i,i+2} = 1$. As a consequence

$$c_{ij} = b_{i-1,j} + b_{ij} = \begin{cases} 1+1 = 2 \equiv 0 (\bmod 2), \text{ if } i = 4l+2 \\ 0+0 = 0, \qquad\qquad\quad \text{ if } i = 4l+4 \end{cases} \tag{39}$$

– Computation of $c_{2k-1,j}$ with $j > 2k-1$, that is, $j = 2k, 2k+1$. For the sake of simplicity we will distinguish two subcases:
  - If $j = 2k$ then

$$\begin{aligned} c_{2k-1,2k} &= \sum_{h=1}^{2k+1} a_{2k-1,h} b_{h,2k} \\ &= b_{2k-3,2k} + b_{2k-1,2k} + b_{2k,2k} + b_{2k+1,2k} \\ &= \begin{cases} 1+0+1+0 = 2 \equiv 0 (\bmod 2), \text{ if } k \text{ is even} \\ 0+1+0+1 = 2 \equiv 0 (\bmod 2), \text{ if } k \text{ is odd} \end{cases} \end{aligned} \tag{40}$$

  - If $j = 2k+1$ then:

$$\begin{aligned} c_{2k-1,2k+1} &= \sum_{h=1}^{2k+1} a_{2k-1,h} b_{h,2k+1} \\ &= b_{2k-3,2k+1} + b_{2k-1,2k+1} + b_{2k,2k+1} + b_{2k+1,2k+1} \\ &= \begin{cases} 1+0+0+1 = 2 \equiv 0 (\bmod 2), \text{ if } k \text{ is even} \\ 0+1+1+0 = 2 \equiv 0 (\bmod 2), \text{ if } k \text{ is odd} \end{cases} \end{aligned} \tag{41}$$

– Computation of $c_{2k,j}$ with $j > 2k$, that is, $j = 2k+1$: since $a_{2k,h} = 0$ for $1 \leq h \leq 2k-2$, $a_{2k,2k-1} = a_{2k,2k} = 1$ and $a_{2k,2k+1} = 1$ then

$$c_{2k,2k+1} = \sum_{h=1}^{2k+1} a_{2k,h} b_{h,2k+1} = b_{2k-1,2k+1} + b_{2k,2k+1}. \tag{42}$$

As

$$b_{2k-1,2k+1} = \begin{cases} 1, \text{ if } 2k-1 = 4l+1 \iff k \text{ is odd} \\ 0, \text{ if } 2k-1 = 4l+3 \iff k \text{ is even} \end{cases} \tag{43}$$

$$b_{2k,2k+1} = \begin{cases} 1, \text{ if } 2k = 4l+2 \iff k \text{ is odd} \\ 0, \text{ if } 2k = 4l+4 \iff k \text{ is even} \end{cases} \tag{44}$$

then

$$c_{2k,2k+1} = \begin{cases} 1+1 = 2 \equiv 0 (\bmod 2), \text{ if } k \text{ is odd} \\ 0+0 = 0, \qquad\qquad\quad \text{ if } k \text{ is even} \end{cases} \tag{45}$$

Using a similar argument, we can also prove that $B \cdot A = Id$.

4 Conclusions and Further Work

In this work the notion of linear cellular automaton on full trees with loops is introduced and its reversibility problem is studied. Specifically, it is shown that some linear cellular automata of this type are reversible and the inverse cellular automaton is explicitly computed.

Future work aimed at solving the complete reversibility problem for all $\mathcal{A}_T$ with $T \in \mathcal{T}_k$, and studying the applications of $\mathcal{A}_T$ in different fields such as f-reversible processes on graphs or digital image processing.

Acknowledgements. This research has been partially supported by Ministerio de Ciencia, Innovación y Universidades (MCIU, Spain), Agencia Estatal de Investigación (AEI, Spain), and Fondo Europeo de Desarrollo Regional (FEDER, UE) under projects with references TIN2017-84844-C2-2-R (MAGERAN) and MTM2017-86042-P, and the project with reference SA054G18 supported by Consejería de Educación (Junta de Castilla y León, Spain).

References

1. Augustynowicz, A., Baetens, J.M., De Baets, B., et al.: A note on the reversibility of 2D cellular automata on hexagonal grids. J. Cell. Autom. **13**, 521–526 (2018)
2. Bhattacharjee, K., Das, S.: Reversibility of d-state finite cellular automata. J. Cell. Autom. **11**, 213–245 (2016)
3. Chang, C.H., Chang, H.: On the Bernoulli automorphism of reversible linear cellular automata. Inf. Sci. **345**(1), 217–225 (2016)
4. Chang, C.H., Su, J.Y., Akin, H., et al.: Reversibility problem of multidimensional finite cellular automata. J. Stat. Phys. **168**(1), 208–231 (2017)
5. Di Gregorio, S., Trautteur, G.: On reversibility in cellular automata. J. Comput. Syst. Sci. **11**(3), 382–391 (1975)
6. Hoekstra, A.G., Kroc, J., Sloot, P.M.A.: Simulating Complex Systems by Cellular Automata. Springer, Berlin (2010)
7. Morita, K.: Reversible computing and cellular automata-a survey. Theor. Comput. Sci. **395**, 101–131 (2008)
8. Richardson, D.: Tessellations with local transformations. J. Comput. Syst. Sci. **6**, 373–388 (1972)
9. Sarkar, P.: A brief history of cellular automata. ACM Comput. Surv. **32**, 80–107 (2000)
10. Sarkar, P., Barua, R.: The set of reversible 90/150 cellular automata is regular. Discret. Appl. Math. **84**, 199–213 (1998)
11. Seck-Tuoh-Mora, J.C., Martínez, G.J., Alonso-Sanz, R., Hernández-Romero, N.: Invertible behavior in elementary cellular automata with memory. Inf. Sci. **199**, 125–132 (2012)
12. Toffoli, T., Margolus, N.H.: Invertible cellular automata: a review. Physica D **45**, 229–253 (1990)
13. Wang, X., Luan, D.: A novel image encryption algorithm using chaos and reversible cellular automata. Commun. Nonlinear Sci Numer. Simul. **18**, 3075–3085 (2013)

Virtual Reality Tool for Learning Sign Language in Spanish

Amelec Viloria[1](✉), Isabel Llerena[1], and Omar Bonerge Pineda Lezama[2]

[1] Universidad de la Costa, Barranquilla, Colombia
{aviloria7,illerena1}@cuc.edu.co
[2] Universidad Tecnológica Centroamericana (UNITEC), San Pedro Sula, Honduras
omarpineda@unitec.edu

Abstract. Language is the means of human access to the world. Languages have the virtue of opening up alternative ways of thinking and understanding the place people inhabit, relating to it, expanding it and modifying it. As a possibility of communication, languages open up opportunities to relate to other people, to get closer to them and to develop a broader understanding of the social and the human elements [1]. This research presents a visual tool designed to allow the learning of multiple words that are part of Spanish Sign Language (SSL) through an anthropomorphic model that is completely manipulable and programmable.

Keywords: Virtual reality tool · Learning sign · Language

1 Introduction

Learning a language is a way to access, in a different way, the world we all shape, and it is part of the diversity, the ideals, the concepts through which people recognize themselves, relate to each other and finally reinvent themselves [2]. Language has the virtue of allowing to transform the conceptions of ourselves and of others. People know each other fundamentally through communicative processes; therefore, the broader these processes are, the more possibilities there are to diversify the world [3].

However, there are many people with communication problems, including deafness which is one of the most common. People who present this problem use sign language (SL) to communicate and translation systems have been developed (Voice/Text to SL) to assist in this task [4]. But, because SL is dependent on countries and cultures, there are differences between grammars, vocabularies and signs, even if they come from places with similar spoken languages [5].

1.1 State of the Art

Due to the lack of interest on the part of system specialists in addressing SSL, no programs or similar systems were found that cover the need for SSL learning [1, 6–10], as there are some similar systems that seek learning through a video game [11] or visual

Y. Dong et al. (Eds.): DCAI 2020, AISC 1237, pp. 251–257, 2021.
https://doi.org/10.1007/978-3-030-53036-5_27

recognition of human gestures [12], but are mostly focused on American Sign Language, so this work is proposed to cover the needs of this sector, and the use of virtual reality is proposed to support people interested in learning SSL.

Recently, an increasing number of people who are not hearing-impaired have sought to learn SSL in order to communicate with this group [12].

1.2 Virtual Reality

Virtual Reality (VR) is a representation of reality through electronic means, which gives the sensation of experiencing a real situation, in which one can interact with the surroundings [13].

This study deals with non-immersive VR, also known as desktop virtual reality, which is limited to the use of a computer and does not require the use of other devices such as helmets or VR viewers. Due to its low cost, this type of virtual reality has been widely disseminated and accepted [14].

2 Methodology

Rapid Application Development model (RAD) is a software development process initially developed by James Martin in 1980 [15]. The method comprises iterative development, prototyping and the use of CASE utilities. Traditionally, rapid application development tends to include usability, utility and speed of execution.

By understanding the requirements and limiting the scope of the project, the process allows the development team to create a "fully functional system" within short periods of time. When used primarily for information system applications, the approach comprises the following phases [16–18]:

Management modeling. The flow of information between management functions is modeled to answer the following questions: What information does the management process handle? What is the information generated? Who generates it? Where does the information go? Who processes it?

Data modeling. The information flow defined as part of the management modeling phase is refined as a set of data objects needed to support the business. The characteristics (called attributes) of each of the objects and the relationships between these objects are defined.

Process modeling. The data objects defined in the data modeling phase are transformed to achieve the information flow needed to implement a management function. Process descriptions are created to add, modify, delete, or retrieve a data object. This is the communication between the objects.

Generation of applications. The RAD assumes the use of fourth generation techniques. Instead of creating software with third generation programming languages, the RAD process works to reuse existing program components (where possible) or to create reusable components (where necessary). In all cases, automatic tools are used to facilitate the construction of the software.

Proof of delivery. Because the RAD process emphasizes reuse, many of the program components have already been tested. This reduces testing time. However, all new components must be tested and all interfaces must be thoroughly exercised.

3 Design

The design of this tool consists of a VR model to imitate the movements made with the body by hearing impaired people to communicate.

Among the necessary actions to carry out this project, it was necessary to know and perform the movements in VR of some words of the LSE to test the functionality of the model, as well as to program the time and position interpolators, to later relate them to the touch sensors, for the functions of playback, forward, backward, delay and pause of the model animation (Fig. 1).

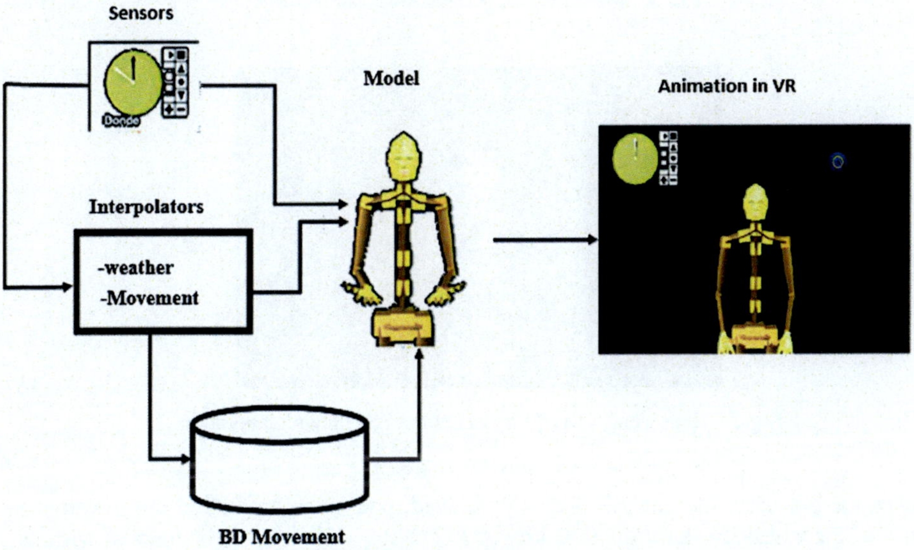

Fig. 1. Tool design.

4 Development

The present Virtual Reality (VR) tool for learning Spanish Sign Language (SSL) aims to allow any interested person to learn different words in SSL, through a visual medium and allowing the manipulation of different aspects of the animation, such as the speed of gestures, their meaning, the words to be represented, as well as being able to pause and resume the animation, among others [19].

For example, to correctly learn the movements necessary to represent the word "Rojo" it can be observed that the speaker must bring the index finger to the chin and make a vertical movement on it (Fig. 2), and to represent a glass, the speaker must make the gesture of taking a glass or cup with one hand over the other (Fig. 3).

Parallel animation allows to see both the word that is represented and the animation of the human model as shown in Fig. 4, as well as to manipulate the animation through the upper controls (Fig. 5) with which one can manipulate the speed at which the model

Fig. 2. Model saying ROJO. Own elaboration.

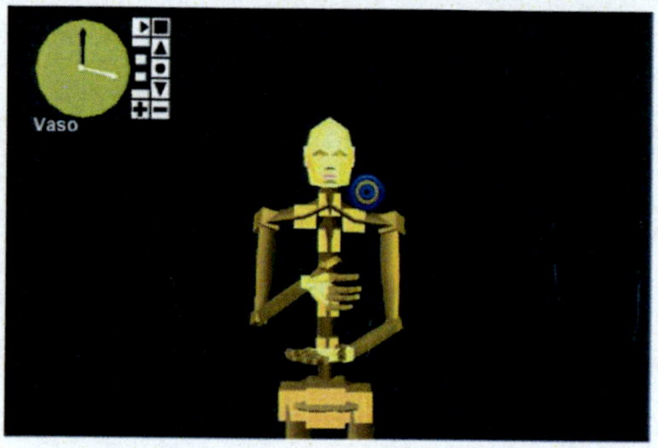

Fig. 3. Representation of a vessel at the SSL. Own elaboration.

makes the gestures, the gesture that is perceived, pause the animation and invert the sense of the words, because the structure of a sentence often differs in contrast with its representation in SSL.

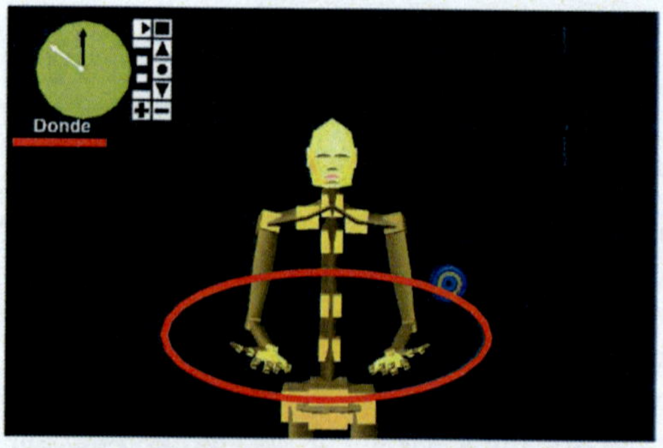

Fig. 4. Parallel operation. Own elaboration.

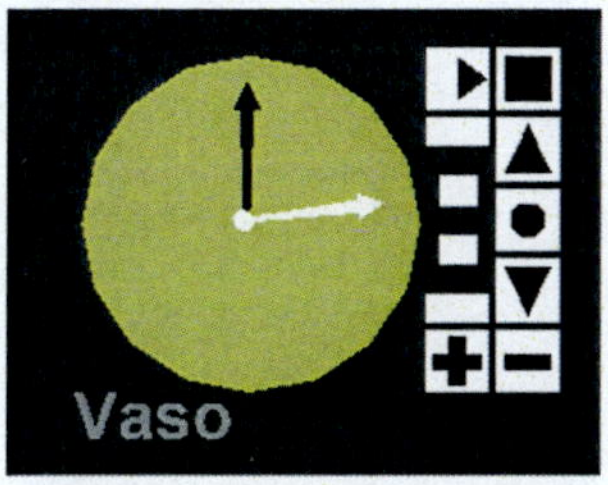

Fig. 5. Application controls and timer. Own elaboration.

The development of this tool was carried out using VRML (Virtual Reality Modeling Language) [20], due to the simplicity and ease that this language offers to represent virtual worlds with reduced computational resources.

A proprietary database, created from the movements and positions established by SSL, was also used, based mainly on [21], and available online.

For those who wish to learn, practice and understand SSL, this tool is a didactic, interactive, dynamic and user-friendly alternative, since it allows to learn and work at their own pace [22].

5 Conclusions

The developed prototype combines the technology and tools that exist today, to help learning a new language through the use of virtual models. This type of tool could be used to close the gap that exists in terms of communication, allowing anyone to access or use new technologies and tools, as well as being used for implementing artificial intelligence systems [23] or deep learning [24] that can be operated by means of SSL gestures.

It is considered that the use of virtual reality improves the learning of a new language, because unlike a video or a set of images, the interaction with the virtual environment is greater, due to the fact that the proposed VR system has speed controls, which makes it possible to see the gestures as many times as necessary and controlling the time they are played.

6 Future Research

In order to extend this study, it is proposed to implement a more extensive database that will make it possible to create a complete compendium of the words and expressions of SSL, and subsequently apply it to other LS in a different region or culture, as well as to evaluate whether using this method presents any significant improvement in learning compared to the standard method of teaching this language.

When the system contains a more complete database, its usability will be evaluated, as an important criterion in the performance and quality of the VR system. Although its evaluation may have a subjective component, the study of norms, guidelines, standards and different proposals will allow the relevant aspects in this process to be unified and serve as a basis for structuring a guide for improving the SSL system.

There are multiple fields of current computing that could take advantage of tools like this for developing programs or translators for impaired people.

References

1. Rodríguez-Ortiz, I.R., Pérez, M., Valmaseda, M., Cantillo, C., Díez, M.A., Montero, I., Moreno-Pérez, F.J., Pardo-Guijarro, M.J., Saldaña, D.: A Spanish Sign Language (LSE) adaptation of the communicative development inventories. J. Deaf Stud. Deaf Educ. **25**(1), 105–114 (2020)
2. Singleton, J.L., Quinto-Pozos, D., Martinez, D.: Typical and atypical sign language learners. In: The Routledge Handbook of Sign Language Pedagogy (2019)
3. Gago, J.J., Victores, J.G., Balaguer, C.: Sign language representation by teo humanoid robot: end-user interest, comprehension and satisfaction. Electronics **8**(1), 57 (2019)
4. Villameriel, S., Costello, B., Dias, P., Giezen, M., Carreiras, M.: Language modality shapes the dynamics of word and sign recognition. Cognition **191**, 103979 (2019)
5. Li, D., Rodriguez, C., Yu, X., Li, H.: Word-level deep sign language recognition from video: a new large-scale dataset and methods comparison. In: The IEEE Winter Conference on Applications of Computer Vision, pp. 1459–1469 (2020)
6. Beal, J.: University American Sign Language (ASL) second language learners: receptive and expressive ASL performance. J. Interpret. **28**(1), 1 (2020)
7. Malaia, E.A., Krebs, J., Roehm, D., Wilbur, R.B.: Age of acquisition effects differ across linguistic domains in sign language: EEG evidence. Brain Lang. **200**, 104708 (2020)
8. Stoll, S., Camgöz, N.C., Hadfield, S., Bowden, R.: Sign language production using neural machine translation and generative adversarial networks. In: 29th British Machine Vision, Northumbria University, Newcastle Upon Tyne, UK (2018)
9. Chung, S., Lim, J.Y., Ju Noh, K., Kim, G., Jeong, H.: Sensor Data Acquisition and Multimodal Sensor Fusion for Human Activity Recognition Using Deep Learning (2019)
10. Gutierrez-Sigut, E., Vergara-Martínez, M., Marcet, A., Perea, M.: Automatic use of phonological codes during word recognition in deaf signers of Spanish Sign Language. FEAST. Formal Exp. Adv. Sign Lang. Theory **1**, 1–15 (2018)
11. Gutierrez-Sigut, E., Costello, B., Baus, C., Carreiras, M.: LSE-sign: a lexical database for Spanish Sign Language. Behav. Res. Methods **48**(1), 123–137 (2016)
12. Parcheta, Z., Martínez-Hinarejos, C.D.: Sign language gesture recognition using HMM. In: Iberian Conference on Pattern Recognition and Image Analysis, pp. 419–426. Springer, Cham, June 2017
13. Goldin-Meadow, S., Brentari, D.: Gesture, sign, and language: the coming of age of sign language and gesture studies. Behav. Brain Sci. **40**, 1–60 (2017)
14. Ebling, S., Camgöz, N.C., Braem, P.B., Tissi, K., Sidler-Miserez, S., Stoll, S., Hadfield, S., Haug, T., Bowden, R., Tornay, S., Razavi, M.: SMILE Swiss German sign language dataset. In: Proceedings of the Eleventh International Conference on Language Resources and Evaluation (LREC 2018), May 2018
15. Torres Samuel, M., Vásquez, C., Viloria, A., Hernández Fernandez, L., Portillo Medina, R.: Analysis of patterns in the university Word Rankings Webometrics, Shangai, QS and SIRScimago: case Latin American. Lecture Notes in Computer Science (Including subseries Lecture Notes in Artificial Intelligent and Lecture Notes in Bioinformatics (2018)
16. Mulík, S., Carrasco-Ortiz, H., Amengual, M.: Phonological activation of first language (Spanish) and second language (English) when learning third language (Slovak) novel words. Int. J. Bilingualism **23**(5), 1024–1040 (2019)

17. Viloria, A., Lis-Gutiérrez, J.P., Gaitán-Angulo, M., Godoy, A.R.M., Moreno, G.C., Kamatkar, S.J.: Methodology for the design of a student pattern recognition tool to facilitate the teaching-learning process through knowledge data discovery (big data). In: International Conference on Data Mining and Big Data, pp. 670–679. Springer, Cham, June 2018
18. Lu, X., Zheng, Y., Ren, W.: Motivation for learning Spanish as a foreign language: the case of Chinese L1 speakers at university level. Círculo de Lingüistica Aplicada a la Comunicación **79**, 79–99 (2018)
19. Lonigan, C.J., Allan, D.M., Goodrich, J.M., Farrington, A.L., Phillips, B.M.: Inhibitory control of Spanish-speaking language-minority preschool children: measurement and association with language, literacy, and math skills. J. Learn. Disabil. **50**(4), 373–385 (2017)
20. Zink, D.N., Reyes, E., Kuwabara, H., Strauss, G.P., Allen, D.N.: Factorial validity of the emotional verbal learning test-spanish (EVLT-S). Arch. Clin. Neuropsychol. **34**(7), 1283 (2019)
21. Ahmed, M.A., Zaidan, B.B., Zaidan, A.A., Salih, M.M., Lakulu, M.M.B.: A review on systems-based sensory gloves for sign language recognition state of the art between 2007 and 2017. Sensors **18**(7), 2208 (2018)
22. Fitzpatrick, E.M., Hamel, C., Stevens, A., Pratt, M., Moher, D., Doucet, S.P., Neuss, D., Bernstein, A., Na, E.: Sign language and spoken language for children with hearing loss: a systematic review. Pediatrics **137**(1), e20151974 (2016)
23. Lane, H.: A chronology of the oppression of sign language in France and the United States. In: Recent perspectives on American Sign Language, pp. 119–161. Psychology Press (2017)
24. Viloria, A., Parody, A.: Methodology for obtaining a predictive model academic performance of students from first partial note and percentage of absence. Indian J. Sci. Technol. **9**, 46 (2016)

Data Augmentation Using Gaussian Mixture Model on CSV Files

Ashish Arora[1(✉)], Niloufar Shoeibi[2(✉)], Vishwani Sati[3], Alfonso González-Briones[2,4,5], Pablo Chamoso[2,5], and Emilio Corchado[2]

[1] Indian Institute of Technology Dharwad, Dharwad, India
160010008@iitdh.ac.in
[2] BISITE Research Group, University of Salamanca. Edificio I+D+i, Calle Espejo 2, 37007 Salamanca, Spain
{Niloufar.Shoeibi,alfonsogb,chamoso,escorchado}@usal.es
[3] Amity School of Engineering and Technology, Amity University, Noida, India
Vishwani.sati@gmail.com
[4] Research Group on Agent-Based, Social and Interdisciplinary Applications (GRASIA), Complutense University of Madrid, Madrid, Spain
[5] Air Institute, IoT Digital Innovation Hub (Spain), Carbajosa de la Sagrada, 37188 Salamanca, Spain

Abstract. One of the biggest challenges in training supervised models is the lack of amount of labeled data for training the model and facing overfitting and underfitting problems. One of the solutions for solving this problem is data augmentation. There have been many developments in data augmentation of the image files, especially in medical image type datasets, by doing some changes on the original file such as Random cropping, Filliping, Rotating, and so on, in order to make a new sample file. Or use Deep Learning models to generate similar samples like Generative Adversarial Networks, Convolutional Neural Networks and so on. However, in numerical dataset, there have not been enough advances. In this paper, we are proposing to use the Gaussian Mixture Models (GMMs) to augment more data very similar to the original Numerical dataset. The results demonstrated that the Mean Absolute Error decreases meaning that the regression model became more accurate.

Keywords: Big data · Numerical data augmentation · Gaussian Mixture Model (GMM) · Machine learning · Regression supervised models

1 Introduction

Due to the tremendous growth of Artificial Intelligence (AI) and Internet of Things (IoT) [12–20] in recent years; the importance of data has grown immensely. Data Scarcity has become the bottleneck for AI in today's world. Lack of Data is still a persistent problem in many fields where AI can be implemented. In some cases, even if the dataset is available and large enough, then annotation is a big problem whenever you are dealing

Y. Dong et al. (Eds.): DCAI 2020, AISC 1237, pp. 258–265, 2021.
https://doi.org/10.1007/978-3-030-53036-5_28

with a supervised learning task. Manual annotation is possible but is highly cumbersome when involved with a large dataset. In one of the notable works [1], authors have tried to automatically annotate an image using a proposed label propagation framework based on Kernel Canonical Correlation Analysis (KCCA). It builds a semantic space where the correlation of visual features is well-preserved in the semantic embeddings.

To avoid the need for a large dataset, researchers have used the concept of Transfer Learning. In the case of Transfer Learning, the model is already trained on a different dataset and it is retrained on the dataset, associated with the problem by fixing some of the weights of the trained network. But this concept completely fails when your dataset is not related to the dataset on which model is trained.

Recently, the notion of Zero-shot Learning has also evolved. It solves the problem of the need for labeled data in the training set. In this case, the model is trained on a set of classes, but the predictions are made on a new set of classes, which is disjoint from the classes on which the model has been trained. This technique is good but still, it is ineffective in many of the AI-related problems.

Data Augmentation is the way of generating synthetic data using the given dataset. There are different techniques through which you can do data augmentation. In the case of images, simple augmentation techniques can be like rotation, flip, scale, crop, translation, adding Gaussian noise, etc. Moreover, advanced techniques can also be performed in some cases. One such example is conditional GANs. With the help of conditional GANs, augmentation can be done. But these tasks are computationally intensive.

In our methodology, we have used Gaussian Mixture Models (GMM) to generate synthetic data with CSV files. GMM is a probabilistic model which learns the distribution of the data, and then samples out the artificial examples based on this distribution. Our detailed methodology is described in Sect. 3.

2 Related Work

The biggest problem for training machine learning models [10] is the lack of having the biggest problem for training machine learning models is the lack of having enough labeled data. In some cases, like medical datasets [11], having labels for each data is so costly or there are not enough expert resources to do the labeling appropriately. If the amount of data for training the model not be enough, the model will face the problems of overfitting and underfitting. For dealing with this problem, there are many solutions, one of them is to extend more data is data augmentation.

Zhong, Z., et al. introduced Random Erasing as a novel data augmentation method that deletes some pixels from the original data randomly and by applying Random Erasing at different levels, it makes more data for training the Convolutional Neural Networks (CNNs). This method is so easy to implement, and it is complementary to other data augmentation techniques like random cropping, flipping, and yields [2].

Data augmentation has been advanced a lot in computer vision and image processing. Therefore, there are many improvements in the medical sciences. For instance, Bowles, C. et al., proposed to use Generative Adversarial Networks (GANs) as a novel way to extract more information from a medical dataset, by generating synthetic samples very similar in appearance to the real images [3].

Also, APAFrid-Adar, M. et al. proposed using a GAN-based model for synthetic medical image augmentation for increasing the performance of the Convolutional Neural Network (CNN) in liver lesion classification [4].

Mariani, G. et al. proffered Balancing GAN (BAGAN) [5] as a tool for augmentation that returns balance to imbalanced data. In some cases, the data is even not enough to generate more by training GANs. They demonstrated that BAGAN generates more realistic images in imbalanced datasets in comparison to other stated GANs.

In [6] Nguyen, P., et al. proposed to use EmbNum+, a numerical embedding for learning both discriminant representations and a similarity metric from numerical columns, to do the attribute augmentation. Attribute augmentation generates samples by changing the size of the attributes and randomly choose the numerical values in the original attributes. However, in numerical datasets, there has not been enough advancement and there are still more rooms to work on.

3 Data and Proposed Method

We have used the dataset related to PLATINUM project which is the dataset of density of woods for our experiment. In this dataset there are 1157 examples in total. Each example is contained in a row and each row is attributed with 71 entries. 7 entries correspond to the densities of different regions. These 7 regions are: Density Extreme Left, Density Left, Density Extreme center, Density center, Density Extreme Right, Density Right, and the mean is given as the average density. The rest 64 values in a row correspond to different features like humidity, pressure difference, thickness, etc. In our proposed approach, we are trying to predict the average density of the wooden table based on the given features. These features exclude regional densities.

3.1 Data Filtration

Since there are too many features involved in our dataset, it is better to remove the redundant ones. In our dataset, there are too many features that belong to a specific type like different variables corresponding to Pressure difference. So, we clubbed all the features of a specific type and plotted the correlation map for the analysis. One of the correlation maps of a specific type (Frame specific feature) is shown in Fig. 1. In Fig. 1, the correlation map is plotted with 13 different features of the same category. It can be easily seen that many of these features are highly correlated with each other. In our model, there is no need to incorporate these features because they are redundant in nature. In total, we selected 4 features out of these 13 features in this category. These 4 features are selected based on their correlation values calculated respecting to average density.

Subsequently, these analyses are performed on all the variables belonging to different categories. After preprocessing the data, we are left with only 26 variables in total.

We split 70% of the data for the training set and the rest 30% for the testing set. After this splitting, we had only 821 rows in the training set. We tried different models to train our model, but it was overfitting as shown in our Results section. The primary reason for our model getting overfitting was the availability of a smaller number of training

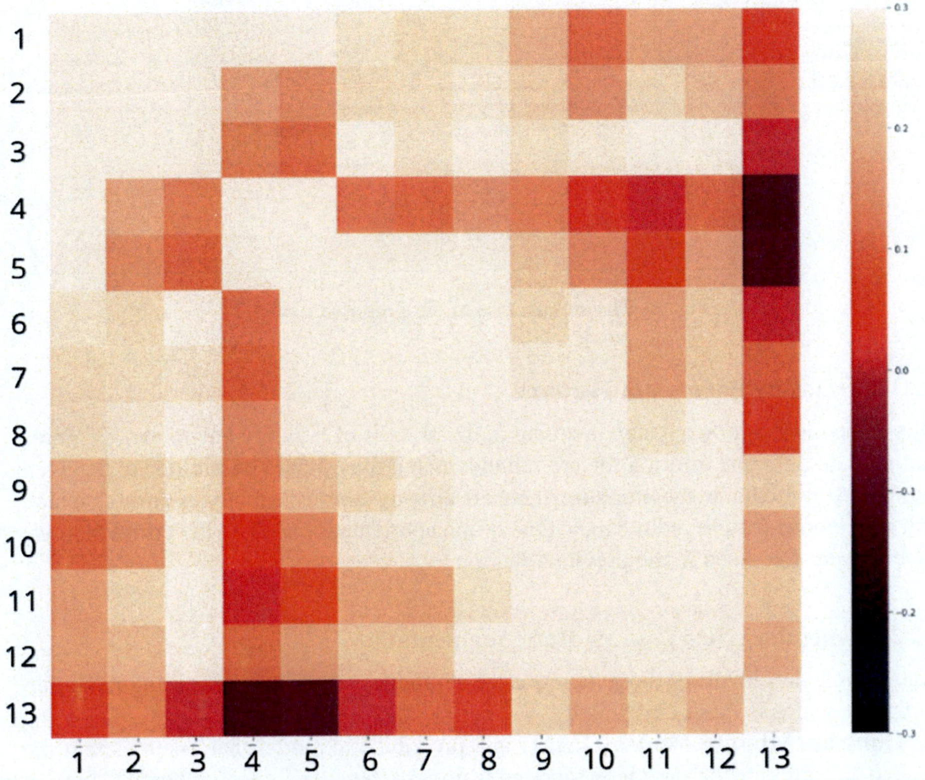

Fig. 1. The correlation values of each feature calculated respect to other features. (1 to 13 are the labels corresponding to different features of type "frame specific pressure")

examples. We even tried to do it with a single layer model, but in this case, the model was not even getting trained properly and thus resulting in less test accuracy.

To increase the diversity of the data for our training model, we aimed to generate synthetic data from our training data. As already discussed, this technique of generating more data is Data Augmentation. In our case, we trained a Gaussian Mixture Model (GMM) using the training data for our data augmentation process. A brief description of our GMM model is given in the next section. The synthetic data generated from this model is then combined with the original training data for our prediction model.

3.2 Proposed Method

Our proposed methodology consists of 3 steps (Fig. 2):

1. Removing redundant features
2. Generating new data via data augmentation
3. Regression Model

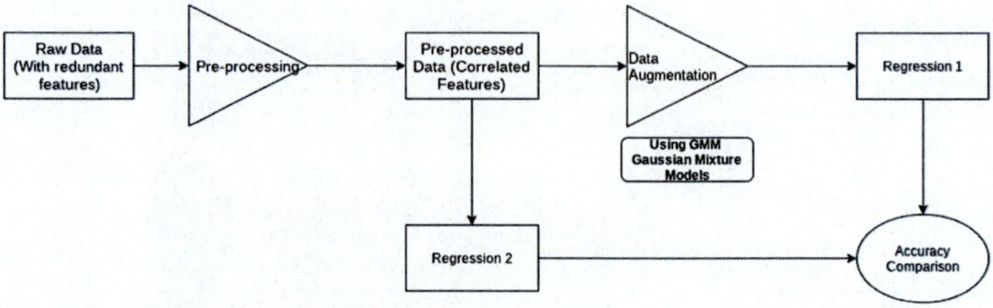

Fig. 2. The architecture of our proposed model

3.2.1 Removing Redundant Features

This is the most significant step involved in the process of building any model. Features that provide the same information are redundant and thus should be avoided to be given as input to the model. In the literature, there are already various techniques through which you can remove feature redundancy. One of the approaches based on the correlation we have already discussed in the previous section.

3.2.2 Generating New Data via Data Augmentation

The second step involves the generation of synthetic data using a data augmentation technique which is described below:

Gaussian Mixture Model (GMM) is a probabilistic model that assumes that the given data points are the samples generated from a mixture of a finite number of Gaussian densities. The equation corresponding to this mixture is written in (1). If there are n Gaussian densities, then the mixture is a convex combination of these densities.

$$f(x) = \sum_{i=1}^{n} \lambda_i * \mathcal{N}\left(x, u_i, \sigma_i^2\right) \tag{1}$$

$$\sum_{i=1}^{n} \lambda_i = 1 \tag{2}$$

In (1), different are the associated weights of normal densities. In this case, a normal density is represented with mean and as the variance. It should be noted here that the sum of all is equal to 1.

We have used the python Scikit-learn library [7, 8] for the implementation of this GMM model. Our GMM model is trained using the Expectation-maximization Algorithm. [9] It is a well-founded statistical iterative algorithm to estimate and of the gaussian mixture.

Since in our task, we have a 26-dimensional feature vector instead of 1, so we will be dealing with multivariate Gaussian densities. The number of estimators in (1) depends on the number of components(n). We have experimented with our model with different values of n, as described in the next section. After training our GMM model using the training set, we sample out a few examples from this trained model. In total, we generated a total of 7000 synthetic examples. Depends on the number of components(n). We have experimented with our model with different values of n, as described in the next section.

After training our GMM model using the training set, we sample out a few examples from this trained model. In total, we generated a total of 3000 synthetic examples.

3.2.3 Regression Model

These synthetic examples are then combined with the original dataset and subsequently fed to our regression model. We have used different algorithms for our regression problem. Linear Regressor, Decision Tree Regressor, Random Forest Regressor, K-Neighbors Regressor, and Ridge Regressor are trained on our dataset. The results are described in the next section.

4 Results

In this section, we have experimented with our proposed architecture with 5 different regressors. These 5 regressors are Linear Regressor, Decision Tree Regressor, Random Forest Regressor, K-Neighbors Regressor, and Ridge Regressor. All the regressors have been implemented using the python Scikit-learn library. Since it is a regression problem, we have used Mean Squared Error (MSE) as the loss function. We have also made a comparison using the R2-squared metric, which is a statistical measure of how good the data is fitted to the regression model. The R2-squared value lies between 0 and 1.

Table 1 corresponds to the statistics when the model is trained without incorporating GMM samples. In total, 817 examples were considered for training and 348 for testing. Testing Mean Absolute Error (MAE) and R2-squared are shown for each of the 5 regressors.

Table 1. Training without GMM samples

Regressor	Training MAE	Testing MAE	R2_SCORE
Linear regressor	11.14	9.66	0.385
Decision tree regressor	0.0	15.94	−0.182
Random forest regressor	3.76	11.12	0.379
K-neighbours regressor	8.66	12.27	0.276
Ridge regressor	9.72	11.11	0.401

Table 2 corresponds to the entries when the model is trained with GMM samples. After fitting the GMM on training data, 7000 more examples were sampled from the fitted GMM. In total, 7817 examples were considered for training and the testing set remains the same.

It can be easily seen that all the 5 regressors are performing better with the augmented data when incorporated in the training set. Better performance can be attributed in terms of lower MAE and higher R2-squared value. Random Forest Regressor is performing the best amongst all the regressors. It has achieved 6.81 MAE when trained on the augmented data in comparison to 11.12 when trained only on the given training data.

Table 2. Training with GMM samples

Regressor	Training MAE	Testing MAE	R2_SCORE
Linear regressor	10.06	10.05	0.51
Decision tree regressor	1.23 e−06	9.46	0.43
Random forest regressor	2.44	6.81	0.74
K-neighbors regressor	5.55	7.06	0.71
Ridge regressor	10.05	10.06	0.512

5 Conclusion and Future Work

In this paper, we have proposed a methodology of augmenting the data of CSV files using the Gaussian Mixture Model, which is a probabilistic model. For experimentation purposes, we have used only one dataset and evaluated the performance of our model using 5 different algorithms. Results demonstrated that this proposed framework is performing better for all the 5 algorithms when trained on augmented data.

So, for our future work, we can evaluate our framework on different datasets, where data is structured in CSV files, like in this case. Moreover, we can also endeavor to model the distribution of data using densities other than gaussian.

Acknowledgement. This work was developed as part of "PLATINUM: Plataforma horizontal de smart data y deep learning para la industria y aplicación al sector manufacturero", ID RTC-2017-6401-7, project financed by Ministry of Science, Innovation and Universities (Retos-Colaboración 2017).

References

1. Burdescu, D., Mihai, G., Stanescu, L., Brezovan, M.: Automatic image annotation and semantic based image retrieval for medical domain. Neurocomputing **109**, 33–48 (2016). https://doi.org/10.1016/j.neucom.2012.07.030
2. Zhong, Z., Zheng, L., Kang, G., Li, S., Yang, Y.: Random erasing data augmentation. arXiv preprint arXiv:1708.04896 (2017)
3. Bowles, C., Chen, L., Guerrero, R., Bentley, P., Gunn, R., Hammers, A., Dickie, D.A., Hernández, M.V., Wardlaw, J., Rueckert, D.: GAN augmentation: augmenting training data using generative adversarial networks. arXiv preprint arXiv:1810.10863 (2018)
4. Frid-Adar, M., Diamant, I., Klang, E., Amitai, M., Goldberger, J., Greenspan, H.: GAN-based synthetic medical image augmentation for increased CNN performance in liver lesion classification. Neurocomputing **321**, 321–331 (2018)
5. Mariani, G., Scheidegger, F., Istrate, R., Bekas, C., Malossi, C.: BAGAN: data augmentation with balancing GAN. arXiv preprint arXiv:1803.09655 (2018)
6. Nguyen, P., Nguyen, K., Ichise, R., Takeda, H.: EmbNum+: effective, efficient, and robust semantic labeling for numerical values. New Gener. Comput. **37**(4), 393–427 (2019)
7. Pedregosa, F., Varoquaux, G., Gramfort, A., Michel, V., Thirion, B., Grisel, O., Blondel, M., Prettenhofer, P., Weiss, R., Dubourg, V., Vanderplas, J.: Scikit-learn: machine learning in Python. J. Mach. Learn. Res. **12**, 2825–2830 (2011)

8. Buitinck, L., Louppe, G., Blondel, M., Pedregosa, F., Mueller, A., Grisel, O., Niculae, V., Prettenhofer, P., Gramfort, A., Grobler, J., Layton, R.: API design for machine learning software: experiences from the scikit-learn project (2013)
9. Moon, T.K.: The expectation-maximization algorithm. IEEE Signal Process. Mag. **13**(6), 47–60 (1996)
10. Shoeibi, N., Shoeibi, N.: Future of smart parking: automated valet parking using deep Q-learning. In: International Symposium on Distributed Computing and Artificial Intelligence, pp. 177–182. Springer, Cham, June 2019
11. Shoeibi, N., Karimi, F., Corchado, J.M.: Artificial intelligence as a way of overcoming visual disorders: damages related to visual cortex, optic nerves and eyes. In: International Symposium on Distributed Computing and Artificial Intelligence, pp. 183–187. Springer, Cham, June 2019
12. Sánchez, S.M., Vara, R.C., Criado, F.J.G., González, S.R., Tejedor, J.P., Corchado, J.M.: Smart PPE and CPE platform for electric industry workforce. In: International Workshop on Soft Computing Models in Industrial and Environmental Applications, pp. 422–431. Springer, Cham, May 2019
13. Chimeno, S.G., Fernández, J.D., Sánchez, S.M., Ramón, P.P., Ospina, Ó.M.S., Muñoz, M.V., Hernández, A.G.: Domestic violence prevention system. In: International Symposium on Distributed Computing and Artificial Intelligence, pp. 10–14. Springer, Cham, June 2018
14. Marquez, S., Casado-Vara, R., González-Briones, A., Prieto, J., Corchado, J.M.: SiloMAS: a MAS for smart silos to optimize food and water consumption on livestock holdings. In: International Symposium on Distributed Computing and Artificial Intelligence, pp. 27–37. Springer, Cham, June 2018
15. González-Briones, A., Casado-Vara, R., Márquez, S., Prieto, J., Corchado, J.M.: Intelligent livestock feeding system by means of silos with IoT technology. In: International Symposium on Distributed Computing and Artificial Intelligence, pp. 38–48. Springer, Cham, June 2018
16. Sánchez, S.M.: Electronic textiles for intelligent prevention of occupational hazards. In: International Symposium on Distributed Computing and Artificial Intelligence, pp. 217–220. Springer, Cham, June 2019
17. Alonso, R.S., Sittón-Candanedo, I., García, Ó., Prieto, J., Rodríguez-González, S.: An intelligent Edge-IoT platform for monitoring livestock and crops in a dairy farming scenario. Ad Hoc Netw. **98**, 102047 (2020)
18. Alonso, R.S., Sittón-Candanedo, I., Rodríguez-González, S., García, Ó., Prieto, J.: A survey on software-defined networks and edge computing over IoT. In: International Conference on Practical Applications of Agents and Multi-Agent Systems, pp. 289–301. Springer, Cham, June 2019
19. Candanedo, I.S., Nieves, E.H., González, S.R., Martín, M.T.S., Briones, A.G.: Machine learning predictive model for industry 4.0. In: International Conference on Knowledge Management in Organizations, pp. 501–510. Springer, Cham, August 2018
20. González-Briones, A., Hernández, G., Corchado, J.M., Omatu, S., Mohamad, M.S.: Machine learning models for electricity consumption forecasting: a review. In: 2019 2nd International Conference on Computer Applications & Information Security (ICCAIS), pp. 1–6. IEEE, May 2019

Sentiment Analysis in Twitter: Impact of Morphological Characteristics

Jesús Silva[1](✉), Juan Manuel Cera[2], Jesús Vargas[2], and Omar Bonerge Pineda Lezama[3]

[1] Universidad Peruana de Ciencias Aplicadas, Lima, Peru
jesussilvaUPC@gmail.com
[2] Universidad de la Costa, Barranquilla, Colombia
{jcera7,jvargas41}@cuc.edu.co
[3] Universidad Tecnológica Centroamericana (UNITEC), San Pedro Sula, Honduras
omarpineda@unitec.edu

Abstract. This paper presents a series of experiments aimed at the sentiment analysis on texts posted in Twitter. In particular, several morphological characteristics are studied for the representation of texts in order to determine those that provide the best performance when detecting the emotional charge contained in the Tweets.

Keywords: Sentiment analysis · Twitter · Morphological characteristics · Weka

1 Introduction

Analyzing the emotional charge in texts is a task of great importance nowadays. There is a multitude of applications that can benefit from computer procedures that allow to automatically detect if the author's intention was expressed in a "positive", "negative", "objective" or "neutral" way. Let's consider, for instance, the case of a political personality that requires to know if the community has a positive or negative appreciation about him/her. Another example could be the determination of the reputation for a public or private institution [1]. In either of the two examples, there is a need to analyze the point of view of the people (users) about their target entities. Although in the past it was common to apply questionnaires to users, nowadays this practice is rarely used mainly due to the following inconveniences [2]:

1. The application of questionnaires is a costly process, both in terms of time and money.
2. The concentration of such questionnaires requires time and further analysis.
3. The selection of candidates on which the questionnaire is applied must be decided carefully to ensure that the analysis results are appropriate (quantity and quality).
4. The analysis of the data has to be completed promptly to avoid making the conclusions obsolete.

Y. Dong et al. (Eds.): DCAI 2020, AISC 1237, pp. 266–273, 2021.
https://doi.org/10.1007/978-3-030-53036-5_29

In this way, it is much more practical and convenient to use fresh data obtained directly from social networks. People tend to express themselves freely on the topics that are of interest to them. The only problem is that these data are expressed in natural language, and therefore require automatic computational methods for their treatment [3]. Even so, this approach is much more attractive for companies and has resulted in a very active research area within the community related to automatic natural language processing.

The aim of this research is to evaluate the impact on the use of various morphological characteristics on the task of detecting emotional charge in Tweets (positive, negative, neutral and objective). The problem has been solved from the perspective of text classification using automatic learning methods. This perspective was considered based on the existence of a supervised corpus (with manual classification).

1.1 Related Studies

There are studies related to the identification of emotions in Twitter, but few of them pay attention to the analysis of the partial contribution generated by the morphological characteristics. For example, [4] calculates the probability of A priori polarity associated with part of speech (PoS) tags. Up to 100 additional characteristics are used, including the use of emoticons and dictionaries of positive and negative words. The results reported show up to 60% accuracy. On the other hand, [5] proposes a strategy that makes use of few lexical resources; in particular, it uses discursive relationships such as connectivity and conditionals to include them in the classic word bag models with the intention of improving the accuracy values over the classification process.

One of the greatest advances in the task of sentiment analysis has been in a competition proposed in the framework of SemEval 2019 [6–10]. Some of the studies provided a broad overview of various methods and characteristics used in the aforementioned task. There is no doubt that this is an important task that will get attention by the linguistic computer community in next years.

2 Experiments

This section describes the experiments carried out, then a description of the data set, the characteristics evaluated to represent the texts, the type of classifier, and finally, the results obtained.

2.1 Data Set

A set of training and test data provided by [11–13] was used in the experiments. This corpus presents 9,584 tweets written in the English language, which were manually labeled with the following classes: positive, negative, neutral and objective, and can be used as training data. The set of test texts contains 1,547 tweets. A description of its general characteristics shown in Table 1.

The vocabulary of the test corpus shares 3,347 terms with the vocabulary of the training corpus. This means that 74% of its vocabulary is present in the vocabulary used

Table 1. Characteristics of the evaluation corpus.

Characteristic	Training corpus	Test corpus
No. of tweets	9,584	1547
No. of words	110,852	16,997
Vocabulary	19,300	4,365
Average length	20.47	16.87
Positive tweets	3,015	368
Negative tweets	865	214
Neutral tweets	2,772	450
Target tweets	2,359	201

in the training. However, given the size of the training vocabulary (20,000) it is noted that only 16% of the vocabulary is common between the two corpora, which clearly shows that there are many words that will not be useful for the classification task.

2.2 Description of the Characteristics

As mentioned above, the objective of this study is to evaluate the impact that morphological characteristics have on the process of textual representation, when carrying out a task of identifying emotional charge in Tweets. Thus, every word of the Tweets of the training and test corpus are filtered, leaving just those words that comply with one and only one morphological label. The tagging process was carried out using the TreeTagger [14]. The tags used are shown in Table 2.

Since Tweets are extremely short texts, in some cases, the number of words that comply with a PoS tag is zero, so it was decided to select the first five words in those cases. This allows to homogenize the results making them comparable across all morphological tags [15], especially considering that, in the short term, an assembly of the results by morphological category will be made. In this case, each instance must have, at least, one characteristic.

2.3 Classifier

Three classifiers of different nature were selected in order to evaluate the process properly [2, 15]. In particular, a decision tree-based classifier known as J48 was used, which is an implementation of the C4.5 algorithm, one of the most used algorithms for data mining. An implementation of the Naïve Bayes algorithm was also used, which calculates the probability of each characteristic, given the previous characteristic. Finally, an implementation of the Support Vector Machines (SVM) known as SMO was used. An SVM builds a hyperplane or set of hyperplanes in a very high dimensionality space that can be used in classification or regression problems. The goal is to find a good separation between the classes, which will allow a correct classification [14, 16].

The implementations of J48, Naïve Bayes and SMO present in the WEKA3 tool were used, using the existing default parameters in each classifier [4, 8].

Table 2. Morphological labels used in the classification process [16].

PoS label	Description
JJ	Adjective
NN	Noun in singular
VBN	Verb in past participle
VB	Verb in its base form
RB	Adverb
IN	Intersection
NP	Propper name in singular
PP	Preposition
RBR	Comparative adverb
RBS	Superlative adverb
PR	Particle
VBG	Verb in gerund or past participle
JJR	Comparative adjective
JJS	Superlative adjective
MD	Modal
NPS	Propper name in plural
PDT	Predeterminer
VBZ	Present tense verb, third person singular
VBP	Present tense verb, not third person singular
WDT	Wh-type determiner
WP	Wh-type pronoun
WPS	Wh-type possessive pronoun
WRB	Wh-type adverb
NNS	Noun in plural

3 Results

This section presents the results obtained on each of the corpora (training and testing), using the characteristics mentioned above. It is important to mention that the experiments were performed on each of the PoS tags independently. Table 3 shows the percentage of data that were classified correctly (C) and the percentage of those that were classified incorrectly (I). the highest values were highlighted, which are obtained mainly by the Support Vector Machines (SMO).

Figure 1 shows that the behavior obtained by SMO is superior, regardless of the type of morphological label used. In particular, the order of importance in the classification process is as follows: WDT, NNS, RP, NP, VBZ, MD, JJ, PP, RB, VB, VBP, NN, IN,

Table 3. Results of the classification process using only the morphological characteristics

PoS label	J48				Naïve Bayes				SMO			
	Training		Test		Training		Test		Training		Test	
	C	I	C	I	C	I	C	I	C	I	C	I
JJ	0.458	0.572	0.29	0.709	0.417	0.554	0.307	0.683	0.453	0.493	0.41	0.64
NN	0.44	0.59	0.293	0.706	0.406	0.565	0.298	0.692	0.438	0.508	0.413	0.637
VBN	0.444	0.586	0.299	0.7	0.397	0.574	0.288	0.702	0.429	0.517	0.395	0.655
VB	0.436	0.594	0.285	0.714	0.392	0.579	0.293	0.697	0.441	0.505	0.392	0.658
RB	0.455	0.575	0.303	0.696	0.409	0.562	0.315	0.675	0.446	0.5	0.407	0.643
IN	0.426	0.604	0.294	0.705	0.392	0.579	0.291	0.699	0.438	0.508	0.388	0.662
NP	0.433	0.597	0.301	0.698	0.398	0.573	0.28	0.71	0.427	0.519	0.402	0.648
PP	0.447	0.583	0.297	0.702	0.4	0.571	0.291	0.699	0.45	0.496	0.382	0.668
RBR	0.443	0.587	0.302	0.697	0.399	0.572	0.285	0.705	0.434	0.512	0.389	0.661
RBS	0.449	0.581	0.293	0.706	0.399	0.572	0.285	0.705	0.432	0.514	0.391	0.659
PR	0.438	0.592	0.285	0.714	0.389	0.582	0.292	0.698	0.428	0.518	0.391	0.659
VBG	0.442	0.588	0.3	0.699	0.397	0.574	0.28	0.71	0.43	0.516	0.389	0.661
JJR	0.426	0.584	0.298	0.701	0.399	0.572	0.285	0.705	0.429	0.517	0.392	0.658
JJS	0.432	0.578	0.29	0.709	0.4	0.571	0.287	0.703	0.437	0.509	0.399	0.651
MD	0.416	0.594	0.279	0.72	0.398	0.573	0.295	0.695	0.426	0.52	0.394	0.656
NPS	0.424	0.586	0.293	0.706	0.397	0.574	0.284	0.706	0.431	0.515	0.386	0.664
PDT	0.425	0.585	0.295	0.704	0.398	0.573	0.284	0.706	0.429	0.517	0.392	0.658
VBZ	0.425	0.585	0.285	0.714	0.399	0.572	0.297	0.693	0.427	0.519	0.393	0.657
VBP	0.425	0.585	0.287	0.712	0.402	0.569	0.303	0.582	0.438	0.508	0.408	0.642
WDT	0.419	0.591	0.294	0.705	0.398	0.573	0.287	0.598	0.428	0.518	0.389	0.661
WP	0.424	0.586	0.294	0.671	0.398	0.573	0.285	0.6	0.431	0.515	0.394	0.656
WPS	0.424	0.586	0.293	0.672	0.397	0.574	0.284	0.601	0.43	0.516	0.391	0.659
WRB	0.422	0.588	0.291	0.674	0.396	0.575	0.285	0.6	0.43	0.516	0.395	0.655
NNS	0.423	0.587	0.3	0.665	0.399	0.572	0.29	0.595	0.428	0.518	0.392	0.658

JJS, RBR, RBS, WP, NPS, VBG, WRB, WPS, PDT, JJR, VBN. That is, according to the results obtained in the training corpus, the five characteristics with the greatest weight are: adjective, preposition, adverb, verb in its base form and verb in present (not third person singular) [17, 18].

For the case of the test corpus (see Fig. 2), it is observed that there are only two cases in which the SMO classification algorithm does not obtain the best results (preposition and adverb). In these two cases, the Naïve Bayes algorithm obtains better results than SMO. The order of importance of each morphological label varies a lot with respect to the one obtained on the training corpus. This order is as follows: NN, JJ, VBP, RB, NP,

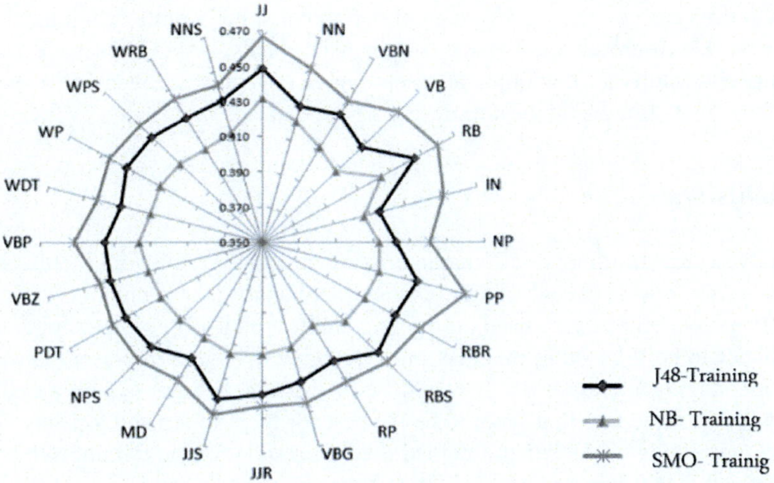

Fig. 1. Results obtained for each morphological label on the training corpus.

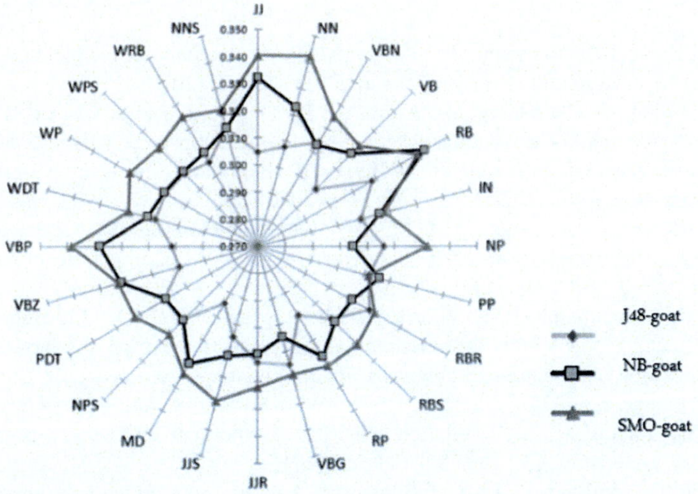

Fig. 2. Results obtained for each morphological label on the test corpus.

JJS, WRB, VBN, WP, MD, VBZ, VB, PDT, JJR, NNS, RBS, WPS, RP, RBR, VBG, WDT, IN, NPS, PP. That is, according to the results obtained in the corpus of evidence, the five characteristics with the greatest weight are: noun in singular, adjective, verb in present tense (not third person singular), adverb and proper noun in singular.

Thus, it can be noted that the following characteristics are the most important in both corpora (training and test): adjective, present tense verb (not third person singular) and adverb. It is curious that the preposition is found as an important characteristic in the training corpus, while in the test corpus it was the less important characteristic.

These results can be improved by searching for an assembly of morphological characteristics that best represent the Tweets. An exhaustive analysis should be made to avoid considering characteristics that duplicate their degree of representativeness and to look only for those that, as a whole, best represent all the data.

4 Conclusions

In this study, several morphological characteristics were analyzed in order to determine the degree of importance of each of them in the task of analysis of sentiments in Twitter. Three different classifiers were used on training and test data, observing that the stable characteristics to be selected in this task must be: adjective, verb in present tense (not third person singular) and adverb. It would be interesting to observe how a classifier behaves with the three tags that seem to be the best for the training and test set, as well as the five morphological tags that obtained the best results for the training set, on one hand, and the five that behaved best for the test set on the other hand.

References

1. Zahra, K., Imran, M., Ostermann, F.O.: Automatic identification of eyewitness messages on Twitter during disasters. Inf. Process. Manag. **57**(1), 102107 (2020)
2. Kaul, A., Mittal, V., Chaudhary, M., Arora, A.: Persona classification of celebrity Twitter users. In: Digital and Social Media Marketing, pp. 109–125. Springer, Cham (2020)
3. Motamedi, R., Jamshidi, S., Rejaie, R., Willinger, W.: Examining the evolution of the Twitter elite network. Soc. Netw. Anal. Mining **10**(1), 1 (2020)
4. Rodríguez-Ruiz, J., Mata-Sánchez, J.I., Monroy, R., Loyola-González, O., López-Cuevas, A.: A one-class classification approach for bot detection on Twitter. Comput. Secur. **91**, 101715 (2020)
5. Vásquez, C., Torres-Samuel, M., Viloria, A., Borrero, T.C., Varela, N., Lis-Gutiérrez, J.P., Gaitán-Angulo, M.: Visibility of research in universities: the triad product-researcher-institution. Case: Latin american countries. In: International Conference on Data Mining and Big Data, pp. 225–234. Springer, Cham, June 2018
6. Burnap, P., Williams, M.L.: Us and them: identifying cyber hate on Twitter across multiple protected characteristics. EPJ Data Sci. **5**(1), 11 (2016)
7. Luo, F., Cao, G., Mulligan, K., Li, X.: Explore spatiotemporal and demographic characteristics of human mobility via Twitter: a case study of Chicago. Appl. Geogr. **70**, 11–25 (2016)
8. Kabakuş, A.T., Şimşek, M.: An analysis of the characteristics of verified Twitter users. Sakarya Univ. J. Comput. Inf. Sci. **2**(3), 180–186 (2019)
9. Nguyen, Q.C., Brunisholz, K.D., Yu, W., McCullough, M., Hanson, H.A., Litchman, M.L., Li, F., Wan, Y., VanDerslice, J.A., Wen, M., Smith, K.R.: Twitter-derived neighborhood characteristics associated with obesity and diabetes. Sci. Rep. **7**(1), 1–10 (2017)
10. Gurajala, S., White, J.S., Hudson, B., Voter, B.R., Matthews, J.N.: Profile characteristics of fake Twitter accounts. Big Data Soc. **3**(2), 2053951716674236 (2016)
11. Chu, K.H., Majmundar, A., Allem, J.P., Soto, D.W., Cruz, T.B., Unger, J.B.: Tobacco use behaviors, attitudes, and demographic characteristics of tobacco opinion leaders and their followers: Twitter analysis. J. Med. Internet Res. **21**(6), e12676 (2019)
12. Agarwal, A., Toshniwal, D.: Face off: travel habits, road conditions and traffic city characteristics bared using Twitter. IEEE Access **7**, 66536–66552 (2019)

13. Kim, Y.H., Woo, H.J.: Exploring Spatiotemporal Characteristics of Twitter data Using Topic Modelling Techniques. Abstracts of the ICA, 1 (2019)
14. Jamison, A.M., Broniatowski, D.A., Quinn, S.C.: Malicious actors on Twitter: a guide for public health researchers. Am. J. Public Health **109**(5), 688–692 (2019)
15. Torres-Samuel, M., Vásquez, C., Viloria, A., Lis-Gutiérrez, J.P., Borrero, T.C., Varela, N.: Web visibility profiles of top100 Latin American universities. In: International Conference on Data Mining and Big Data, pp. 254–262. Springer, Cham, June 2018
16. Saeidi, M., Venerandi, A., Capra, L., Riedel, S.: Community Question Answering Platforms vs. Twitter for Predicting Characteristics of Urban Neighbourhoods. arXiv preprint arXiv: 1701.04653 (2017)
17. Silva, J., Varela, N., Ovallos-Gazabon, D., Palma, H.H., Cazallo-Antunez, A., Bilbao, O.R., Llinás, N.O., Lezama, O.B.P.: Data mining and social network analysis on Twitter. In: International Conference on Communication, Computing and Electronics Systems, pp. 401–408. Springer, Singapore (2020)
18. Silva, J., Naveda, A.S., Suarez, R.G., Palma, H.H., Núñez, W.N.: Method for collecting relevant topics from Twitter supported by big data. In: Journal of Physics: Conference Series, vol. 1432, no. 1, p. 012094. IOP Publishing, January 2020

The Use of Artificial Intelligence for Clinical Coding Automation: A Bibliometric Analysis

A. Ramalho[1,2](✉), J. Souza[2], and A. Freitas[1,2]

[1] MEDCIDS – Department of Community Medicine, Information and Health Decision Sciences, Faculty of Medicine, University of Porto, Porto, Portugal
andrelcramalho@gmail.com

[2] CINTESIS – Centre for Health Technology and Services Research, Porto, Portugal

Abstract. In hospital settings, all information concerning the patient's diseases and medical procedures are routinely registered in free-text format to be further abstracted and translated into standard clinical codes. The derived coded data is used for several purposes, from health care management and decision-making to billing and research. However, clinical coding is mostly performed manually, is very time-consuming, is inefficient and prone to error task. We conducted a bibliometric analysis of the scientific production on automated clinical coding using AI methods in the context of the International Classification of Diseases (ICD) classification system. The study aims to provide an overview and evolution of the research on automated clinical coding through Artificial Intelligence (AI) techniques. We did not consider time or language restrictions. Our analyses focused on characteristics of the retrieved literature, scientific collaboration and main research topics. A total of 1611 publications on automated coding during the last 46 years were retrieved, with significant growth occurring after 2009. The top 10 most productive publication sources are related to medical informatics, even though no common publication source or author was identified. The United States had by far the highest number of publications in this field. We found that natural language processing and machine learning were the main AI methodological areas explored for automated coding applications. Automated clinical coding using AI techniques is still rising and will undoubtedly face several challenges in the coming years. The results of the bibliometric analysis can assist the conduction of a more in-depth review to assess the variation of specific techniques and compare the performance of different methodological approaches regarding automated coding applications.

Keywords: Clinical coding · Artificial intelligence · International classification of diseases

1 Introduction

There is a growing interest in the reuse of coded clinical data for research, decision-making and health care quality improvement. However, direct analysis of coded clinical

Y. Dong et al. (Eds.): DCAI 2020, AISC 1237, pp. 274–283, 2021.
https://doi.org/10.1007/978-3-030-53036-5_30

data is highly dependent on how clinical coding is performed within hospital settings. This process consists in the translation of medical text containing information regarding the patients' diseases and procedures into a numeric (or alphanumeric) standardized code [1, 2]. Initially applied to mortality data, clinical coding has been used to classify health conditions, morbidity, hospital stays and medical procedures, providing a valuable starting line for clinical audit. Therefore, it has great utility for epidemiological planning, health actions and programs, measuring treatment effectiveness and for economic and process management of health care services.

The International Classification of Diseases (ICD) is a base document prepared by the World Health Organization (WHO) that directs the identification of health trends and statistics worldwide. Besides, it determines an international standard list of codes for reporting and classifying diagnoses of diseases, health conditions and procedures, for all clinical and research purposes [3]. Several countries have implemented and updated their ICD version for morbidity reporting. The ninth (ICD-9) and tenth (ICD-10) revisions, along with their respective clinical modifications (i.e. ICD-9-CM and ICD-10-CM in the United States) and other variations are currently the most adopted ICD versions, even though the implementation of these versions have not been able to keep up with the pace of updates given many limiting factors. ICD-10 was endorsed in May 1990, and this version is cited by over 20,000 scientific articles and is used by more than 100 countries around the world [3]. Despite this, there are still many countries and health services that have not implemented the tenth revision, mainly because of the difficulty to convert ICD-9 datasets into ICD-10 datasets automatically. ICD-9 has 6,969 codes, whereas ICD-10 present 12,420 (14,199 with four-character place codes in Chap. 20 (External Causes of Morbidity and Mortality). The eleventh revision (ICD-11) was already launched on June 18, 2018 and following its endorsement by the Forty-Second World Health Assembly in May 2019, WHO Member States could adopt ICD-11 from January 1, 2022.

The fact is that ICD coding is hard work for health services. As an adjunct in supporting clinical coding, artificial intelligence (AI) has contributed to several data treatment methodologies for automated code assignment, as the volume of medical information to be coded would be bulky enough to be a lost cause. Several coding support systems have been proposed across the literature [4–7]. However, even though AI applications are regarded as essential support in this matter, there is still no consensus on which method would be the most suitable to obtain a more accurate and up-to-date clinical coding.

In this article, we present a bibliometric analysis with the primary objective to identify the scientific production related to the automation of clinical coding in the context of the ICD classification system and to provide an overview and evolution of the research in this field. Furthermore, this study also aimed to gather information on techniques and approaches for the implementation of automated coding applications.

2 Methods

2.1 Bibliographic Database

The choice of the bibliographic database to be used in this study considered the broadest possible coverage regarding publications linked to AI methods for automated clinical coding. In this sense, we opted for a peer-reviewed database, Scopus, which has more than 22,800 titles and approximately 5,000 international publishers. Moreover, Scopus is a broader and more generalized database, comprising all research fields, and allows the use of practical and economic tools for data extraction and aggregation in several formats, providing more detailed information for the analyses.

2.2 Study Design

A bibliometric method was implemented to assess the conducted research on automated ICD coding using AI techniques from inception to February 21, 2020. No publication language restriction was considered.

A search expression was defined and calibrated [8] through test rounds for individual and combined terms from an initial set of studies. After the calibration, the most relevant search expression was performed, using **TITLE-ABS-KEY** filter, as follows:

> ("artificial intelligence" OR "machine learning" OR "deep learning" OR "natural language processing" OR "text mining" OR "text classification" OR "automated coding" OR "auto-coding" OR "algorithmic approaches" OR "automatic encoding" OR "automatic coding" OR "automatic assignment" OR "diagnosis code assignment" OR "clinical coding" OR "hospital information system") AND ("ICD-9*" OR "ICD-10*" OR "ICD10*" OR "ICD9*" OR "international classification of diseases")

2.3 Software and Data Analysis

We analyzed all included articles by index and author keywords, publication year, document by source, country, type, top-cited articles, authors and co-authorships. Index and author keywords were analyzed by frequency, and a ranking of publication sources was performed by relevance.

The total and the average number of citations were analyzed by country. Authorship analysis was carried out by computing the number of single-authored and multi-authored publications. Overall collaboration and the frequency of publications that were cited together, i.e., co-citations, were also analyzed using a software tool designed specifically for bibliometric analysis – VOSviewer [9]. Microsoft Excel software gave us support to data preparation and formatting in order to perform the analyses of the data file exported (i.e., comma-separated values' table) from Scopus.

3 Results

The search on Scopus was performed on February 21, 2020 and a total of 1611 documents were retrieved.

3.1 Keywords

The study of keywords is essential to give a more unobstructed view of the performance and assist us in calibrating the search expression. More general terms such as "human", "female" and "male", which presented high frequencies - 1129, 471 and 440 occurrences, respectively - were not considered for assessment. The top 15 most relevant index or author keywords and their frequencies were "international classification of diseases"(901), "clinical coding"(739), "coding"(648), "classification"(275), "natural language processing"(250), "electronic health records"(232), "icd-10"(225), "hospital information systems"(206), "machine learning"(179), "icd-9"(169), "hospitalization"(169), "algorithm"(167), "artificial intelligence"(141), "icd-9-cm"(123), "information processing"(105).

The co-occurrence of terms and words related to the study shows the relevance of the search expression used. The relationship analysis of the co-occurring keywords was performed using the VosViewer software tool [9], designed specifically for bibliometric analysis. The figure is a density map in which color represents clusters indicating the strength of occurrence and the lines connecting them show the relationships between them. Figure 1 shows the relationships between the terms of the highest appearance in the search expression. The highest co-occurrence appears between the terms: "coding", "clinical coding" and "international classification of diseases".

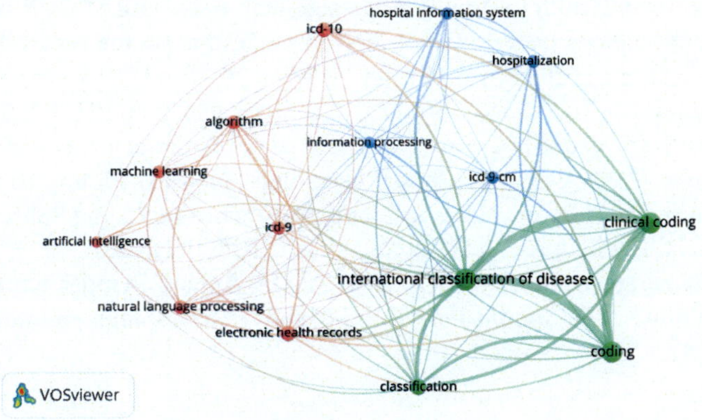

Fig. 1. Keyword co-occurrence.

3.2 Publication Year

Research on automated clinical coding begun in the early 1970 s, when the first publication came out [10]. The topic started to have a greater expression in terms of number of publications mainly after 2000, even though a considerable the number of publications were already observed in the mid-1990s, namely in 1995 ($n = 15$) and 1998 ($n = 12$), denoting the increasing concern and interest on automated coding systems shortly after the beginning of the utilization of the tenth revision (ICD-10) by WHO member states, in 1994 [7, 11–14]. Since 2009, publications gained even more attention from researchers, reaching exponential peaks in consecutive years (Fig. 2), with a record of publications in 2015 (n = 185). There were only four publications until 1990.

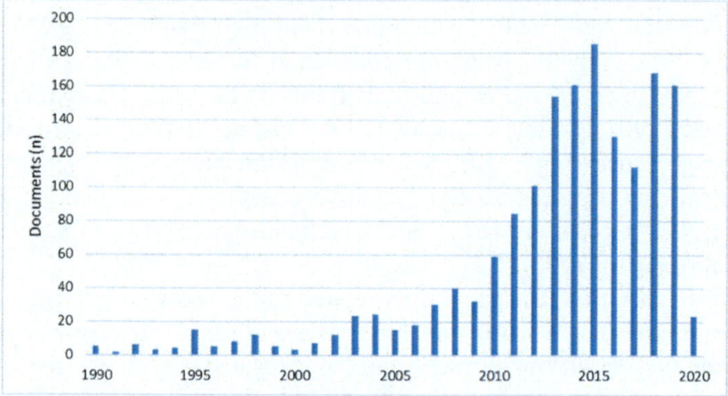

Fig. 2. Documents by publication year, 1990–2020

3.3 Documents by Source

The source with the highest number of publications in this research field was the "Journal of AHIMA American Health Information Management Association" (SJR impact factor 0.08), whose publications occurred after 2010. In fact, this source reached the highest number of publications in 2013 (n = 18) and, between 2010 and 2016, achieved a total of 76 indexed documents. Top 10 most productive sources also included "Studies in Health Technology and Informatics" (n = 61), "Journal of the American Medical Informatics Association" (n = 41), "CEUR Workshop Proceedings" (n = 33), "Journal of Biomedical Informatics" (n = 33), "Lecture Notes in Computer Science Including Subseries Lecture Notes in Artificial Intelligence and Lecture Notes in Bioinformatics" (n = 28), "Health Information Management Journal" (n = 27), "AMIA Annual Symposium Proceedings AMIA Symposium AMIA Symposium" (n = 26) and "Pharmacoepidemiology and Drug Safety" (n = 24).

3.4 Documents by Country

Notably, most of the publications were from the United States (n = 683, 59%), with the highest number of publications concentrated in the period 2013-2015, with a total of 232 documents. American publications in the three most frequent years strained to address various methodological aspects of ICD coding and the transition from ICD-9-CM to ICD-10-CM/PCS for medical records, which started in October 1, 2015 in the United States [15–18]. Australia, France and United Kingdom were the following countries with highest number of indexed documents, reflecting their movement on automated coding by means of AI [19–22]. Top 10 countries by number of publications are the United States (n = 683), Australia (n = 96), France (n = 90), United Kingdom (n = 87), Germany (n = 64), Canada (n = 52), Brazil (n = 47), Spain (n = 44), China (n = 43) and Italy (n = 40).

3.5 Documents by Type

The three most frequent types of documents in the results were the "articles" category (n = 1235), followed by "conference papers" (n = 196) and "reviews" (n = 86). Other publication categories included Notes (27), Editorials (22), Letters (22), Short Survey (16), Book Chapters (4), Conference Reviews (2) and Undefined (2). There were only three reviews related to the transition or automation of coding involving ICD-9, ICD-10 and their variations [6, 23, 24].

Conference papers had little expression until 2014 and, after this period, more papers were added to the indexed evidence, especially in 2018 and 2019 (n = 27 and n = 30, respectively), in events such as the "International Conference on Information and Knowledge Management", "World Congress on Medical and Health Informatics", "Conference on Artificial Intelligence in Medicine", among others. Figure 3 shows the evolution of the number of publications in the three main types of studies resulting from the search, highlighting the publication peaks for both articles and conference papers in 2015, and a low but constant frequency for reviews over time.

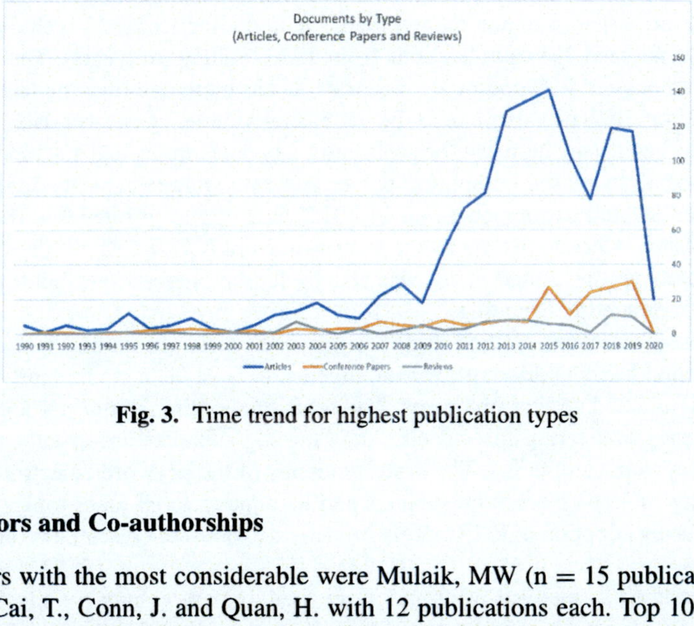

Fig. 3. Time trend for highest publication types

3.6 Authors and Co-authorships

The authors with the most considerable were Mulaik, MW (n = 15 publications), followed by Cai, T., Conn, J. and Quan, H. with 12 publications each. Top 10 authors in this field also included Chute, C.G, (10), Murphy, S.N. (10), Barta, A. (9), Liao, K.P. (9), Fung, K.W. (8) and Pathak, J. (8). In the co-authorship evaluation, the analysis was performed by the VosViewer software and considered a maximum number of 20 authors per document and a minimum number of 5 documents by author. The analysis resulted in three co-authorship interconnected clusters, and the authors with the highest straight link were Cai, T, Liao, K.P. and Murphy, S.N. (Fig. 4).

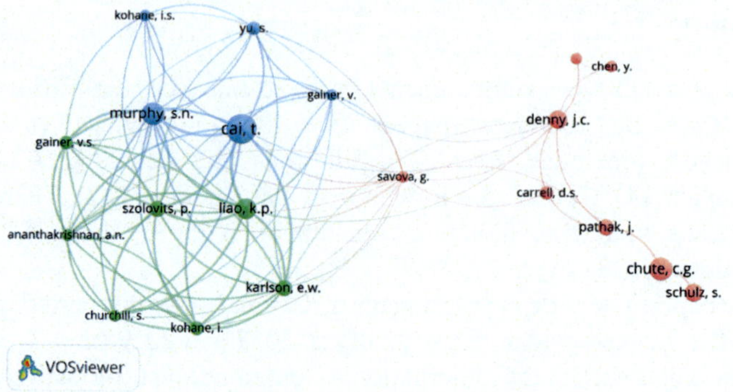

Fig. 4. Co-authorships by cluster of relationship (represented by colors). The size of circles is proportional to the number of documents.

4 Discussion

To the best of our knowledge, this study can be considered the first bibliometric analysis to map scientific research trends on automated ICD coding using AI techniques. The present study analyzed 1611 publications from 1973 to 2019 (46 years). The scientific literature on automated ICD coding by means of AI has increased over the last decades, particularly after 2009, possibly due to the increasing volume of clinical data generated by the growing health care industry, the proliferation of data mining and machine learning techniques and an increased computing power and data storage capacity [25]. A peak of the number of publications occurred in 2015, mostly due to increased research in the United States, which experienced the transition from ICD-9-CM to ICD-10-CM in that year. Moreover, the United States was also by far the biggest contributor to overall research on automated ICD coding.

The proposed research methodology identified the fields of natural language processing, machine learning and artificial intelligence to have been significantly used for developing automated coding processes. The topic "data mining" did not appear with a high frequency, indicating that specific data mining subareas are usually identified by authors' keywords and terms. The high frequency of the keyword "electronic health record" among the top 15 research subjects and its relation to all other topics suggests that the increasing adoption of EHRs might be critical to boost the research on automated coding.

Journals related to medical informatics are predominant among the most productive sources. However, we found that the top 10 most productive sources account for nearly 24% of the total number of publications and the number of publications by journal is relatively sparse, suggesting that there are no hegemonic sources in this field. Furthermore, empirical data show that there is no influential author in this field and no strong co-authorship association was observed for the author with the highest number of publications.

The time span and continuous grown of publications on automated clinical coding indicate that researchers have been trying to solve the several challenges related this task. The subjective nature of medical language itself is an important barrier to automated clinical coding, as free-text clinical notes often present a non-standard syntax, ambiguous abbreviations, negating expressions, spelling and grammatical errors and several synonyms for the same clinical concepts [22, 26–28]. Another critical barrier to develop automated coding applications has been the scarcity of hospitals using EHRs, which makes impossible the use of automated coding at hospitals that still use paper records and limits the amount of available training data [29]. An even more long-standing challenge is the dimension of existence of tens of thousands of possible medical labels, which present a frequency distribution that is highly imbalanced in most hospital datasets, especially for rare diseases, resulting in many absent labels in training data [22, 26].

Regarding the limitations of this study, we mention the restriction of our search to peer-reviewed publications. Also, the database was chosen based on the authors' access and knowledge and thereby other scientific databases that were not searched might have presented relevant publications within the scope of the bibliometric analysis. Despite the mentioned limitations, however, a bibliometric analysis of a large volume of publications and a summary of research topics can be a helpful proxy for the overall techniques and state of art in this field. Therefore, future studies may benefit from investigating how different techniques are being employed to solve specific automated coding problems in specific medical domains, select groups of methods and compare the impacts of different methodological approaches in the context of automated coding.

5 Conclusion

Considering the results and scope of this bibliometric analysis, researchers should be aware that the developments of automated clinical coding is still rising and will certainly face several challenges in the coming years. It is difficult to conclude whether one specific methodological area will prevail, but it seems that machine learning has been gaining importance in the studied field. This study provides a contextual outline which may be useful for conducting a more in-depth and refined systematic review of techniques for automated coding, in which specific AI methods could be described in more detail, including their performance and the medical fields in which they were applied. Based on these results, suggestions for future research can be made, namely adding more scientific databases in order to find more relevant and current publications, perform an in-depth assessment of the variation of specific techniques and medical fields over the years and compare different methodological domains, such as machine learning, natural language processing and deep learning, in order to discover which approaches have provided the best results so far and thus guide future research approaches to obtain a better performance of different automated coding applications.

Acknowledgments. The project "Clikode - Automatic Processing of Clinical Coding, (3I) Innovation, Research of AI models for hospital coding of Procedures and Diagnoses", POCI-05-5762-FSE-000230, is financed by Portugal 2020, through the European Social Fund, within the scope of COMPETE 2020 (Operational Programme Competitiveness and Internationalization of Portugal 2020).

References

1. Tatham, A.: The increasing importance of clinical coding. Br. J. Hosp. Med. **69**(7), 372–373 (2008)
2. Stanfill, M.H., Williams, M., Fenton, S.H., Jenders, R.A., Hersh, W.R.: A systematic literature review of automated clinical coding and classification systems. J. Am. Med. Inf. Assoc. **17**(6), 646–651 (2010)
3. WHO International Classification of Diseases (ICD) Information Sheet. https://www.who.int/classifications/icd/factsheet/en/. Accessed 22 Feb 2020
4. Ferrão, JC., Oliveira, MD., Janela, F., Martins, HMG.: Clinical coding support based on structured data stored in electronic health records. In: 2012 IEEE International Conference on Bioinformatics and Biomedicine Workshops, pp. 790–797 (2012)
5. Chen, D., Zhang, R., Qiu, R.G.: Leveraging semantics in wordnet to facilitate the computer-assisted coding of ICD-11. IEEE J. Biomed. Health Inf. **24**(5), 1469–1476 (2019)
6. Chute, C.G., Cohn, S.P., Campbell, K.E., et al.: The content coverage of clinical classifications for the computer-based patient record institute's work group on codes & structures. J. Am. Med. Inf. Assoc. **3**, 224–233 (1996)
7. Lussier YA., Shagina L., Friedman C.: Automating ICD-9-CM encoding using medical language processing: a feasibility study. J. Am. Med. Inf. Assoc. 1072e2 (2000)
8. Hutson, M.: AI glossary: artificial intelligence, in so many words. Science **357**(6346), 19 (2017)
9. van Eck, N.J., Waltman, L.: Software survey: VOSviewer, a computer program for bibliometric mapping. Scientometrics **84**(2), 523–538 (2010)
10. Mau, G.: Problems of a pediatric diagnostic key as part of a hospital documentation system. Klin. Padiatr. **185**(5), 400–402 (1973)
11. Chaux, R., Treussier, I., Audeh, B., Pereira, S., Hengoat, T., Paviot, B.T., Bousquet, C.: Automated control of codes accuracy in case-mix databases by evaluating coherence with available information in the electronic health record. Stud. Health Technol. Inf. **264**, 551–555 (2019)
12. Li, M., Fei, Z., Zeng, M., Wu, F.X., Li, Y., Pan, Y., Wang, J.: Automated ICD-9 coding via a deep learning approach. IEEE/ACM Trans. Comput. Biol. Bioinf. **16**(4), 1193–1202 (2019)
13. Becker, B.F.H., et al.: ADVANCE consortium.: CodeMapper: semiautomatic coding of case definitions. A contribution from the ADVANCE project. Pharmacoepidemiol. Drug Saf. **26**(8), 998–1005 (2017)
14. McMaster, C., Liew, D., Keith, C., Aminian, P., Frauman, A.: A machine-learning algorithm to optimise automated adverse drug reaction detection from clinical coding. Drug Saf. **42**(6), 721–725 (2019)
15. Rubenstein, J.N.: How will the transition to ICD-10 affect urology coding? An analysis of ICD-9 code use from a large group practice. Urol. Pract. **2**(6), 312–316 (2015)
16. Fant, C., Theiss, M.A.: Transitioning to ICD-10. Nurse Pract. **40**(10), 22–31 (2015)
17. Outland, B., Newman, M.M., William, M.J.: Health policy basics: implementation of the international classification of disease, 10th revision. Ann. Intern. Med. **163**(7), 554 (2015)
18. Kavuluru, R., Rios, A., Lu, Y.: An empirical evaluation of supervised learning approaches in assigning diagnosis codes to electronic medical records. Artif. Intell. Med. **65**(2), 155–166 (2015)
19. Nguyen, AN., Truran, D., Kemp, M., et al.: Computer-assisted diagnostic coding: effectiveness of an NLP-based approach using SNOMED CT to ICD-10 mappings. In: AMIA Annual Symposium Proceedings, vol. 2018, pp. 807–816 (2018)
20. Kaur, R., Ginige, J.A.: Comparative analysis of algorithmic approaches for auto-coding with ICD-10-AM and ACHI. Stud. Health Technol. Inf. **252**, 73–79 (2018)

21. Tsopra, R., Peckham, D., Beirne, P., Rodger, K., Callister, M., White, H., et al.: The impact of three discharge coding methods on the accuracy of diagnostic coding and hospital reimbursement for inpatient medical care. Int. J. Med. Inf. **115**, 35–42 (2018)
22. Catling, F., Spithourakis, G.P., Riedel, S.: Towards automated clinical coding. Int. J. Med. Inf. **120**, 50–61 (2018)
23. Soo, I.H.-Y., Lam, M.K., Rust, J., Madden, R.: Do we have enough information? how ICD-10-am activity codes measure UP. Health Inf. Manage. J. **38**(1), 22–34 (2009)
24. American Health Information Management Association: Destination 10: healthcare organization preparation for ICD-10-CM and ICD-10-PCS. J AHIMA, vol. 75, pp.56A–D (2004)
25. Islam, M.S., Hasan, M.M., Wang, X., Germack, H.D., Noor-E-Alam, M.: A systematic review on healthcare analytics: application and theoretical perspective of data mining. Healthc. (Basel Switz.) **6**(2), 54 (2018)
26. Brasil, S., Pascoal, C., Francisco, R., Dos Reis Ferreira, V., Videira, P.A., Valadão, A.G.: Artificial Intelligence (AI) in rare diseases: is the future brighter? Genes (Basel) **10**(12), 978 (2019)
27. Campbell, S., Giadresco, K.: Computer-assisted clinical coding: a narrative review of the literature on its benefits, limitations, implementation and impact on clinical coding professionals. Health Inf. Manage. **49**(1), 5–18 (2020)
28. Alonso, V., Santos, J.V., Pinto, M., Ferreira, J., Lema, I., Lopes, F., Freitas, A.: Problems and barriers during the process of clinical coding: a focus group study of coders' perceptions. J. Med. Syst. **44**(3), 62 (2020)
29. Rinkle, V.A.: Computer assisted coding - a strong ally, not a miracle aid. J. Health Care Compliance **17**(1), 55–58–67–68 (2015)

A Feature Based Approach on Behavior Analysis of the Users on Twitter: A Case Study of AusOpen Tennis Championship

Niloufar Shoeibi[1](✉), Alberto Martín Mateos[1], Alberto Rivas Camacho[1], and Juan M. Corchado[1,2]

[1] Bisite Research Group, University of Salamanca, Salamanca, Spain
{Niloufar.shoeibi,alberto_martin}@usal.es
[2] Air Institute, IoT Digital Innovation Hub, Salamanca, Spain

Abstract. Due to the advancement of technology, and the promotion of smartphones, using social media got more and more popular. Nowadays, it has become an undeniable part of people's lives. So, they will create a flow of information by the content they share every single moment. Analyzing this information helps us to have a better understanding of users, their needs, their tendencies and classify them into different groups based on their behavior. These behaviors are various and due to some extracted features, it is possible to categorize the users into different categories. In this paper, we are going to focus on Twitter users and the AusOpen Tennis championship event as a case study. We define the attributions describing each class and then extract data and identify features that are more correlated to each type of user and then label user type based on the reasoning model. The results contain 4 groups of users; Verified accounts, Influencers, Regular profiles, and Fake profiles.

Keywords: Social media analytics · Behavior analysis · User behavior mining · Feature extraction · Twitter · Verified · Influencers · Regular and fakes

1 Introduction to Social Media

Social networks have consolidated as a source of communication and transmission of the information at a global level over the last few years. An infinite number of topics can be dealt with, so there is a huge amount of information spreading by users.

In some cases, the objective may be to detect certain profiles, for example, with behaviors that induce unethical thinking or activities, such as sexist ideologies [1]. Other times, the aim may be to detect relevant or current issues in real-time [2]. Because of these mentioned reasons and many more, the importance of behavior analysis on social networks is undeniable.

Y. Dong et al. (Eds.): DCAI 2020, AISC 1237, pp. 284–294, 2021.
https://doi.org/10.1007/978-3-030-53036-5_31

1.1 Behavior Analysis

Each creature has a behavior towards the environment and others. By analysis of the behavior, it is possible to discover the patterns of thinking in different situations. However, social media can be considered as an environment for human beings so they can express themselves through their interactions. The analysis of the data extracted from the user's behavior is an important block in this type of case study knowing well the environment in which the events occur to extract information and knowledge [3]. Besides, different studies try to identify possible types of profiles in the networks: bots, fakes, and so. [4] is an example, but since there is no robust concept about what characteristics each profile has, it is complicated to reach a unique solution.

1.2 Data Extraction on Social Media

Each social network has different casuistry, and these different types of profiles that have been discussed behave differently. Thus, it is important to make a good analysis; for instance, finding similar users, topics, or arguments in different social networks [4]. The challenge is unstructured data extracted from different platforms [5]. Therefore, data mining techniques will play a decisive role.

In this paper, we are going to focus on Twitter, which is a platform that allows two-way communication in which any user can interact with another quickly and easily, so we don't have the problem of different sources. Besides, messages can be spread through "tweets", related to any daily aspect and in which only users are identified as verified or not verified. To go one step further, each account must be analyzed in more detail. Therefore, in this research, we are going to detect different types of profiles according to their behavior on the Twitter social network, with the hope that in the future, we will be able to address this study in a more detailed way and detect possible profiles with different purposes.

This paper organized as follows: Sect. 2, related work. Section 3, pertains to data and proposed method which describes the data extraction and feature selection and the technique that has been used for labelling the users. In Sect. 4, the results will be more explained. In Sect. 5, the conclusion of the results and future work has been presented. And finally, Sect. 6 is indicated to the references.

2 Related Work

Most of the research in this area is related to taking advantage of social network data, either through machine learning algorithms (supervised and unsupervised), deep learning, graph theory, etc.

Rashidi et al. studied opportunities and challenges for exploring the capacity of modelling travel behavior. They used the data extracted from social networks to obtain information based on features like trips, their purpose, mode of transport, duration, etc. through surveys. However, the processing time is very slow [6].

Large organizations try to influence choices in social networks, which is going to cause a lack of freedom of expression. In this article, Subrahmanian et al. developed an algorithm for detecting bots based on tweet parameters, profiles, and environment [7].

Erşahin et al. create a twitter fake account detection based on supervised discretization techniques by the reason for the increase of the exposure of incorrect information through fakes profiles has increased [8].

A method that tries to group different profiles according to influential words with the complexity of semantics was developed by Sundararaman et al. [9].

One study has the main objective to characterize the behavior of cancer patients. In this article, Crannell et al. match different types of cancer-patient with several sentiments [10].

NLP-based word embedding grouping method for publishing health surveillance was published by Dai et al. This method is tested versus other bags of words methods [11]. Lastly, Kaneko et al. presented a method based on using keyword burst and image clustering instead of only text analysis for event photo mining [12].

3 Data Extraction from Twitter API

The analysis of the data is an important block in this type of case study, knowing well the environment in which the events occur to extract information and knowledge.

Tables 1 and 2 show the list of variables considered from the tweets of each of the users and about the information of their accounts.

Table 1. Features obtained about the tweets of each Twitter profile.

Features (tweets)	Definition
Text	Tweet text
Favorites	Number of favorites that have a tweet
Retweet	Number of retweets that have a tweet
Created at	Date of the publication of a tweet

Table 2. Features obtained about each Twitter profile.

Features (user profile)	Definition
Name	Name the user, as they have defined it
Screen name	Name of the twitter account, that it is unique
Listed count	Number of public lists that a user is a member
Biography profile	Biography profile text
Followings	Number of followings that have an account
Followers	Number of followers that have an account
Favourites count	Number of favorite tweets that have an account
Statuses count	Number of tweets (RT + own tweets) that have an account
Creation_at	Date of the creation of an account

From all the default variables allowed by the twitter API (Tables 1 and 2) and the possible characteristics that can identify different types of user profiles, the variables of Table 3 have been added to the study and, consequently, those of Table 4 that are directly related.

Table 3. New variables from the data analysis (metadata).

New features	Definition
Number of own Tweets	Number of tweets published by a user in the last week
Number of Retweets	Number of tweets retweeted by a user in the last week
Number of own Tweets	Number of tweets (published and retweeted) by a user in the last week
Favorited Tweets count	Number of favorites that have a tweet published
Retweet & Tweets count	Number of retweets that have a tweet published
Mentions	Number of mentions inside tweets published by a user in the last week
Tweets URL	Number of tweets with a URL in it text inside tweets published by a user in the last week
Time between Tweets	Minutes between tweet publications
Twitter years	Number of years that have an account

Table 4. Rates from new variables (metadata).

Rates	Definition
Tweets per Retweets rate	The ratio of tweets published per tweets retweeted in the last week
Tweets year ratio	The ratio of tweets (published and retweeted) per year
Time between Tweets	Minutes between tweets (mean) in the last week
Followings per Follower	The ratio of followings per followers' rate

4 Feature Extraction Using Graph Theory and Analysis

In this paper, we proposed analyzing the content of the tweet to find the *mentioned users* or the *retweet* as a means of defining the relationship between the users. After extracting 5000 number of tweets about our case study subject, "AusOpen", which is the Australian tennis championship, we explore the content, and if the tweet is a *retweet*, our proposed method extracts the screen names of the profiles whose tweets have been retweeted by others. If not, we will extract the profiles that have been tagged in the tweet. Then define the graph of relations from the source, which is the user who is doing retweets to the target which his/her content has been retweeted. After building the graph, we do the graph analysis and measure new features extracted from the graph and defined relationship. These features are represented in Table 5.

Table 5. Features Extracted from Graph Analysis [13, 14].

Rates	Definition
Eccentricity	The maximum shortest distance of one node from others. The less Eccentricity, the more influencing the power of the node
Clustering Coefficient Centrality	Which nodes in a network are tending to be in the same cluster based on the degree of the nodes. $cc = \frac{n}{t}$
Closeness Centrality	Indicates how close a node is to other nodes in a network by capturing the average distance based on one vertex to another. $cl = \frac{1}{\sum_{v \neq u} d(u, v)}$
Betweenness Centrality	Shows how influential is the node. The more the value of betweenness centrality is, the more important that node would be to the shortest paths through the network. So, if that node is removed, so many connections will be lost. $b = \sum_{s \neq v \neq t} \frac{\delta_{st}(u)}{\delta_{st}}$
Harmonic Closeness Centrality	This measure is so similar to closeness centrality, but it can be used in networks that are not connected to each other. It means that, when two nodes are not connected, the distance will be infinity, and Harmonic Closeness is able to handle infinity just by replacing the average distance between the nodes with the harmonic mean
In-Degree Centrality	This centrality indicates the importance via the number of edges entering the node
Out-Degree Centrality	This centrality indicates the importance via the number of edges going out of the node
Degree Centrality	This measures how many connections a node has. In other words, it's the summation of the In-Degree and Out-Degree of the node and shows how important a node is, by the number of connections. $Deg(v) = InDeg(v) + OutDeg(v)$

n: No of connection between neighbors of a particular node
t: Total number of possible connections among all the neighbors of the node
d(u, v): the geodesic distance between u, v.
s: source, t: destination
st: number of shortest paths between (s, t)
st(u): number of shortest paths between (s, t) that pass-through u.

5 Different User Groups Base on Behaviors

Based on the behavior of users, we have defined four different categories; "Verified Accounts", "Influencer Users", "Regular Users" and "Fake Accounts".

- **"Verified Accounts"**, which are the ones that have been verified by Twitter, and they have the verified blue mark beside their names on twitter panel. We call these users "Verified" people who are politicians, famous artists, TV shows, special sports events, etc. It is possible to detect this class by checking the "Verified" term in the JSON file extracted from the tweet object.
- **"Influencer Users"** are the users that have a great influence on other users, and others believe in the content they are sharing feelings, thoughts, or expressions. However, they are not verified by twitter. It isn't very easy to detect this category of users because they are regular users, but sometimes, they have the same behavior as the famous accounts, like verified accounts. To detect this category, in addition to the features appearing in the JSON file extracted from Twitter API, we need to go deeper and analyze the behavior using graph analysis, extracting more complicated features and analyze the content. Betweenness Centrality, In-Degree Centrality, and many more. An Influencer account should have too many retweets, so we can say that In-Degree is a feature that defines how important that specific node is. Then more value of in-Degree shows the more times the tweet of that profile has been retweeted. And after applying this filter, so many of the nodes have been dropped because they didn't have retweets.
- **"Regular Users"** are the ones that express their thoughts, opinion, and feelings, and they are not aiming to make an influence on other people's thoughts and opinions. They are not having too many followers and followings, and their interaction ratio is not high.
- **"Fake Accounts"** which are the accounts that usually have an irregular behavior with their content, like fake news, spam, incoherent tweets, etc. So, they usually aim to put false influence or to change statistics in society. This type of profile can be defined with several different features, and it is hard to tell them apart from regular profiles and make a strong definition of it. For example, some of them can retweet so much, others can have a default image profiles, and others can have both features. From this study, we created new variables that can help as a good filter to identify these profiles, like the number of tweets (published and retweeted) per day, based on the features that appear in the JSON file [15, 16].

In the Table 6, we can see a summary of the characteristics of each of the profiles:

Table 6. Different categories of users based on their behavior.

Profile type	Characteristics	Related features
Verified	- Celebrities - Famous Sport Players - Politics	- *Verified* = True
Influencers	People who are not verified by twitter but their content influence other thoughts	- A high number of their tweets - A high number of followers - The low time between tweets - The high number of interactions with their own tweets - High in-degree centrality
Regulars	People who are no verified by twitter, publish a few contents, favorites, with a balanced number of followers and followings, not in large numbers	
Fakes	People who are not verified by twitter, but their contents are fake news, spam, incoherent tweets, etc.	- The high number of Retweets - The low time between tweets - Default image profile - No biography - Numbers on its account name - The small number of followers - 2001 followings - Tweets duplicated - Self-Loop - High outdegree number - Low indegree number

6 Results

With the application of several filters based on the behavior analysis, we can label the data. We apply two different filters to our dataset based on features from the graph analysis and then, from the metadata. Graph analysis is leading us to find the filters

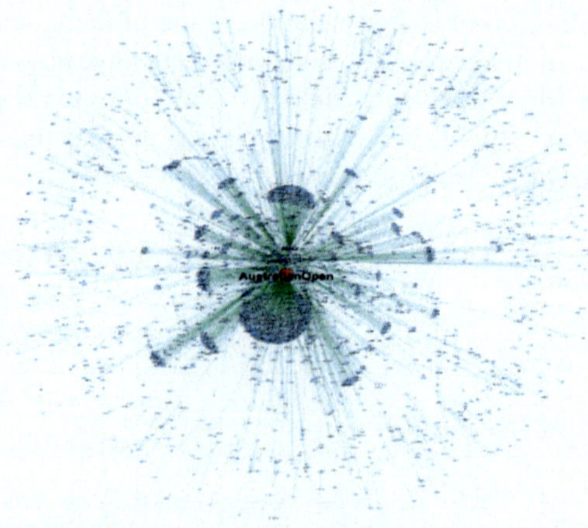

Fig. 1. The graph of relationship of people who tagged each other or retweeted other's tweets in the concept of AusOpen Tennis Championship

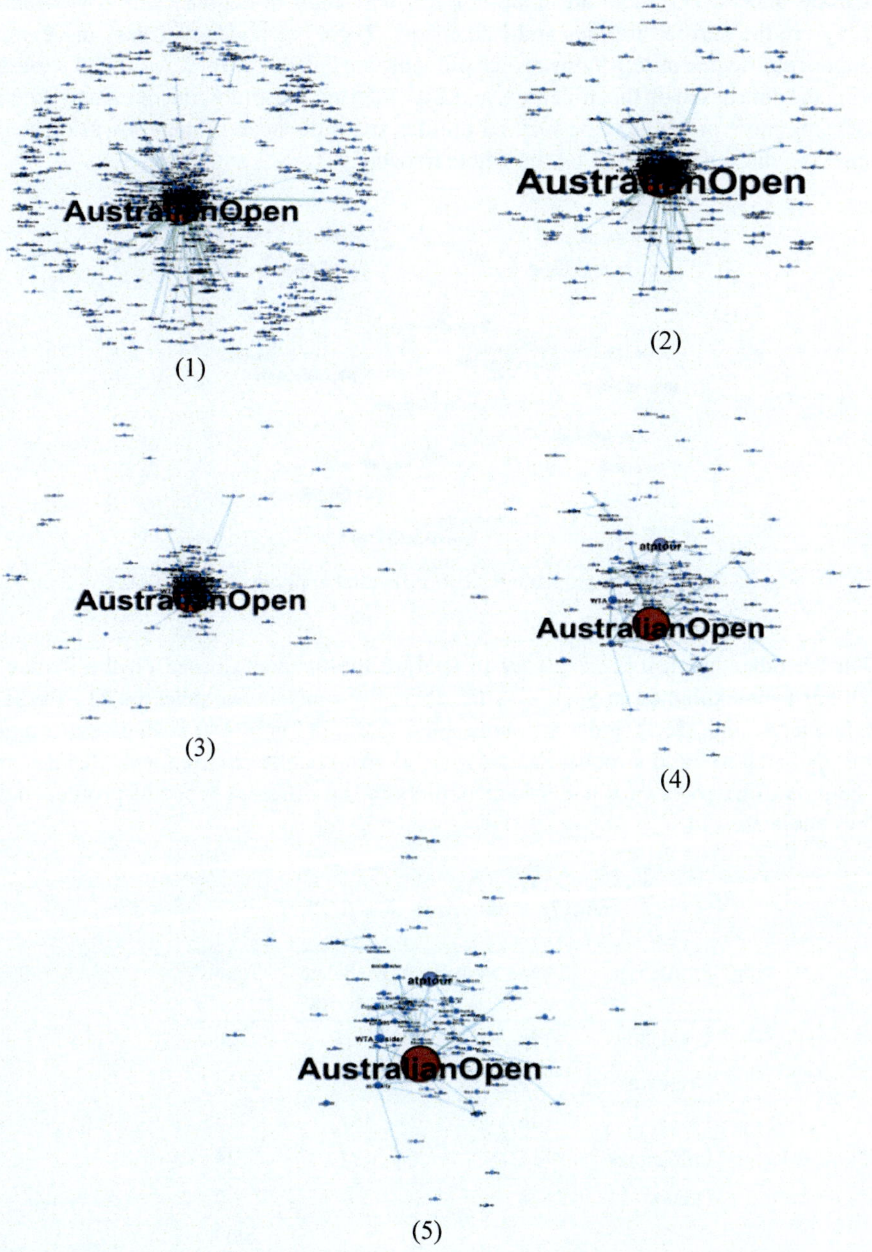

Fig. 2. Graph revolution of finding *Influencers* and *Verified Profiles* after applying in-degree centrality filters in order to remove *Fakes* and *Regulars*.

helping us to have more accurate labeling. Figure 1 is representing the graph of the relationship between all the users who tweeted about the AusOpen tennis championship. As can be understood, there are nodes (users) with high in-degree centrality, which possibly are the verified profiles and influencers. There are nodes with less in-degree and higher out-degree or self-loops, which probably are Fakes. In Fig. 2. we demonstrated filtering the data based on the in-degree, and after filtering five times, the remained group of users are more probable to be verified profiles and influencers. Figure 3 was created by applying the self-loop filter on the whole network.

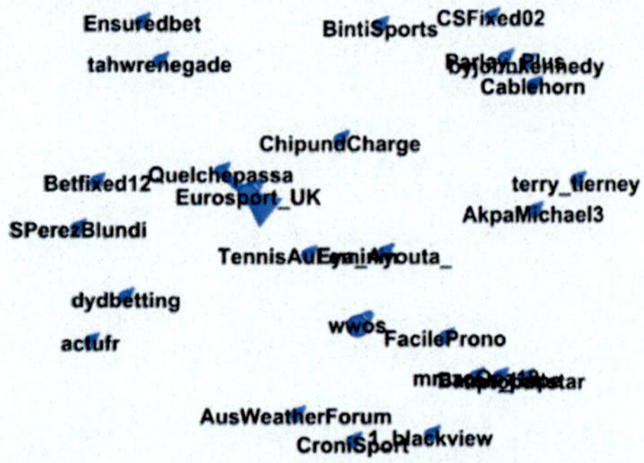

Fig. 3. The group of probable Fake Profiles after applying self-loop filter

On the other hand, it is necessary to analyze the metadata based on the feature, which has been explained in Sect. 3. After doing the analysis, we detected 415 Fakes, 49 Influencers, and 2266 regular accounts, such shows in Table 7. It is an essential and useful approach to build a robust dataset in an almost chaotic environment. Besides, it can help us to discover new patterns or outliers in the different types of profiles and various subjects.

Table 7. Profile types classification

Profile type	Graph analysis	Metadata analysis	Total
Verified	179	0	179
Regulars	9	2257	2266
Fakes	26	389	415
Influencers	17	32	49
Others	2678	0	0
			2909

7 Conclusion and Future Work

In this paper, we defined the graph of relationship based on the interaction of users called "retweet". This relationship, which is from the person (source) who retweeted another person's tweet (target), has been demonstrated in a directed graph in which the nodes are users. The edges are which are arrows from the source to target, showing retweets. By using graph analysis, we generated more features then added them to the original ones that have been extracted from Twitter and each profile of the user. The results show that adding these new features helped the behavior analysis of the users.

In the future, we are looking forward to trying other social media like Facebook, Instagram, and so many more and making the various machine learning models for training them with the dataset generated by the method this paper proposed in this paper and then measure the type of the account.

References

1. Waseem, Z., Hovy, D.: Hateful symbols or hateful people? Predictive features for hate speech detection on twitter. In: Proceedings of the NAACL Student Research Workshop, pp. 88–93, June 2016
2. Xie, W., Zhu, F., Jiang, J., Lim, E.P., Wang, K.: TopicSketch: real-time bursty topic detection from twitter. IEEE Trans. Knowl. Data Eng. **28**(8), 2216–2229 (2016)
3. Oh, C., Roumani, Y., Nwankpa, J.K., Hu, H.F.: Beyond likes and tweets: consumer engagement behavior and movie box office in social media. Inf. Manag. **54**(1), 25–37 (2017)
4. El, A., Azab, A.M.I., Mahmoud, M.A., Hefny, H.: Fake account detection in twitter based on minimum weighted feature set. World Acad. Sci. Eng. Technol. Int. J. Comput. Inf. Eng. **10**(1) (2016)
5. Injadat, M., Salo, F., Nassif, A.B.: Data mining techniques in social media: a survey. Neurocomputing **214**, 654–670 (2016)
6. Rashidi, T.H., Abbasi, A., Maghrebi, M., Hasan, S., Waller, T.S.: Exploring the capacity of social media data for modelling travel behaviour: opportunities and challenges. Transp. Res. Part C: Emerg. Technol. **75**, 197–211 (2017)
7. Subrahmanian, V.S., Azaria, A., Durst, S., Kagan, V., Galstyan, A., Lerman, K., Menczer, F.: The DARPA Twitter bot challenge. Computer **49**(6), 38–46 (2016)
8. Erşahin, B., Aktaş, Ö., Kılınç, D., Akyol, C.: Twitter fake account detection. In: 2017 International Conference on Computer Science and Engineering (UBMK), pp. 388–392. IEEE, October 2017
9. Sundararaman, D., Srinivasan, S.: Twigraph: discovering and visualizing influential words between Twitter profiles. In: International Conference on Social Informatics, pp. 329–346. Springer, Cham, September 2017
10. Crannell, W.C., Clark, E., Jones, C., James, T.A., Moore, J.: A pattern-matched Twitter analysis of US cancer-patient sentiments. J. Surg. Res. **206**(2), 536–542 (2016)
11. Dai, X., Bikdash, M., Meyer, B.: From social media to public health surveillance: word embedding based clustering method for twitter classification. In: SoutheastCon 2017, pp. 1–7. IEEE, March 2017
12. Kaneko, T., Yanai, K.: Event photo mining from twitter using keyword bursts and image clustering. Neurocomputing **172**, 143–158 (2016)
13. Perez, C., Germon, R.: Graph creation and analysis for linking actors: application to social data. In: Automating Open Source Intelligence, pp. 103–129. Syngress (2016)

14. Deverashetti, M., Pradhan, S.K.: Identification of topologies by using harmonic centrality in huge social networks. In: 2018 3rd International Conference on Communication and Electronics Systems (ICCES), pp. 443–448. IEEE, October 2018
15. Bovet, A., Makse, H.A.: Influence of fake news on Twitter during the 2016 US presidential election. Nat. Commun. **10**(1), 1–14 (2019)
16. Gurajala, S., White, J.S., Hudson, B., Matthews, J.N.: Fake Twitter accounts: profile characteristics obtained using an activity-based pattern detection approach. In: Proceedings of the 2015 International Conference on Social Media & Society, pp. 1–7, July 2015

S-COGIT: A Natural Language Processing Tool for Linguistic Analysis of the Social Interaction Between Individuals with Attention-Deficit Disorder

Jairo I. Vélez[1,2,3](✉), Luis Fernando Castillo[2,3](✉), and Manuel González Bedia[2,3](✉)

[1] University of Caldas, Manizales, Colombia
{jairo.velez,luis.castillo}@ucaldas.edu.co
[2] University of Zaragoza, Zaragoza, Spain
{790853,mgbedia}@unizar.es
[3] National University of Colombia, Manizales, Colombia
lfcastilloos@unal.edu.co

Abstract. This paper describes the design and implementation of a computer platform aimed at monitoring social interaction in subjects with the attention-deficit disorder as a special cognitive condition. Applying Natural Language Processing (NLP) algorithms it is intended to support the monitoring and intervention of individuals with special needs through language analysis such as attention-deficit disorder.

This tool allows the interaction of people through a web platform that can be accessed directly from a terminal computer or a mobile client for exchanging text messages as well as the loading of text where not necessarily corresponds to a conversation but a text written by an individual. The analysis of the texts is carried out through the integration of different algorithms for the application of mathematical techniques, indexes and models related to the processing, obtaining and visualization of the information.

Keywords: Natural language processing · Attention-deficit disorder · Text analytics

1 Introduction

The use of technology and computer science for the collection and analysis of data referring to different areas of knowledge has been seen as an important area of study in recent years with a large number of contributions [8]. The application of different techniques and methodologies has helped to contribute significantly in the approach to various pathologies for their diagnosis and as a means of encouraging different skills that although they have been taken as deficiencies in the development of the individual [3], it has been observed that those individuals may be hidden capacities that require various means for their exploitation. However, in studies such as [13] results were obtained

Y. Dong et al. (Eds.): DCAI 2020, AISC 1237, pp. 295–302, 2021.
https://doi.org/10.1007/978-3-030-53036-5_32

based on statistical calculations using averaged measures such as the mean or standard deviation, immediately eliminating the information obtained for each trial and obviating the correlation between the information of different interactions. These averaged measures lead to dispensing with data on variability and temporality that can be of great importance for daily monitoring in the development of the individual.

Different studies have demonstrated the relationship between different mental pathologies and the development of language, these two fields affecting each other in the individual when one of the difficulties is present [25]. Various tools have been implemented to assist patients under some mental condition with the requirements of their communicative needs as in the case of Augmented and Collaborative Communications (ACC) [4, 9], which correspond to generally pictographic systems through which the individual can make their needs known.

According to the above, we start from a baseline established by [2] to implement a tool that allows to the user to know several things about an individual with Attention Deficit Disorder by processing a text that can be extracted from several means like message exchange between two individuals or from a text document which contains the transcription of an interaction with an individual that presents this condition. Such processing is made through the use of natural language algorithms that provides, among others, to identify subjective predicates, the use of expressive structures as well as parenthetic, confirmational particles and to register the frequency of factual verbs Vs non-factual verbs.

2 Theoretical Review

There are varied concepts and theoretical foundations that support the development of this research based on technologies, computer science and the cognitive study of human beings, specifically those with special conditions and needs. The integration of these components allows the neuropsychological and cognitive study to be approached from other perspectives; as these are commonly based on traditional methods of test and observation, agglomerating the results under averaged measures that influence the reading of the conditions of the treated individual. Different conditions can be analyzed through language, such as autism, dyslexia, schizophrenia, syndromes such as Williams or Rett, aphasia, among others. Although there are numerous methodologies for natural language processing, techniques for Information Retrieval and Search have been adopted as they provide measures of interest to assess the general development of the patient over time, in addition to allowing to identify patterns in linguistic constructions and compare these regularities with elements of different corpora being able to reflect in this way patterns between different individuals.

According to [5] Text corpora have been manually annotated with such data structures to compare the performance of various systems. The availability of standard benchmarks has stimulated research in Natural Language Processing (NLP) and effective systems have been designed for all these tasks. Such systems are often viewed as software components for constructing real-world NLP solutions. As mentioned in [19] four standard NLP tasks: POS, CHUNK, NER, and SRL are listed below:

Part-Of-Speech Tagging (POS)
POS aims at labeling each word with a unique tag that indicates its syntactic role, for example, plural noun, adverb, etc.

CHUNK
Chunking aims at labeling segments of a sentence with syntactic constituents such as a noun or verb phrases (NP or VP). Each word is assigned only one unique tag, often encoded as a begin-chunk (e.g., B-NP) or inside-chunk tag (e.g., I-NP).

Named Entity Recognition (NER)
NER labels atomic elements in the sentence into categories such as "PERSON" or "LOCATION". As in the chunking task, each word is assigned a tag prefixed by an indicator of the beginning or the inside of an entity.

Semantic Role Labeling (SRL)
SRL aims at giving a semantic role to a syntactic constituent of a sentence.

To apply natural language processing is necessary to have a source from which we can extract text, that source is a corpus and to facilitate the NLP processing we can use a programming language like Python and a specific library called NLTK.

Language Processing Tools
Corpora is a collection of written or spoken natural language material, stored on a computer, and used to find out how language is used. So more precisely, a corpus is a systematic computerized collection of authentic language that is used for linguistic analysis as well as corpus analysis. If you have more than one corpus, it is called corpora [21], Python is an interpreted programming language whose philosophy emphasizes the readability of its code. It is a multiparadigm programming language since it supports object-orientation, imperative programming and, to a lesser extent, functional programming. It is an interpreted, dynamic and multiplatform language [22], and the way of processing a corpus employing Python we use the Natural Language Toolkit (NLTK) which is a library that has four types of corpora: isolated corpus, categorized corpus, overlapping corpus, and temporal corpus.

Computer science has then contributed important ideas about representation (e.g. data structures) and the processes that operate in these structures, so the idea that the processing of human information as a sequence of computational processes operating in mental representations is maintained as the basis of modern cognitive psychology. A third source that influenced this movement was the study of linguistics and generative grammar [24] because as such the rules governing sentences and sentences are mentalistic hypotheses about the cognitive processes responsible for the verbal behaviors that are commonly observed [18]. Finally, some of the most popular document representations and similarity measures used in the literature to enable certain systems to leverage knowledge sources in a textual format such as word count vector, term frequency-inverse document frequency, latent semantic indexing, latent dirchlet allocation, and random projection in which its use has progressively gained popularity in this context and it is currently present in some of the most popular toolboxes for text-topic modeling, and it is often used to approximate TF-IDF distances between documents in a compressed representation [14].

3 Materials and Methods

3.1 System Architecture

As has been said, our implementation starts from a baseline in which there were elicited some of the requirements, developed - in java using a Model-View-Controller pattern - several routines to analyze text corpora and using a plotting library to visualize the results of those routines.

Our proposed architecture, as can be seen in Fig. 1, has several layers, according to the architectural pattern provided by the Django framework that corresponds to the Model View Template controller (MVT) which proposes a layer for the data model, another for the views, another for templates and one more for persistence services whenever models are migrated to an Object Relational Model (ORM). Thus, in the Templates layer there is a base template from which the other artifacts for data acquisition, transformation and analysis inherit; These GUIs are manifested through the views of the Apps package in which the different control modules rest on the flow between the data server in which the artifacts related to the storage of the data mainly rest (text of the interactions) and the messaging service that is managed by the XMPP server (Openfire) that, eventually, can be accessed through a mobile device if the acquisition of data is done through a chat app.

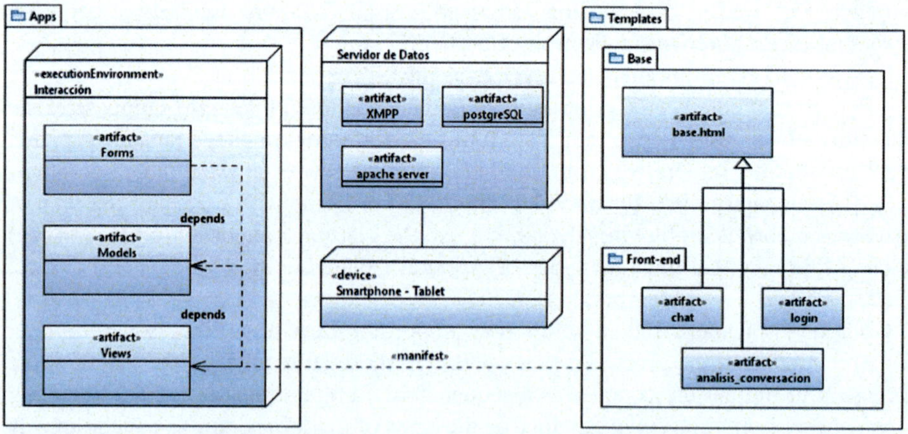

Fig. 1. The software architecture of S-COGIT

3.2 NLP Processing Techniques

The main techniques used to process text are related to accessing text corpora, processing raw text to use regular expressions for detecting word patterns, normalizing, tokenizing and segmenting text. Once the text has been normalized, we classify it employing decision trees, Naive Bayes, and maximum entropy classifiers to model linguistic patterns and build feature-based grammars to finally analyze the meaning of sentences.

Finally, when the text has been processed, the user can visualize information related to it.

Following the architecture shown previously, the S-COGIT components diagram (Fig. 2) shows the component diagram detailing, at a lower level of abstraction, the process carried out on the data obtained, which is extracted from a persistent model using a structured query language, and submitted to a set of operations defined in the NLP interface, each of which is implemented in its component, being consistent with modularity as one of the proposed architectural principles.

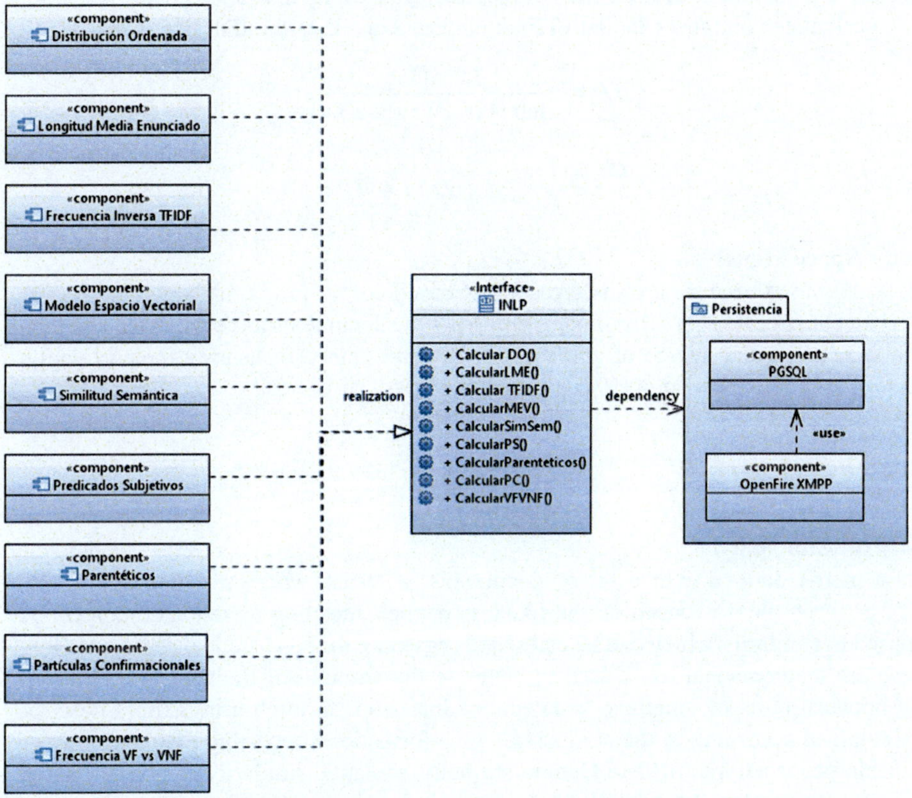

Fig. 2. The component diagram of S-COGIT

Ordered Distribution

It is a space where the number of messages is related to an interaction scene which can be defined as the number of exchanged messages during a time range defined by a change of subject or a change of paragraph.

Mean Length Utterance

According to [1] MLU was described originally by Roger Brown as a global index that allows measuring syntax development. It is calculated based on the number of terms in a statement [20]. The speech is divided into statements, then for each statement, you have to apply Eq. (1) to calculate its MLU

$$MLU = \frac{Number\ of\ words\ in\ statement}{Number\ of\ statements} \tag{1}$$

Term Frequency (TFIFD)

According to [15] it is a statistic representation of data that allows quantifying present symbols in a document by allocating values according to its relevancy [23].

To calculate you must calculate TF like in Eq. 2 and then the IDF like in Eq. 3.

$$TF = \frac{f(t, d)}{\max\{f(t, d) : t \in d\}} \tag{2}$$

$$idf = log \frac{|D|}{|\{d \in D : t \in d\}|} \tag{3}$$

Vector Space Model

It is widely used because it transforms a text-based corpus into a mathematical format with you can operate [11]. This model represents the documents as vectors [16] and then evaluates the similar degree of a document against a query, measuring the correlation between the vectors by Eq. 4 [17].

$$\cos(d_j, q) = \frac{d_j \cdot q}{|d_j| \cdot |q|} \tag{4}$$

Semantic Similarity

It is a metric defined over a set of documents or terms, where the idea of the distance between them is based on the likeness of their meaning or semantic content as opposed to similarity which can be estimated regarding their syntactical representation. These are mathematical tools used to estimate the strength of the semantic relationship between units of language, concepts or instances, through a numerical description obtained according to the comparison of information supporting their meaning or describing their nature [10] and Computationally, semantic similarity can be estimated by defining a topological similarity, by using ontologies to define the distance between terms/concepts.

Subjective Predicates

According to [6] Subjective Predicates can be defined as predicates that exist within an already identified subjective sentence.

Subjective predicates can be computationally identified comparing the terms of every sentence in the document with a list of most common transitive verbs -transitive verbs require one or two compliments.

Use of Parenthetics

Parentheticals are defined as any text sequence between parentheses. They have been approached from isolated perspectives, like translation pairs extraction. It is commonly stated that parentheticals provide optional information: they can be removed without affecting understanding [7].

Frequency of Factual Verbs vs No Factual Verbs

Factual or non-factual expression depends on the assertion of the speaker's degree of commitment to the truth of the proposition. However, the assertion has truth value: it is true or false. Factivity expresses a strong commitment of the speaker to the true proposition. It means that the speaker commits the state of affairs represented in the proposition is true. Non-factivity expresses a weak commitment of the speaker to the true proposition. The speaker has no enough evidence, consequently, the speaker gives a different hypothesis about the state of affairs mentioned [12].

4 Results

As a main result, we developed a tool that allows processing several algorithms on a text extracted from a corpus composed of several documents related to interaction, through message exchange or transcripted conversations, of individuals with attention deficit disorder (ADD) condition.

This tool is useful for studying such condition, providing a visual and friendly interface to identify patterns and measure the use of the language of the individuals that are part of the study, and it facilitates to the scientist the processing of all the information, related to the analysis of the language used, through the natural language processing using the NLTK library and the computing power - that Python provides - based on the statistics for the respective analysis.

5 Conclusions

For a software development that involves solving a problem of such dimensions, a robust and sustained design is required in an appropriate architecture and flexible enough to generate a framework that, in an agile and expeditious manner, provides quality attributes such as usability, portability, maintainability, security, among others.

The visualization is an essential characteristic to show the results of the analyzes and thus, in an innovative way, to represent, using graphs and highlights in the text, the operations carried out so that its interpretation is easier.

References

1. Ayala Flores, C.L.: Evaluación de la sintaxis durante el desarrollo del lenguaje., Parole: revista de creación literaria y de filología, pp. 127–138 (1988)
2. Arroyave, M.: Representación de la interacción en términos de la información para condiciones cognitivas especiales: Caso de Estudio Asperger y TDAH, Universidad de Caldas, Colombia (2017)
3. Boucenna, S., et al.: Interactive technologies for autistic children: a review. Cogn. Comput. 722–740 (2014)

4. Campigotto, R., McEwen, R., Demmans Epp, C.: Especially social: Exploring the use of an iOS application in special needs classrooms. Comput. Educ. **60**(1), 74–86 (2013)
5. Collober, R., et al.: Natural language processing (almost) from scratch. J. Mach. Learn. Res. **12**, 2493–2537 (2011)
6. Banchs, E., et al.: Information Retrieval Technology, pp. 196–199. Springer (2013)
7. El Maarouf, I., Villaneau, J.: Parenthetical classification for information extraction. In: Proceedings of COLING 2012: Posters, Mumbai, pp. 297–308 (2012)
8. Farrar, F.C.: Transforming home health nursing with telehealth technology. Nurs. Clin. N. Am. **50**(2), 269–281 (2015)
9. Fernández-López, Á., Rodríguez-Fórtiz, M.J., Rodríguez-Almendros, M.L., Martínez-Segura, M.J.: Mobile learning technology based on iOS devices to support students with special education needs. Comput. Educ. **61**, 77–90 (2013)
10. Harispe, S., Ranwez, S., Janaqi, S., Montmain, J.: Semantic similarity from natural language and ontology analysis. Synth. Lect. Hum. Lang. Technol. **8**(1), 1–25 (2015)
11. Hassanpour, S., O'Connor, M.J., Das, A.K.A.: A semantic-based method for extracting concept definitions from scientific publications: evaluation in the autism phenotype domain. J. Biomed. Semant. **4**(1), 14 (2013)
12. Ho, V.H., et al.: Factual and non-factual expression. Int. J. Eng. Lang. Lit. Humanit. (2015)
13. Kalman, Y.M., Geraghty, K., Thompson, C.K., Gergle, D.: Detecting linguistic HCI markers in an online Aphasia Support Group(2012)
14. Lopez-S, D., Revuelta, J., González, A., Corchado, J.: Hybridizing metric learning and case-based reasoning for adaptable clickbait detection, pp. 3–5. Springer Nature (2017)
15. Mahdhaoui, A., Chetouani, M.: Understanding parent-infant behaviors using non-negative matrix factorization. In: Toward Autonomous, Adaptive, and Context-Aware Multimodal Interfaces. Theoretical and Practical Issues, pp. 436–447 (2011)
16. Manning, C.D., Raghavan, P., Schutze, H.: An Introduction to Information Retrieval. Cambridge University Press, Cambridge (2009)
17. Manwar, A.B., Hemant, S., Mahalle, K.D., Chinchkhede, D.V.C.: A vector space model for information retrieval: a Matlab approach. Ind. J. Comput. Sci. Eng. **3**(2), 222–229 (2012)
18. Miller, G.A.: The cognitive revolution: a historical perspective. Trends Cogn. Sci. **7**, 141–144 (2003)
19. Bird, S., Klein, E., Loper, E.: Natural Language Processing with Python. O'Reilly, Newton (2009)
20. Pavez, M.M.: Presentación del indice de desarrollo del lenguaje "promedio de longitud de los enunciados" (PLE). (2002)
21. Thanaki, J.T.: Python Natural Language Processing. Packt Publishing, Birmingham (2017)
22. Python Documentation by Version. https://www.python.org/doc/versions/ Accessed 13 June 2019
23. Regneri, M., King, D.: Automatically evaluating atypical language in narratives by children with autistic spectrum disorder. In: Natural Language Processing and Cognitive Science: Proceedings, p. 11 (2014)
24. Smith, E.: Cognitive psychology: history. In: International Encyclopedia of the Social & Behavioral Sciences, pp. 2140–2147. Elsevier Ltd. (2005)
25. Stephanie, D., Julie, F.: Exploring links between language and cognition in autism spectrum disorders: complement sentences, false belief, and executive functioning. J. Commun. Disord. **54**, 15–31 (2015)

A Machine Learning Platform for Stock Investment Recommendation Systems

Elena Hernández-Nieves[1(✉)], Álvaro Bartolomé del Canto[1], Pablo Chamoso-Santos[1], Fernando de la Prieta-Pintado[1], and Juan M. Corchado-Rodríguez[1,2,3,4]

[1] Bisite Research Group, University of Salamanca, Salamanca, Spain
elenahn@gmail.com, alvarob96@gmail.com, chamoso@gmail.com, fer@gmail.com, corchado@gmail.com

[2] Air Institute, IoT Digital Innovation Hub, Salamanca, Spain

[3] Department of Electronics, Information and Communication, Faculty of Engineering, Osaka Institute of Technology, Osaka, Japan

[4] Pusat Komputeran dan Informatik, Universiti Malaysia Kelantan, Kelantan, Malaysia

Abstract. This research aims to create an investment recommendation system based on the extraction of buy/sell signals from the results of technical analysis and prediction. In this case it focuses on the Spanish continuous market. As part of this research, different techniques have been studied for data extraction and analysis. After having reviewed the work related to the initial idea of the research, it is shown the development carried out, together with the data extraction and the machine learning algorithms for prediction used. The calculation of technical analysis metrics is also included. The development of a visualization platform has been proposed for high-level interaction between the user and the recommendation system. The result is a platform that provides a user interface for both data visualization, analysis, prediction and investment recommendation. The platform's objective is not only to be usable and intuitive, but also to enable any user, whether an expert or not in the stock market, to abstract their own conclusions from the data and evaluate the information analyzed by the system.

Keywords: Artificial intelligence · Machine learning · Investment recommendation system, and data visualization

1 Introduction

High potential returns can derive from well-defined trading strategies. This motivates stock market forecasters to develop approaches that successfully predict index values or stock prices. A stock market prediction is considered to be successful if it achieves the best results using the minimum data input and the least complex stock market model [2]. The rise of Machine Learning and the growing computing capacity have made it possible to elaborate new products on the basis of traditional financial services, developing economic-financial tools that offer greater flexibility and speed [18].

Y. Dong et al. (Eds.): DCAI 2020, AISC 1237, pp. 303–313, 2021.
https://doi.org/10.1007/978-3-030-53036-5_33

This research aims to create an investment recommendation system based on the extraction of buy/sell signals from the results of technical analysis and prediction. In this case it focuses on the Spanish continuous market. As part of this research, different techniques have been studied for data extraction and analysis. The algorithms to be used by the recommender system have been determined and the factors that should be analyzed in the technical analysis. A visualization platform has been proposed for high-level interaction between the recommendation system and the user. The article is structured as follows: Sect. 2 reviews the existing solutions for forecasting stock ratings, Sect. 3 presents the implementation (the data extraction and the machine learning algorithms for prediction). Section 4 outlines the results of the whole research process. In Sect. 5, conclusions are drawn, and the future lines of research are discussed to further improve the proposed machine learning platform for stock investment recommendation systems.

2 Related Work

This section examines the state of the art and its contributions. The different techniques and solutions for stock prediction are carefully reviewed. Stock investment recommendations are mainly based on ANN techniques. The authors in [2], compiled the applications of intelligent techniques for forecasting stock ratings and quotes through soft-computing methods. They classified the articles according to the initial data processing, the sample size, the type of technique and its characteristics (number of ANN layers or functions belonging to a fuzzy set), the validation of the datasets and the training method. The results of the authors' research indicate that neural networks and neuro-fuzzy models are suitable for predicting stock market values. Furthermore, the experiments demonstrate that soft-computing techniques outperform conventional models in most cases. Although the described study provided an exhaustive analysis, it only covered research published in the period from 1992 to 2006, which is perhaps a little outdated. Another research that examines the literature on predicting stock market movements has been carried out [23]; it compiles researches that use automatic learning techniques and artificial intelligence. The research of this author identified artificial neural networks (ANNs) as the dominant machine learning technique in the area of stock market prediction. The authors in [25] concluded their research by stating that it is necessary to build a database with information on historical events in the stock market. Their study compiled machine learning techniques for stock prediction.

Other authors observed that predicting stock values is challenging due to the lack of certainty in making the prediction. This is due to the many factors that may be involved. The authors discussed the analysis procedure performed by investors, including the approach to improving prediction efficiency which combines Machine Learning algorithms with fundamental and technical analyses [19].

In [24], the authors provided a preliminary approach to a multi-agent recommendation system for computer investment. It is based on a hybrid technique that integrates collaborative and content-based filtering. This content-based model considers the investor's preferences, macroeconomic factors, stock profiles and expected trends in order to give optimal recommendations on the most suitable securities. The collaborative filter evaluates investment behaviors and actions of the investor pairs that typically dominate the economic market in order to suggest similar actions to the target investor. In [17] the authors proposed an intelligent recommendation system based on a technical indicator decision tree optimized by GA-SVM. The system is capable of learning the patterns of stock price fluctuations so as to recommend the appropriate trading strategy one day in advance. The authors validated the system's performance according to fifteen different measures. Its performance was compared with traditional trading based on technical indicators, and with the traditional buy and hold strategy.

3 Proposal

After having reviewed the work related to the initial idea of the research, this section shows the development carried out, together with the data extraction and the machine learning algorithms for prediction used. The calculation of technical analysis metrics is also included. In this section the most relevant aspects of the developed recommendation platform are explained.

3.1 The Data: Spanish Continuous Market

To extract data, the system receives a request from the user through an API (Application Programming Interface) endpoint or through a call to a method in the package developed in Python for version 3.x. In the case of such a request the system builds the headers with relevant data. If a data history has been requested, the headers include the name of the company and the range of dates, on the contrary only the name of the company is extracted. As shown in Fig. 1, a Python package has been created that supports different versions. The design of the software system package consisted in the creation of a Python package uploaded to PyPI (Python Package Indexer). It was decided to create a Python package for data extraction from Investing.com, a reliable and consistent source of historical data from the Spanish continuous market. First, a Web Scrape has been performed for automatic information retrieval from Investing.com. Since it is not always legally permitted to extract information from a website using this technique, the company was asked for an authorization.

Figure 2 shows the graphic representation of the combination of possible Python package times for each of the different phases involved in Web Scraping. The combinations are shown in best to worst scaling, as follows: request-lxml, request-bs4, urllib3-lxml and urllib3-bs4. Therefore, to send the request to Investing.com and extract the HTML, either GET or POST type requests are optimal, while lxml is optimal for historical data extraction and parsing.

Finally, the resulting scripts give form to an extensible and open Python package called investpy, intended for data extraction from investing.com. The package facilitates the extraction of data from various financial products such as: stocks, funds, government bonds, ETFs, certificates, commodities, etc.

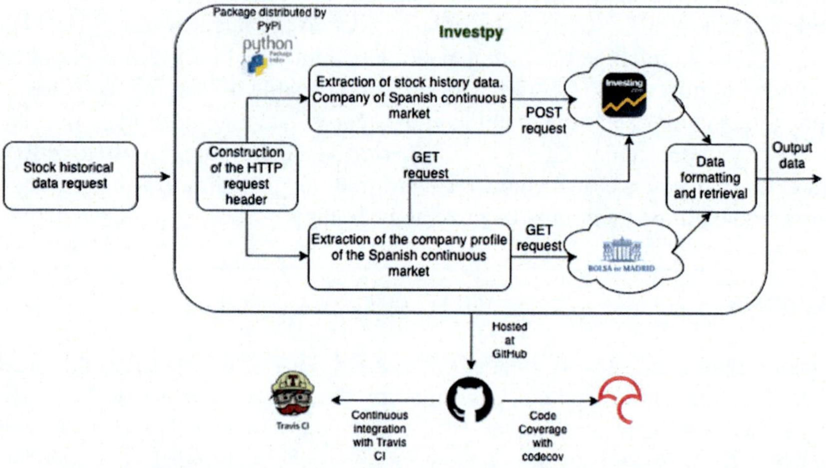

Fig. 1. Data extraction package design specification

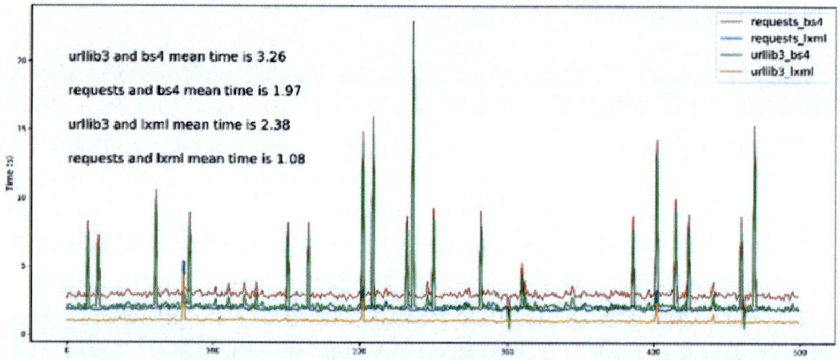

Fig. 2. Best web scraping combination

3.2 Machine Learning Algorithms for Prediction

To predict the future behavior of a stock, Machine Learning regression algorithms [1, 15, 21, 23] are applied and technical factors are analyzed.

Since the prediction aims to determine the closing stock market price, the set of opening values is defined as the input variables, and the set of closing values is defined as the output variables, i.e. the closing values are the objective variable of the algorithm. Given the nature of the problem, regression algorithms must be applied. This is because when working with continuous data, regression algorithms can indicate the pattern in a given set of data. Consequently, these algorithms are applied in cases where the relationship between a scalar dependent variable or objective variable Y and one or more input variables X is to be modelled.

The following section describes the algorithms used by the system to predict the last (unknown) closing value based on historical market data, from the last (known) opening value:

- Random Forest Regressor: These algorithms are an automated learning method for classification, regression and other tasks. A random forest is a meta-stimulus that fits a series of classification decision trees into various subsamples of the data set and uses the means to improve productive accuracy and control over fit.
- Gradient Boosting Regressor: It is an automated learning technique that builds the model in a scenic way, just like methods based on reinforcement. It generalizes models allowing for the optimization of an arbitrary and differentiable loss function.
- SMV-LinearSVR: learning models that analyze data for classification and regression analysis. An SVM training algorithm builds a model that assigns new examples to one or another category, making it a non-probabilistic binary linear classifier. In SVR we try to adjust the error within a certain threshold. In this case, it is similar to SVR with the kernel = linear parameter.
- MLP Regressor: a kind of artificial feedback neural network. MLP uses a supervised learning technique called backpropagation for the construction of the network. In addition, its multiple layers and non-linear activation distinguish MLP from a linear perceptron. It also allows to distinguish data that are not linearly separable.
- KNNeighbors Regressor: Non-parametric method used for classification and regression.

To calculate the factors for the technical analysis, the TA-Lib library has been used through the wrapper written in Python with the same name. Pandas' utilities have been used to calculate moving averages. Technical Analysis is an analysis used to weight and evaluate investments. It identifies opportunities to acquire or sell stocks based on market trends. Unlike fundamental analysis which attempts to determine the exact price of a stock, Technical Analysis focuses on the detection of trends op patterns in market behaviour for the identification of signals to buy or sell assets, along with various graphical representations that help evaluate the safety or risk of a stock [8].

This type of analysis can be used in any financial product as long as historical data is available. It is required to include both share prices and volume. Technical analysis is very often employed when a short-term analysis is required, thus, it can help to adequately address the problem presented in this research, where it predicts the closing value of a share in a day.

In [8] we have found the following indicators to be considered in the analysis:

- Relative Strength Index (RSI): It is a Momentum Indicator (these indicators reflect the difference between the current closing price and the closing price of the previous N days), which measures the impact of frequent changes in the price of a stock to identify signs of overbuying or overselling. The representation of the RSI is shown on an oscillator, Eq. 1, that is, a line whose value oscillates between two extremes, which in this case is between 0 and 100.

$$RSI_{step\ one} = 100 - \left[\frac{100}{1 + \frac{Average\ gain}{Average\ loss}} \right] \quad (1)$$

- Stochastic Oscillator (STOCH): It is a Momentum Indicator that compares the closing price of a stock on a given day with the range of closing values of that stock over a certain period of time, defined by the time window, Eq. 2. It also allows to adjust the sensitivity of the oscillator either by adjusting the time window or by calculating the moving average of the STOCH result. Like RSI it identifies signals of overbought or oversold stock within a range of 0 to 100 possible values.

$$\%K = \left(\frac{C - L14}{H14 - L14}\right) \times 100 \tag{2}$$

Where C is the most recent closing price, *L14* is the lowest price traded of the 14 previous trading sessions, *H14* the highest price traded during the same 14-day period and $\%K$ the current value of the stochastic indicator.

- Ultimate Oscillator (ULTOSC): It is a Momentum Indicator used to measure the evolution of a stock over a series of time frames using a weighted average of 3 different windows or time frameworks, Eq. 3. Therefore, it acquires a lower volatility and identifies fewer buy-sell signals than other oscillators that only depend on a single time frame. When the lines generated by ULTOSC diverge from the closing values of a stock, buy and sell signals are identified for it.

$$UO = \left[\frac{(A_7 \times 4) + (A_{14} \times 2) + A_{28}}{4 + 2 + 1}\right] \times 100 \tag{3}$$

Where *UO* is Ultimate Oscillator, *A* is Average. For the calculation of Average is necessary take into account the followings equations and terms like *Prior Close (PC)*, *Buying Pressure (BP)* and *Ture Range (TR)*. Eqs. 4 and 5

$$BP = Close - Min(Low, PC) \tag{4}$$

$$TR = Max(High, Prior\ Close) - Min(Low, Prior\ Close) \tag{5}$$

Once this has been taken into consideration, we can proceed with the calculation of the *Average*.

$$Average_7 = \frac{\sum_{p=1}^{7} BP}{\sum_{p=1}^{7} TR} \tag{6}$$

$$Average_{14} = \frac{\sum_{p=1}^{14} BP}{\sum_{p=1}^{14} TR} \tag{7}$$

$$Average_{28} = \frac{\sum_{p=1}^{28} BP}{\sum_{p=1}^{28} TR} \tag{8}$$

- William %R (WILLR): Also known as the Williams Percent Range, it is a Momentum Indicator that fluctuates between −100 and 0 and measures and identifies levels of stock overbuying or overselling, Eq. 9. WILLR is very similar to the STOCH in its use, as it is used for the same purpose. This indicator compares the closing value of a stock with the range between the maximum and minimum values within a given time frame.

$$Williams\ \%R = \frac{Highest\ High - Close}{Highest\ High - Lowest\ Low} \tag{9}$$

Where *Highest High* is the highest price in the lookback period, typically 14 days, *Close* is the most recent closing price and *Lowest low* is the lowest price in the lookback period, typically 14 days.

Moving averages are also used in Technical Analysis as it also represents the Momentum or value change in a timeframe N. Hence, moving averages help to understand the market trend and, like Momentum Indicators, allow to identify buy and sell signals from the historical data of a stock in a previously mentioned timeframe N. In this research, we will apply the simple moving average (SMA) and the exponential moving average (EMA) for timeframes of 5, 10, 20, 50, 100 and 200 days, so we will have indicators in different periods.

- Simple Moving Average (SMA): It's an arithmetic moving average. It is calculated by adding the recent closing values of an action for a window of size N and dividing that sum by the size of the window. Thus, when the size of the timeframe N is low, it responds quickly to changes in the value of the stock; while if the size of the window N is high, it responds more slowly.

$$SMA = \frac{A_1 + A_2 + \cdots + A_n}{n} \tag{10}$$

  Where A_n is the price of an asset at period n and n is the number of total periods.
- Exponential Moving Average (EMA): Also called Exponentially Weighted Moving Average since it weights recent observations, i.e. closing prices of a stock closer to the current one. It can be said that EMAs respond better than SMAs to recent changes in a share's price.

$$EMA_{today} = \left(Value_{Today} \times \left(\frac{Smoothing}{1 + Days} \right) \right) + EMA_{Yesterday} \times \left(1 - \left(\frac{Smoothing}{1 + Days} \right) \right) \tag{11}$$

  Where *EMA* is the Exponential moving average.

Since both the algorithmic predictions and the results of the technical factor and moving average calculations result in the next closing value of a stock, the recommendation will be based on identifying buy and sell signals based on the comparison of the predicted value to the value that the stock has at the current time.

4 Results

The result is a platform (Fig. 3) that provides a user interface for both data visualization, analysis, prediction and investment recommendation. The platform's objective at a functional user level is not only to be usable and intuitive, but also that the user, whether an expert or not in the stock market, be able to abstract his own conclusions from the data and evaluate the information analyzed by the system. The created platform depends completely on the Python package developed for data extraction: investpy.

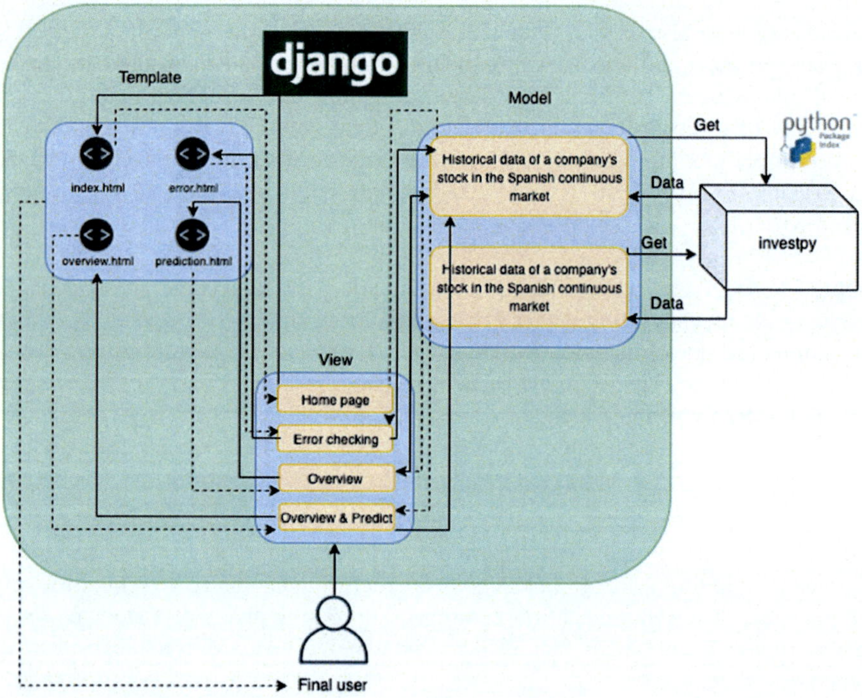

Fig. 3. Django design architecture

The web platform initially shows a screen where the overview option is given on one side and the overview and recommendation option on the other (Fig. 4). The overview functionality covers the extraction and basic visualization of the data. The system retrieves the company profile and the historical data for the last 5 years of the stock. On the basis of those data, it produces a series of representations:

- Time series: offers a graphic representation of the retrieved historical data, where the X and Y axes represent, the value of the stock in euros, and the date on which the stock reached that value, respectively.
- Candlestick chart: this representation shows the opening and closing values for each date and the difference between the maximum and minimum values for the same date.
- Data table: represents the available values. They are called OHLC (Open-High-Low-Close).

Fig. 4. Main view of the web platform

The Overview & Recommendation functionality is the same as the user input check in that it too extracts both company profile and historical data. However, this functionality also includes technical factors and moving averages with the consequent buy/sell recommendation. The generated graphs are visualized on the platform, among them are graphs that compare the different algorithms that the system has applied to make the prediction. This enables the user to identify those that have had a better precision. The platform presents the conclusions abstracted from the resulting values. It shows the buy/sell recommendation based on those values. The process of prediction and recommendation made by the system is transparent to the user.

5 Conclusion

The conducted research provides an initial approach to data analysis and to the combined used of Machine Learning algorithms and techniques, with traditional market analysis. This has made it possible to draw conclusions about future market behaviour. Thus, it can be concluded that when Machine Learning algorithms are trained with a sufficiently large amount of data, it is possible to successfully predict the closing value on the basis of the current opening value of the market. Thus, after identifying buy and sell signals, it has been possible to create a system that recommends the user to buy, hold or sell a stock at a certain time of day, according to the prediction obtained by the regression algorithms.

In addition to the inclusion of NLP (Natural Language Processing) techniques in the recommendation system, other methods are considered [5–7, 9, 13–16, 20]. Furthermore, the proposed model has the potential to enter the global market. It is considered viable given that all historical stock data previously go through a GridSearchCV, which consists in cross validating the optimal hyperparameters to be used by an algorithm from a specific dataset. To this end, it will be necessary to add functionalities to the developed Python package in the future. Additionally, it would be valuable to study the algorithms applied on other markets.

Acknowledgments. This research has been supported by the project "Intelligent and sustainable mobility supported by multi-agent systems and edge computing (InEDGEMobility): Towards Sustainable Intelligent Mobility: Blockchain-based framework for IoT Security", Reference: RTI2018-095390-B-C32, financed by the Spanish Ministry of Science, Innovation and Universities (MCIU), the State Research Agency (AEI) and the European Regional Development Fund (FEDER).

References

1. Arora, N.: Financial Analysis: Stock Market Prediction Using Deep Learning Algorithms (2019)
2. Atsalakis, G.S., Valavanis, K.P.: Surveying stock market forecasting techniques–Part II: soft computing methods. Expert Syst. Appl. **36**(3), 5932–5941 (2009)
3. Corchado, J.M., Lees, B.: A hybrid case-based model for forecasting. Appl. Artif. Intell. **15**(2), 105–127 (2001)
4. Coria, J.A.G., Castellanos-Garzón, J.A., Corchado, J.M.: Intelligent business processes composition based on multi-agent systems. Expert Syst. Appl. **41**(4), 1189–1205 (2014)
5. Dang, N.C., De la Prieta, F., Corchado, J.M., Moreno, M.N.: Framework for retrieving relevant contents related to fashion from online social network data. In: International Conference on Practical Applications of Agents and Multi-Agent Systems, pp. 335–347. Springer, Cham, June 2016
6. Dash, R., Dash, P.K.: A hybrid stock trading framework integrating technical analysis with machine learning techniques. J. Fin. Data Sci. **2**(1), 42–57 (2016)
7. Carneiro, D., Araújo, D., Pimenta, A., Novais, P.: Real time analytics for characterizing the computer user's state. ADCAIJ: Adv. Distrib. Comput. Artif. Intell. J. **5**(4), 1–18 (2016). ISSN 2255-2863
8. Edwards, R.D., Magee, J., Bassetti, W.C.: Technical Analysis of Stock Trends. CRC Press, Boca Raton (2018)
9. Fdez-Riverola, F., Iglesias, E.L., Díaz, F., Méndez, J.R., Corchado, J.M.: Applying lazy learning algorithms to tackle concept drift in spam filtering. Expert Syst. Appl. **33**(1), 36–48 (2007)
10. Fernández-Riverola, F., Diaz, F., Corchado, J.M.: Reducing the memory size of a fuzzy case-based reasoning system applying rough set techniques. IEEE Trans. Syst. Man Cybern. Part C (Appl. Rev.) **37**(1), 138–146 (2006)
11. Frikha, M., Mhiri, M., Gargouri, F.: A semantic social recommender system using ontologies based approach for Tunisian tourism (2015)
12. Fyfe, C., Corchado, J.M.: Automating the construction of CBR systems using kernel methods. Int. J. Intell. Syst. **16**(4), 571–586 (2001)
13. Glez-Bedia, M., Corchado, J.M., Corchado, E.S., Fyfe, C.: Analytical model for constructing deliberative agents. Eng. Intell. Syst. Electr. Eng. Commun. **10**(3), 173–185 (2002)
14. Urbano, J., Cardoso, H.L., Rocha, A.P., Oliveira, E.: Trust and normative control in multi-agent systems. ADCAIJ: Adv. Distrib. Comput. Artif. Intell. J. **1**(1) (2012). ISSN 2255-2863

15. Khaidem, L., Saha, S., Dey, S.R.: Predicting the direction of stock market prices using random forest. arXiv preprint arXiv:1605.00003 (2016)
16. Morente-Molinera, J.A., Kou, G., González-Crespo, R., Corchado, J.M., Herrera-Viedma, E.: Solving multi-criteria group decision making problems under environments with a high number of alternatives using fuzzy ontologies and multi-granular linguistic modelling methods. Knowl. Based Syst. **137**, 54–64 (2017)
17. Nair, B.B., Mohandas, V.P.: An intelligent recommender system for stock trading. Intell. Decis. Technol. **9**(3), 243–269 (2015)
18. Nicoletti, B., Nicoletti, W.: Future of FinTech. Palgrave Macmillan, Basingstoke (2017)
19. Patel, J., Shah, S., Thakkar, P., Kotecha, K.: Predicting stock and stock price index movement using trend deterministic data preparation and machine learning techniques. Expert Syst. Appl. **42**(1), 259–268 (2015)
20. Pawlewski, P., Golinska, P., Dossou, P.-E.: Application potential of agent based simulation and discrete event simulation in enterprise integration modelling concepts. ADCAIJ: Adv. Distrib. Comput. Artif. Intell. J. **1**(1) (2012). ISSN 2255-2863
21. Pimprikar, R., Ramachandran, S., Senthilkumar, K.: Use of machine learning algorithms and twitter sentiment analysis for stock market prediction. Int. J. Pure Appl. Math. **115**(6), 521–526 (2017)
22. Pudaruth, S., Moheeputh, S., Permessur, N., Chamroo, A.: Sentiment analysis from Facebook comments using automatic coding in NVivo 11. ADCAIJ: Adv. Distrib. Comput. Artif. Intell. J. **7**(1), 41–48 (2018)
23. Soni, S.: Applications of ANNs in stock market prediction: a survey. Int. J. Comput. Sci. Eng. Technol. **2**(3), 71–83 (2011)
24. Taghavi, M., Bakhtiyari, K., Scavino, E.: Agent-based computational investing recommender system. In: Proceedings of the 7th ACM Conference on Recommender Systems, pp. 455–458, October 2013
25. Yoo, P.D., Kim, M.H., Jan, T.: Machine learning techniques and use of event information for stock market prediction: a survey and evaluation. In: International Conference on Computational Intelligence for Modelling, Control and Automation and International Conference on Intelligent Agents, Web Technologies and Internet Commerce (CIMCA-IAWTIC 2006), vol. 2, pp. 835–841. IEEE, November 2005
26. Zhou, F., Qun, Z., Sornette, D., Jiang, L.: Cascading Logistic Regression Onto Gradient Boosted Decision Trees to Predict Stock Market Changes Using Technical Analysis (2018)

Applying Machine Learning Classifiers in Argumentation Context

Luís Conceição[1,2(✉)], João Carneiro[2], Goreti Marreiros[2], and Paulo Novais[1]

[1] ALGORITMI Centre, University of Minho, 4800-058 Guimarães, Portugal
pjon@di.uminho.pt

[2] GECAD – Research Group on Intelligent Engineering and Computing for Advanced Innovation and Development, Institute of Engineering, Polytechnic of Porto, 4200-072 Porto, Portugal
{msc,jrc,mgt}@isep.ipp.pt

Abstract. Group decision making is an area that has been studied over the years. Group Decision Support Systems emerged with the aim of supporting decision makers in group decision-making processes. In order to properly support decision-makers these days, it is essential that GDSS provide mechanisms to properly support decision-makers. The application of Machine Learning techniques in the context of argumentation has grown over the past few years. Arguing includes negotiating arguments for and against a certain point of view. From political debates to social media posts, ideas are discussed in the form of an exchange of arguments. During the last years, the automatic detection of this arguments has been studied and it's called Argument Mining. Recent advances in this field of research have shown that it is possible to extract arguments from unstructured texts and classifying the relations between them.

In this work, we used machine learning classifiers to automatically classify the direction (relation) between two arguments.

Keywords: Argument mining · Machine learning classifiers · Argumentation-based dialogues

1 Introduction

Group decision making is an area that has been studied over the years. In organizations, most decisions are taken in groups, either for reasons of organizational structure, as their organizational organigrams oblige them to do so, or for the associated benefits of group decision making, such as: sharing responsibilities, greater consideration of problems and possible solutions, and also allow less experienced elements learn during the process [1–3].

Group Decision Support Systems (GDSS) emerged with the aim of supporting decision makers in group decision-making processes. They have been studied over the past 30 years and have become a very relevant research topic in the field of Artificial Intelligence, being nowadays essentially developed with web-based interfaces [4–8].

Y. Dong et al. (Eds.): DCAI 2020, AISC 1237, pp. 314–320, 2021.
https://doi.org/10.1007/978-3-030-53036-5_34

The globalization of markets and the appearance of large multinational companies mean that many of the managers are constantly traveling through different locations and in areas with different time zones [9].

In order to properly support decision-makers these days, it is essential that GDSS provide mechanisms such as automatic negotiation, represent interests of decision-makers, potentiate the generation of ideas, allow the existence of a process, between others.

Our research group has been working over the last few years on methods and tools that support decision-makers that are geographically dispersed, more specifically on argumentation-based frameworks [10]; in the definition of behaviour styles for agents that represent real decision-makers [11, 12]; in the satisfaction of the decision-makers regarding the group decision-making process [13], as well as in the group's satisfaction in relation to the decision taken.

Our dynamic argumentation framework allows GDSS to be equipped with features that allow decision-makers, even when they are geographically dispersed, to benefit from the typical advantages associated to the face-to-face group decision-making processes. This framework accompanies the decision-maker throughout the group decision-making process, through the implementation of a multi-agent system, where each agent represents a human decision-maker in the search for a solution for the problem, proposing one or more alternatives as solution to the problem from the set of initial alternatives, taking into account the preferences and interests of the decision-maker in the decision problem that is being handled. The dynamic argumentation framework allows decision-makers to understand the conversation performed between agents in the negotiation process, but also the agents (Fig. 1) are able to understand the new arguments introduced and exchanged by decision-makers. Agents can use these new arguments to advice decision-makers and to find solutions during the decision-making process [10].

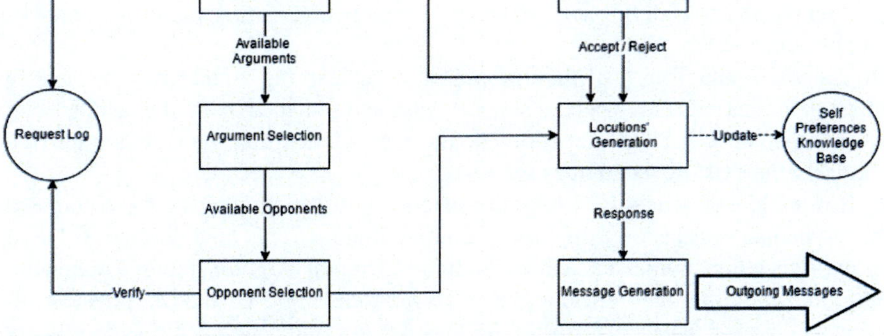

Fig. 1. Agent's communications workflow [11]

The application of Machine Learning (ML) techniques in the context of argumentation has grown over the past few years. Argument Mining (AM) is the automatic identification and extraction of the structure of inference and reasoning expressed in the form of arguments in natural language [14].

In this work our goal is to automatically classify the relation between two arguments introduced by decision-makers in our dynamic argumentation framework. The framework works as a social network where each decision-maker can make a post expressing his/her opinion regarding one (or more) alternative(s) and/or criteria, and the decision-maker has to classify the direction of the argument between "against" (if his/her comment is attacking the idea that he/she is responding) or "in favour" (if his/her comment is supporting the idea that he/she is responding). To create a model to automatically classify the relations between arguments we used the dataset created by Benlamine, Chaouachi, Villata, Cabrio, Frasson and Gandon [15] which consists of a set of arguments 526 arguments exchanged by participants in online debates regarding different topics. The dataset is annotated with the relation between each two arguments in support (if the argument is in favour of the idea expressed in the previous argument) and attack (if the argument is against the idea expressed in the previous argument). With this dataset we ran experiments to train a classifier for the relation between arguments.

The paper is organized as follows. Section 2 we present some of the most relevant works in the literature about AM. Section 3 describes our work and experiences. In Sect. 4 we discuss the obtained results of the trained classifier. Finally, in Sect. 5 we describe the conclusions and some ideas about future work that we intend to follow.

2 Related Work

It is possible to find different approaches in this area, for instance in the work of Rosenfeld and Kraus [16] they used ML techniques in a multi-agent system that supports humans in argumentative discussions by proposing possible arguments to use. To do this, they analysed the argumentative behaviour from 1000 human participants and they proved that is possible to predict human argumentative behaviour using ML techniques [16].

Another interesting work is the one published by Carstens and Toni [17] where they worked in the identification and extraction of argumentative relations. To do that, they classify pairs pf sentences according to whether they stand in an argumentative relation to other sentences, considering any sentence as argumentative that supports or attacks another sentence.

In Cocarascu and Toni [18] they propose a deep learning architecture to identify argumentative relations of attack and support from one sentence to another, using Long Short-Term Memory (LSTM) networks. Authors concluded that the results improved considerably the existing techniques for AM.

In Rosenfeld and Kraus [19] they presented a novel methodology for automated agents for human persuasion using argumentative dialogs. This methodology is based on an argumentation framework called Weighted Bipolar Argumentation Framework (WBAF) and combines theoretical argumentation modelling, ML and Markovian optimization techniques. They performed field experiments and concluded that their agent is able to persuade people no worse than people are able to persuade each other.

Swanson, Ecker and Walker [20] published a paper where they created a dataset by extracting dialogues from online forums and debates. They established two goals: the extraction of arguments from online dialogues and the identification of argument's facet similarity. The first consists in the identification and selection of text excerpts that contain arguments about an idea or topic, and the second is about argument's semantic, an attempt to identify if two arguments have the same meaning about the topic.

In Mayer, Cabrio, Lippi, Torroni and Villata [21] authors performed argument mining in clinical trials in order to support physicians in decision making processes. They extracted argumentative information such as evidence and claims, from clinical trials which are extensive documents written in natural language, using the MARGOT system [22].

3 Training the Classifier of Arguments Relations

In this Section we describe the process we have followed to train our argument relation classifier. As we stated in Sect. 1, the dataset used in this work consists in a set of arguments extracted by Benlamine, Chaouachi, Villata, Cabrio, Frasson and Gandon [15] from 12 different online debates. This dataset consists in a set of 526 arguments, composing 263 pairs of arguments extracted from the 12 online debates about different topics such as abortion, cannabis consumption, bullism, among others. We extracted the data from the xml files and built the initial dataset (represented by an excerpt in Table 1) with 3 columns named: "arg1", "arg2", and "relation", corresponding the first to the first argument, the second to the argument that responds to the first, and the relation which consists in the annotation that characterizes the relation between the

Table 1. Excerpt of dataset sentence pairs, labeled according to the relation from Arg2 with Arg1

Arg1	Arg2	Relation
I think that using animals for different kind of experience is the only way to test the accuracy of the method or drugs. I cannot see any difference between using animals for this kind of purpose and eating their meat	And I think there are alternative solutions for the medical testing. For example, for some research, a volunteer may be invited	Against
I don't think the animal testing should be banned, but researchers should reduce the pain to the animal	Maybe we should create instructions to follow when we make tests on animals. Researchers should use pain killer just not to torture animals	InFavour
I think that using animals for different kind of experience is the only way to test the accuracy of the method or drugs. I cannot see any difference between using animals for this kind of purpose and eating their meat	Animals are not able to express the result of the medical treatment but humans can	Against

arguments ("InFavour"/"Against"). The annotated dataset contains 48% (127) argument pairs classified as "InFavour" and 52% (136) argument pairs classified as "Against".

To build our classifier model we represented each sentence pair with a Bag-of-Words (BOW) and we added some calculated features aiming to improve the results of the classifier as the authors done in [17]. The calculated features consisted in Sentence Similarity [23], Edit Distance [24], and Sentiment Score [25] measures. Sentence Similarity and Edit Distance measures were calculated for each pair of argumentative sentences, while Sentiment Score was calculated individually to each argumentative sentence.

The experiments on building the classifier were carried out using two algorithms: Support Vector Machines (SVM) and Random Forest (RF). We also used a k-fold cross-validation procedure to evaluate our models due to the small size of dataset. The k parameter of k-fold was set to 10 and we obtained the results presented in Table 2.

Table 2. 10-fold CV average results on training dataset

	Accuracy (%)	Precision (%)	Recall (%)	F_1 (%)
RF	56,21	60,04	55,38	53,96
SVM	66,07	65,69	69,22	64,22

4 Discussion

The results obtained are not yet satisfactory enough so that we can consider that these models can already be applied in the GDSS prototype that we have been developing, however as we can see in Table 2, the model generated by the SVM algorithm presents higher values in all measures model evaluation. As next steps we intend, first to improve our classifier performing feature selection analysis on the dataset in order to reduce the BOW to the main features (words) that can have more influence in the model definition and second to test our classifier with other debate datasets.

We also intend to continue working on the creation of the classification models, based on datasets generated with our argumentation framework, in this case we will have more features added to our dataset, such as: data on the decision makers' behavior style in the decision-making process, data on the decision-maker preferences on each of the alternatives and criteria of the problem, among others, which we believe could be usefull to construct more precise classifiers.

5 Conclusions

Supporting and representing decision-makers in group decision making processes is a complex task, but when we consider that the decision-makers may not be in the same place at the same time, the task is even more complex. The utilization of artificial intelligence mechanisms in the conception of GDSS's is boosting the improvement and acceptance of these systems by the decision-makers.

In this work we applied ML supervised classification algorithms to arguments extracted from online debates with the goal of creating an argumentative sentence classifier to perform automatic classification of argumentative sentences exchanged in GDSS's by decision-makers.

Although the results obtained do not have a high accuracy, we believe that it is possible to automatically classify the argumentative sentences exchanged by decision-makers during a group decision-making process with greater precision, when we obtain more features that help to relate the argumentative sentences. Those features can be extracted or calculated from the original sentences and others will be generated by our dynamic argumentation framework. Despite that, we still plan to study deeply the feature selection techniques in the natural language process and run experiences to compare the results.

Acknowledgments. This work was supported by National Funds through the FCT – Fundação para a Ciência e a Tecnologia (Portuguese Foundation for Science and Technology) within the Projects UIDB/00319/2020, UIDB/00760/2020, GrouPlanner Project (POCI-01-0145-FEDER-29178) and by the Luís Conceição Ph.D. Grant with the reference SFRH/BD/137150/2018.

References

1. Bell, D.E.: Disappointment in decision making under uncertainty. Oper. Res. **33**, 1–27 (1985)
2. Huber, G.P.: A theory of the effects of advanced information technologies on organizational design, intelligence, and decision making. Acad. Manag. Rev. **15**, 47–71 (1990)
3. Luthans, F., Luthans, B.C., Luthans, K.W.: Organizational Behavior: An Evidence Based Approach. IAP (2015)
4. Huber, G.P.: Issues in the design of group decision support systems. MIS Q.: Manag. Inf. Syst. **8**, 195–204 (1984)
5. DeSanctis, G., Gallupe, B.: Group decision support systems: a new frontier. SIGMIS Database **16**, 3–10 (1985)
6. Marreiros, G., Santos, R., Ramos, C., Neves, J.: Context-aware emotion-based model for group decision making. IEEE Intell. Syst. **25**, 31–39 (2010)
7. Conceição, L., Martinho, D., Andrade, R., Carneiro, J., Martins, C., Marreiros, G., Novais, P.: A web-based group decision support system for multicriteria problems. Concurr. Comput.: Pract. Exp. e5298 (2019)
8. Carneiro, J., Andrade, R., Alves, P., Conceição, L., Novais, P., Marreiros, G.: A consensus-based group decision support system using a multi-agent MicroServices approach. In: International Conference on Autonomous Agents and Multi-Agent Systems 2020. International Foundation for Autonomous Agents and Multiagent Systems (2020)
9. Grudin, J.: Group dynamics and ubiquitous computing. Commun. ACM **45**, 74–78 (2002)
10. Carneiro, J., Martinho, D., Marreiros, G., Jimenez, A., Novais, P.: Dynamic argumentation in UbiGDSS. Knowl. Inf. Syst. **55**, 633–669 (2018)
11. Carneiro, J., Martinho, D., Marreiros, G., Novais, P.: Arguing with behavior influence: a model for web-based group decision support systems. Int. J. Inf. Technol. Decis. Mak. **18**, 517–553 (2018)
12. Carneiro, J., Saraiva, P., Martinho, D., Marreiros, G., Novais, P.: Representing decision-makers using styles of behavior: an approach designed for group decision support systems. Cogn. Syst. Res. **47**, 109–132 (2018)

13. Carneiro, J., Saraiva, P., Conceição, L., Santos, R., Marreiros, G., Novais, P.: Predicting satisfaction: Perceived decision quality by decision-makers in Web-based group decision support systems. Neurocomputing **338**, 399–417 (2019)
14. Lawrence, J., Reed, C.: Argument mining: a survey. Comput. Linguist. **45**, 765–818 (2020)
15. Benlamine, S., Chaouachi, M., Villata, S., Cabrio, E., Frasson, C., Gandon, F.: Emotions in argumentation: an empirical evaluation. In: Twenty-Fourth International Joint Conference on Artificial Intelligence (2015)
16. Rosenfeld, A., Kraus, S.: Providing arguments in discussions on the basis of the prediction of human argumentative behavior. ACM Trans. Interact. Intell. Syst. (TiiS) **6**, 30 (2016)
17. Carstens, L., Toni, F.: Towards relation based argumentation mining. In: Proceedings of the 2nd Workshop on Argumentation Mining, pp. 29–34 (2015)
18. Cocarascu, O., Toni, F.: Identifying attack and support argumentative relations using deep learning. In: Proceedings of the 2017 Conference on Empirical Methods in Natural Language Processing, pp. 1374–1379 (2017)
19. Rosenfeld, A., Kraus, S.: Strategical argumentative agent for human persuasion. In: Proceedings of the Twenty-Second European Conference on Artificial Intelligence, pp. 320–328. IOS Press (2016)
20. Swanson, R., Ecker, B., Walker, M.: Argument mining: extracting arguments from online dialogue. In: Proceedings of the 16th Annual Meeting of the Special Interest Group on Discourse and Dialogue, pp. 217–226 (2015)
21. Mayer, T., Cabrio, E., Lippi, M., Torroni, P., Villata, S.: Argument mining on clinical trials. In: COMMA, pp. 137–148 (2018)
22. Lippi, M., Torroni, P.: MARGOT: a web server for argumentation mining. Expert Syst. Appl. **65**, 292–303 (2016)
23. Miller, G.A.: WordNet: a lexical database for English. Commun. ACM **38**, 39–41 (1995)
24. Navarro, G.: A guided tour to approximate string matching. ACM Comput. Surv. (CSUR) **33**, 31–88 (2001)
25. Esuli, A., Sebastiani, F.: Sentiwordnet: a publicly available lexical resource for opinion mining. In: LREC, pp. 417–422. Citeseer (2006)

Type-Theory of Parametric Algorithms with Restricted Computations

Roussanka Loukanova[1,2(✉)]

[1] Institute of Mathematics and Informatics, Bulgarian Academy of Sciences, Sofia, Bulgaria
rloukanova@gmail.com
[2] Stockholm University, Stockholm, Sweden

Abstract. The paper extends the formal language and reduction calculus of Moschovakis type-theory of recursion, by adding a restrictor operator on terms with predicative constraints. Terms with restrictions over memory variables formalise inductive algorithms with generalised, restricted parameters. The theory provides restricted computations for algorithmic semantics of mathematical expressions and definite descriptors, in formal and natural languages.

Keywords: Formal language · Algorithms · Recursion · Restrictor · Restricted memory variables · Reduction calculus · Canonical forms · Restricted computations

1 Introduction

For a detailed background on a new theory of the notion of algorithm, see the initial and original work in [4]. The theory of typed, algorithmic, acyclic computations was introduced in [5], by designating it with L^{λ}_{ar}, via a demonstration of its possible applications to computational semantics of natural language. For more recent developments of the language and theory of acyclic algorithms L^{λ}_{ar}, see, e.g., [1–3].

In this paper, we extend the formal language and reduction calculus of L^{λ}_{ar}, by incorporating constraints on objects and algorithmic computations. In Sect. 2, we add a new restrictor operator for formation of terms with propositional restrictions, at the object level of the language. In Sect. 4, to cover (acyclic) algorithms with constraint computations, we extend the reduction calculus of the theory L^{λ}_{ar}. In Sect. 5, we introduce the notion of generalised variables, as restricted memory slots, by using the restrictor operator.

In Sect. 7, we provide possibilities for applications of the restricted algorithms to represent definite descriptors, which are abundant in data of various forms in specific domains. The technique is a new, algorithmic approach, which pertains to semantics of programming languages and other specification languages in Computer Science. Restricted variables designate memory slots for storing

Y. Dong et al. (Eds.): DCAI 2020, AISC 1237, pp. 321–331, 2021.
https://doi.org/10.1007/978-3-030-53036-5_35

information, which is constrained to satisfy propositional conditions. The definite descriptor provides constraints for imposing properties over data and in algorithms with restricted parameters. The algorithmic restrictor is specifically important for computational semantics of formal languages, e.g., in programming, specification languages in data and Artificial Intelligence (AI), and computational syntax-semantics of natural language. In particular, we present alternative possibilities for algorithmic semantics of singular, nominal expressions. For example, a Noun Phrase (NP), which is a definite description, like "the dog", can designate an object in a semantic domain, via identifying it as the unique object satisfying the property denoted by the description "dog". Definite descriptors are abundant in the languages of mathematics, as specialised, subject delimited fragments of human language, which can be intermixed with mathematical expressions, e.g., "the natural number n, such that $n + 1 = 2$".

2 Algorithms with Acyclic Recursion and Restrictor

In this section, we provide its major notions of $\mathrm{L}^{\lambda}_{\mathrm{ar}}$ while extending its formal language and theory. The formal language of this extended $\mathrm{L}^{\lambda}_{\mathrm{ar}}$ has an additional constant for a restrictor operator (such that), for parametric algorithms with restricted computations. The extended L^{λ}_{rar} has the same set of types, constants and variables, as $\mathrm{L}^{\lambda}_{\mathrm{ar}}$.

$\mathsf{Types}_{\mathrm{L}^{\lambda}_{\mathrm{ar}}}$. The set $\mathsf{Types}_{\mathrm{L}^{\lambda}_{\mathrm{ar}}}$ of $\mathrm{L}^{\lambda}_{\mathrm{ar}}$ is defined recursively, e.g., by the rules in the style of Backus-Naur Form $\tau :\equiv \mathsf{e} \mid \mathsf{t} \mid \mathsf{s} \mid (\tau_1 \rightarrow \tau_2)$. The type e is for objects that are entities of the semantic domains, s is for states consisting of context information, e.g., possible worlds, time and space locations, speakers, listeners. t is for truth values. For any $\tau_1, \tau_2 \in \mathsf{Types}$, the type $(\tau_1 \rightarrow \tau_2)$ is for functions from objects of type τ_1 to objects of type τ_2. The abbreviation $\widetilde{\sigma} \equiv \mathsf{s} \rightarrow \sigma$ is for state-dependent types $\widetilde{\sigma}$. We shall use the typical Curry coding, the type $(\tau_1 \rightarrow \cdots \rightarrow (\tau_n \rightarrow \sigma))$, is for functions on n-arguments ($n \geq 0$).

Constants Consts (also abbreviated by K): are given by denumerable sets of typed constants, $\mathsf{Consts} = \bigcup_{\tau} \mathsf{Consts}_{\tau}$, for $\mathsf{Consts}_{\tau} = \{\mathsf{c}^{\tau}_0, \ldots, \mathsf{c}^{\tau}_k, \ldots\}$, for $\tau \in \mathsf{Types}$. *Pure Variables:* by denumerable sets of typed variables, $\mathsf{PureV} = \bigcup_{\tau} \mathsf{PureV}_{\tau}$, for $\mathsf{PureV}_{\tau} = \{\mathsf{v}^{\tau}_0, \mathsf{v}^{\tau}_1, \ldots\}$. *Recursion (Memory) Variables:* by denumerable sets of typed variables $\mathsf{RecV} = \bigcup_{\tau} \mathsf{RecV}_{\tau}$, for $\mathsf{RecV}_{\tau} = \{\mathsf{r}^{\tau}_0, \mathsf{r}^{\tau}_1, \ldots\}$. The vocabulary is without intersections: $\mathsf{Consts} \neq \mathsf{RecV} \neq \mathsf{PureV}$, and the set of all variables is $\mathsf{Vars} = \mathsf{PureV} \cup \mathsf{RecV}$.

We use the typical notations for type assignments, $A : \tau$, and A^{τ}, to express that A is a term of type τ.

Definition 1. *The set* $\mathsf{Terms}_{L^{\lambda}_{rar}}$ *of the* L^{λ}_{rar}*-terms consists of the expressions defined by the following rules:*

$$A :\equiv \mathsf{c}^{\tau} : \tau \mid x^{\tau} : \tau \tag{1a}$$
$$\mid B^{(\sigma\to\tau)}(C^{\sigma}) : \tau \tag{1b}$$
$$\mid \lambda(v^{\sigma})\,(B^{\tau}) : (\sigma \to \tau) \tag{1c}$$
$$\mid \big(A_0^{\sigma_0} \text{ where } \{\, p_1^{\sigma_1} := A_1^{\sigma_1}, \ldots, p_n^{\sigma_n} := A_n^{\sigma_n} \,\}\big) : \sigma_0 \tag{1d}$$
$$\mid \big(\mathsf{A}_0^{\sigma_0} \text{ such that } \{\, \mathsf{C}_1^{\tau_1}, \ldots, \mathsf{C}_m^{\tau_m} \,\}\big) : \sigma_0' \tag{1e}$$

given that $\mathsf{c} \in \mathsf{Consts}$, $x \in \mathsf{PureV} \cup \mathsf{RecV}$, *and, for* $n \geq 0$, $i = 0, \ldots, n$, $\sigma_i \in \mathsf{Types}$, $A_i \in \mathsf{Terms}_{\sigma_i}$, *and, for* $i = 1, \ldots, n$, $p_i \in \mathsf{RecV}_{\sigma_i}$, *are pairwise different recursion variables, such that the sequence of assignments* $\{p_1^{\sigma_1} := A_1^{\sigma_1}, \ldots, p_n^{\sigma_n} := A_n^{\sigma_n}\}$ *satisfies the Acyclicity Constraint (AC),* (3a)–(3c), $\mathsf{C}_1 : \tau_1, \ldots, \mathsf{C}_m : \tau_m \in \mathsf{Terms}_{L^{\lambda}_{rar}}$; *and each* $\tau_i \in \mathsf{Types}$ *is either* $\tau_i \equiv \mathsf{t}$, *i.e., the type* t *of truth values, or* $\tau_i \equiv \widetilde{\mathsf{t}}$, *i.e., the type* $\widetilde{\mathsf{t}}$ *of truth values that may depend on states. The type of the term in* (1e) *is that of* $\mathsf{A}_0^{\sigma_0}$ *lifted to state dependent, in case at least one of components is such, i.e.:*

$$\sigma_0' \equiv \begin{cases} \sigma_0, & \text{if } \tau_i \equiv \mathsf{t}, \text{ for all } i \in \{1, \ldots, n\} \\ \sigma_0 \equiv (\mathsf{s} \to \sigma), & \text{if } \tau_i \equiv \widetilde{\mathsf{t}}, \text{ for some } i \in \{1, \ldots, n\} \\ \widetilde{\sigma_0} \equiv (\mathsf{s} \to \sigma_0), & \text{if } \tau_i \equiv \widetilde{\mathsf{t}}, \text{ for some } i \in \{1, \ldots, n\}, \text{ and} \\ & \quad \text{for all } \sigma \in \mathsf{Types}, \sigma_0 \not\equiv (\mathsf{s} \to \sigma) \end{cases} \tag{2}$$

Acyclicity Constraint (AC). *The sequence of assignments* (3a) *is* acyclic *iff there is a function* rank, *such that, all* $p_i, p_j \in \{p_1, \ldots, p_n\}$ *satisfy* (3c)*:*

$$\{p_1 := A_1, \ldots, p_n := A_n\} \quad (n \geq 0) \tag{3a}$$
$$\mathsf{rank} : \{p_1, \ldots, p_n\} \to \mathbb{N} \tag{3b}$$
$$\text{if } p_j \text{ occurs freely in } A_i, \text{ then } \mathsf{rank}(p_j) < \mathsf{rank}(p_i) \tag{3c}$$

We call the sequence (3a) *acyclic system of assignments.* The terms A of the form (1d) are called *recursion terms.* For each $\tau \in \mathsf{Types}$, Terms_{τ} is the set of the terms of type τ.

The formal language without the acyclicity constraint provides the version of the type-theory with full recursion L^{λ}_{r}.

The sets $\mathsf{FreeV}(A)$ and $\mathsf{BoundV}(A)$, respectively of the free and bound variables of a term A, are defined by structural recursion on A, in the usual way, with the exception of the recursion terms. For any given recursion term A of the form (1d), all occurrences of $p_1, \ldots, p_n \in \mathsf{RecV}$ in A are bound, and all other free (bound) occurrences of variables in $A_0, \ldots, A_n$ are also free (bound) in A. We shall call any term of the form (1e) a *restricted term.*

We shall use the meta-symbol $\equiv$, which is not in the vocabulary of $\mathrm{L}^{\lambda}_{\mathrm{ar}}$, to express literal identities between expressions.

Often, we use abbreviations for sequences, e.g.:

$$\overrightarrow{p} := \overrightarrow{A} \equiv p_1 := A_1,\ \ldots,\ p_n := A_n \quad (n \geq 0) \tag{4}$$

3 Denotational Semantics of $\mathbf{L}^{\lambda}_{\mathrm{ar}}$

A *standard semantic structure* is a tuple $\mathfrak{A}(\mathsf{Consts}) = \langle \mathbb{T}, \mathcal{I} \rangle$ such that:

- $\mathbb{T} = \{\mathbb{T}_\sigma \mid \sigma \in \mathit{Types}\}$ is a frame of typed objects, such that:
  - $\{0, 1, er\} \subseteq \mathbb{T}_\mathsf{t} \subseteq \mathbb{T}_\mathsf{e}$ $(er_\mathsf{t} \equiv er_\mathsf{e} \equiv er \equiv \mathit{error})$
  - $\mathbb{T}_\mathsf{s} \neq \varnothing$ (the *states*); $\mathbb{T}_{(\tau_1 \to \tau_2)} = (\mathbb{T}_{\tau_1} \to \mathbb{T}_{\tau_2}) = \{f \mid f\colon \mathbb{T}_{\tau_1} \to \mathbb{T}_{\tau_2}\}$ (standard frame)
  - $er_{(\tau_1 \to \tau_2)} = h$, s.t. for every $c \in \mathbb{T}_{\tau_1}$, $h(c) = er_{\tau_2}$
- $\mathcal{I}\colon \mathsf{Consts} \longrightarrow \cup\mathbb{T}$ is the *interpretation function*, respecting the types: for every $\mathsf{c} \in \mathsf{Consts}_\sigma$, $\mathcal{I}(\mathsf{c}) \in \mathbb{T}_\sigma$
- $\mathfrak{A}$ has the set G of all variable valuations: $G = \{g\colon \mathsf{PureV} \cup \mathsf{RecV} \longrightarrow \cup\mathbb{T}\}$

There is a unique *denotation function* $\mathsf{den}^{\mathfrak{A}}\colon \mathsf{Terms} \longrightarrow \{f \mid f\colon G \longrightarrow \cup\mathbb{T}\}$ defined by (D1)–(D5). For a given semantic structure $\mathfrak{A}$, we write $\mathsf{den} \equiv \mathsf{den}^{\mathfrak{A}}$.

(D1) $\mathsf{den}(x)(g) = g(x)$; $\mathsf{den}(\mathsf{c})(g) = \mathcal{I}(\mathsf{c})$
(D2) $\mathsf{den}(A(B))(g) = \mathsf{den}(A)(g)(\mathsf{den}(B)(g))$
(D3) $\mathsf{den}(\lambda x(B))(g)(t) = \mathsf{den}(B)(g\{x := t\})$, for every $t \in \mathbb{T}_\tau$
(D4) $\mathsf{den}(A_0 \text{ where } \{p_1 := A_1, \ldots, p_n := A_n\})(g) =$
$\mathsf{den}(A_0)(g\{p_1 := \overline{p}_1, \ldots, p_n := \overline{p}_n\})$
where the values $\overline{p}_i \in \mathbb{T}_{\tau_i}$ are defined by recursion on $\mathsf{rank}(p_i)$:

$$\overline{p_i} = \mathsf{den}(A_i)(g\{p_{k_1} := \overline{p}_{k_1}, \ldots, p_{k_m} := \overline{p}_{k_m}\}) \tag{5}$$

$p_{k_1}, \ldots, p_{k_m}$ are all $p_j \in \{p_1, \ldots, p_n\}$, such that $\mathsf{rank}(p_j) < \mathsf{rank}(p_i)$
(D5) The denotation of the restricted terms is in two cases, with respect to possible state dependent types of the components

Case 1: for all $i \in \{1, \ldots, n\}$, $C_i : \mathsf{t}$
For every $g \in G$:

$$\mathsf{den}\big(A_0^{\sigma_0} \text{ s.t. } \{\overrightarrow{C}\}\big)(g) = \begin{cases} \mathsf{den}(A_0)(g), & \text{if } \mathsf{den}(C_i)(g) = 1, \\ & \quad \text{for all } i \in \{1, \ldots, n\} \\ er_{\sigma_0} & \text{if } \mathsf{den}(C_i)(g) = er \text{ or} \\ & \quad \mathsf{den}(C_i)(g) = 0, \\ & \quad \text{for some } i \in \{1, \ldots, n\} \end{cases}$$

Case 2: for some $i \in \{1, \dots, n\}$, $C_i : \tilde{\mathsf{t}}$ (state dependent proposition)
For every $g \in G$, and every state $s \in \mathbb{T}_{\mathsf{s}}$:

$$\mathsf{den}\big(A_0^{\sigma_0} \text{ s.t. } \{\overrightarrow{C}\}\big)(g)(s) = \begin{cases} \mathsf{den}(A_0)(g)(s), & \text{if } \mathsf{den}(C_i)(g) = 1, \\ & \text{for all } i \text{ s.th. } C_i : \mathsf{t}, \text{ and} \\ & \quad \mathsf{den}(C_i)(g)(s) = 1, \\ & \text{for all } i \text{ s.th. } C_i : \tilde{\mathsf{t}}, \text{ and} \\ & \quad \sigma_0 \equiv (\mathsf{s} \to \sigma) \\ \mathsf{den}(A_0)(g), & \text{if } \mathsf{den}(C_i)(g) = 1, \\ & \text{for all } i \text{ s.th. } C_i : \mathsf{t}, \text{ and} \\ & \quad \mathsf{den}(C_i)(g)(s) = 1, \\ & \text{for all } i \text{ s.th. } C_i : \tilde{\mathsf{t}}, \text{ and} \\ & \quad \sigma_0 \not\equiv (\mathsf{s} \to \sigma), \\ & \text{for all } \sigma \in \mathsf{Types} \\ er_{\sigma_0}, & \text{otherwise} \end{cases}$$

Informally, the denotation of a term A of the form (1e) is the denotation of A_0, in case all C_i are true, otherwise it is error, i.e., it is error in case some of the terms A_0, C_i is error, or some of the terms C_i is false ($i = 1, \dots, m$).

4 Reduction Calculus

The formal language $\mathrm{L}_{\mathrm{ar}}^{\lambda}$ has reduction rules that reduce its terms to their unique, up to congruence, canonical forms. Here, we extend the set of the rules given in [5], by adding two *Restriction Rules*, (st1)–(st2), for reducing terms having occurrences of the restrictor operator.

Restriction Rules: given that: (st1)/(st2)

- each term in $\overrightarrow{R}$ is immediate and has a type of a truth value
- each C_j ($j = 1, \dots, m$, $m \geq 0$) is proper and has a type of a truth value

Case1: A_0 is an immediate term

$$\begin{aligned} &(A_0 \text{ such that } \{C_1, \dots, C_m, \overrightarrow{R}\}) \\ \Rightarrow\ &(A_0 \text{ such that } \{c_1, \dots, c_m, \overrightarrow{R}\}) \\ &\qquad \text{where } \{c_1 := C_1,\ \dots, c_m := C_m\} \end{aligned} \tag{st1}$$

for fresh $c_j \in \mathsf{RecV}$ ($j = 1, \dots, m$)

Case2: A_0 is a proper term

$$\begin{aligned} &(A_0 \text{ such that } \{C_1, \dots, C_m, \overrightarrow{R}\}) \\ \Rightarrow\ &(a_0 \text{ such that } \{c_1, \dots, c_m, \overrightarrow{R}\}) \\ &\qquad \text{where } \{a_0 := A_0, \\ &\qquad\qquad c_1 := C_1,\ \dots, c_m := C_m\} \end{aligned} \tag{st2}$$

for fresh $a_0, c_j \in \mathsf{RecV}$ $(j = 1, \ldots, m)$

5 Algorithmically Restricted Memory Parameters

Theorem 1 (Restricted Memory Variables). *Assume that, for any $n \geq 1$:*

- $\overrightarrow{R_j}$ *are proper terms, and* $\overrightarrow{I_j}$ *are immediate*
- $p_i \in \mathsf{RecV}$ $(i = 2, \ldots, n)$ *and* $r_j \in \mathsf{RecV}$ $(j = 1, \ldots, n)$ *are fresh with respect to* $p_1, \overrightarrow{R_j}, \overrightarrow{I_j}$ $(j = 1, \ldots, n)$

Then:

$$((\ldots(p_1 \text{ s.t. } \{\overrightarrow{R_1}, \overrightarrow{I_1}\})\ldots) \text{ s.t. } \{\overrightarrow{R_n}, \overrightarrow{I_n}\}) \tag{8a}$$

$$\Rightarrow (p_n \text{ s.t. } \{\overrightarrow{r_n}, \overrightarrow{I_1}\}) \text{ where } \{ \tag{8b}$$

$$p_n := (p_{n-1} \text{ s.t. } \{\overrightarrow{r_{n-1}}, \overrightarrow{I_{n-1}}\}), \tag{8c}$$

$$\ldots,$$

$$p_2 := (p_1 \text{ s.t. } \{\overrightarrow{r_1}, \overrightarrow{I_1}\}), \tag{8d}$$

$$\overrightarrow{r_n} := \overrightarrow{R_n}, \ldots, \overrightarrow{r_1} := \overrightarrow{R_1}\} \tag{8e}$$

The proof is by induction on $n \geq 1$, since it's long, we leave it for extended work.

We call the terms of the form (8a) and (8b)–(8e) *restricted memory variables.* The term (8b)–(8e) is the canonical form of (8a), in case $\overrightarrow{R_i}$ are explicit (without the constant where) and irreducible, for all $i = 1, \ldots, n$.

Note 1. The memory, i.e., recursion, variable $p_1 \in \mathsf{RecV}$ is free in (8a) and (8b)–(8e), while restricted by $\overrightarrow{R_1}$. For all $i = 2, \ldots, n$, the memory variables $p_i \in \mathsf{RecV}$ are both restricted by $\overrightarrow{R_i}$ and bound by the recursion operator where.

In the special case of $\overrightarrow{R_i} = \emptyset$, for all $i = 1, \ldots, n$, the memory variables $p_i \in \mathsf{RecV}$ $(i = 1, \ldots, n)$ are called *basic, restricted memory variables.*

Definition 2 (Reduction Relation, ⇒). *For any two terms A and B, A reduces to B, denoted by $A \Rightarrow B$, iff B can be obtained from A by a finite number of applications of reduction rules, including the new rules for restricted terms.*

Note 2. In the extended calculus, the irreducible terms include the restriction operator such that, abbreviated by s.t., applied only to immediate recursion variables, and thus, in the canonical forms (9), including special casses as in (10), when the constant s.t. occurs in the head part. Restricted memory variables, as in (8b)–(8e), are special cases of immediate variables.

Theorem 2 (Canonical Form Theorem). *For every $A \in \mathsf{Terms}$, there is a unique up to congruence, irreducible term* $\mathsf{cf}(A) \in \mathsf{Terms}$, *the* canonical form of *A, such that:*

(1) For some explicit, irreducible $A_0, \dots, A_n \in \mathsf{Terms}$ *($n \geq 0$):*

$$\mathsf{cf}(A) \equiv A_0 \text{ where } \{p_1 := A_1, \dots, p_n := A_n\} \tag{9}$$

(2) Assume the special case of $A \equiv \big(A \text{ such that } \{\overrightarrow{C}, \overrightarrow{R}\}\big)$*, given that:*
- *each term in* $\overrightarrow{R}$ *is immediate, and has a type of a truth value*
- *each* C_j *($j = 1, \dots, m$, $m \geq 0$) is proper, and has a type of a truth value*

Then (9) has the special form (10):

$$\mathsf{cf}(A) \equiv \big(A'_0 \text{ such that } \{\overrightarrow{c}, \overrightarrow{R}\}\big) \text{ where } \{p_1 := A_1, \dots, p_n := A_n\} \tag{10}$$

for some immediate $A'_0 \in \mathsf{Terms}$*, some explicit, irreducible* $A_1, \dots, A_n \in \mathsf{Terms}$ *($n \geq 0$), and memory variables* $c_j, p_i \in \mathsf{RecV}$ *($j = 1, \dots, m$, $m \geq 0$, $i = 1, \dots, n$), such that* $\overrightarrow{c} \subseteq \overrightarrow{p}$ *(i.e., for all* j*,* $c_j \in \{p_1, \dots, p_n\}$*)*

(3) $A \Rightarrow \mathsf{cf}(A)$

6 On Algorithmic Semantics and Memory Constraints

In this section, we emphasise the significance of the algorithmic semantics in the type-theory of acyclic recursion L^{λ}_{rar}, with respect to its denotational semantics.

6.1 Formalization of the Notion of Iterative Algorithm

The type-theory of acyclic recursion is a formalization of the mathematical notion of algorithm, for computing values of recursive functions designated by *recursion terms.* The values of the functions, when denoted by meaningful, i.e., proper, recursive terms, are computed iteratively by algorithms determined by *the canonical forms of the corresponding terms.* In this paper, we have introduced L^{λ}_{rar}, which is an enrichment of $\mathrm{L}^{\lambda}_{\mathrm{ar}}$, by restricted memory variables (to save parametric data). I.e., the concept of algorithm, with memory parameters, which are constrained to store data, which in addition to being typed, can be restricted to satisfy properties. The algorithms are expressed by terms in canonical forms, which carry data components computed and stored in restricted memory variables, i.e., memory slots. The restrictions of the memory variables, are expressed algorithmically by the canonical forms as well.

Thus, the concept of algorithm, with restricted memory parameters, is formalised at the object level of the syntax and reduction calculus of L^{λ}_{rar}.

For a term $A \in L^{\lambda}_{rar}$, that has occurrences of the restrictor operator, the canonical form $\mathsf{cf}(A)$ has occurrences of the restrictor only in sub-terms that represent restricted immediate terms of the form $(P \text{ such that } \{C_1, \dots, C_m\})$, for immediate terms $P, C_1, \dots, C_m$.

Immediate Terms. In case A is an immediate term of $\mathrm{L}^{\lambda}_{\mathrm{ar}}$, and thus of L^{λ}_{rar}, i.e., $A \equiv_{\mathsf{c}} \mathsf{cf}(A) \equiv_{\mathsf{c}} \lambda(u_1) \dots \lambda(u_n)(V(v_1) \dots (v_m))$, for $V \in \mathsf{Vars}$, $u_j, v_i, \in \mathsf{PureV}$, $i = 1, \dots, n$, $j = 1, \dots, m$, $m, n \geq 0$, the value $\mathsf{den}(A)(g)$ is provided immediately by the values $g(V)$ and $g(v_i)$, and abstracting away from the values $g(u_j)$.

Proper Terms. In case A is a proper term, i.e., non-immediate, there is an algorithm $\mathsf{alg}(A)$ for computing $\mathsf{den}(A)(g)$. The algorithm $\mathsf{alg}(A)$ is determined by its canonical form, (9), $\mathsf{cf}(A) \equiv A_0$ where $\{p_1 := A_1, \ldots, p_n := A_n\}$. The denotation $\mathsf{den}(A)(g) = \mathsf{den}(\mathsf{cf}(A))(g)$ is computed stepwise, iteratively, according to the ranking $\mathsf{rank}(p_i)$, $i = 1, \ldots, n$ $(n \geq 0)$, by starting with the lowest rank, the algorithm computes $\mathsf{den}(A_i)(g)$ and saves the values in the corresponding memory variables p_i. Then, these values are used for computing the denotation $\mathsf{den}(A)(g) = \mathsf{den}(A_0)(g)$ (Fig. 1).

$$\underbrace{\text{Syntax of } \mathrm{L}^{\lambda}_{\mathrm{rar}} \Longrightarrow \text{Algorithms for Iterative Computations} \Longrightarrow \text{Denotations}}_{\text{Algorithmic Semantics of } \mathrm{L}^{\lambda}_{\mathrm{rar}}}$$

Fig. 1. Algorithmic semantics for iterative computations with restricted parameters

6.2 Algorithmic Semantics and Constants

Example 1. The terms A, B, C, respectively in (11a), (12a), (13), designate different algorithms for computing the same numerical value 40, which is the denotation of each of them. The algorithms are determined by their corresponding canonical forms $\mathsf{cf}(A)$, $\mathsf{cf}(B)$, $\mathsf{cf}(C)$, in (11b), (12b), (13):

(1) The term $\mathsf{cf}(A)$ in (11b) determines the algorithm for computing $\mathsf{den}(A)$:

$$A \equiv (200 + 40)/6 \tag{11a}$$

$$\Rightarrow_{\mathsf{cf}} \underbrace{n/d \text{ where } \{ n := (a_1 + a_2),}_{\text{algorithmic pattern}} \;\; \underbrace{a_1 := 200,\; a_2 := 40,\; d := 6}_{\text{algorithmic instantiation of memory slots}} \; \}$$

$$\equiv \mathsf{cf}(A) \tag{11b}$$

(2) The term $\mathsf{cf}(B)$ in (12b) determines the algorithm for computing $\mathsf{den}(B)$:

$$B \equiv (120 + 120)/6 \tag{12a}$$

$$\Rightarrow \underbrace{n/d \text{ where } \{ n := (a_1 + a_2),} \;\; \underbrace{a_1 := 120,\; a_2 := 120,\; d := 6} \}$$

$$\equiv \mathsf{cf}(B) \tag{12b}$$

(3) The term $\mathsf{cf}(C)$ in (13) determines the algorithm for computing $\mathsf{den}(C)$:

$$C \equiv \underbrace{n/d \text{ where } \{ n := (a + a),} \;\; \underbrace{a := 120,\; d := 6} \} \equiv \mathsf{cf}(C) \tag{13}$$

The terms A, B, C, respectively in (11a), (12a), (13), compute the same number, which is denoted by the numeral constant 40, as their denotational values, in case of decimal number system (base 10). But, their denotations are computed by different algorithms expressed by $\mathsf{cf}(A)$, $\mathsf{cf}(B)$, $\mathsf{cf}(C)$.

$$\mathsf{den}(A) = \mathsf{den}(B) = \mathsf{den}(C) = 40 \quad \text{(decimal number system)} \tag{14a}$$

$$\mathsf{alg}(A) \neq \mathsf{alg}(B) \neq \mathsf{alg}(C) \tag{14b}$$

6.3 Parametric Algorithms with Restrictions

The restricted term R, in (15a), which is formed with the restrictor operator, $\mathsf{such\,that}$ (abbreviated s.t.), is a subterm of the term D in (15b)–(15d).

If taken alone, without the where assignments in (15c)–(15d), R designates restricted, parametric algorithm, in which $n, d \in \mathsf{RecV}$ are free restricted memory variables: $n, d \in \mathsf{FreeV}(R)$:

$$R \equiv (n/d \;\mathsf{such\,that}\; \{\, n, d \in \mathbb{Z},\; d \neq 0 \,\}) \qquad \text{(restricted term)} \tag{15a}$$

$$D \equiv \underbrace{(n/d \;\mathsf{such\,that}\; \{\, n, d \in \mathbb{Z},\; d \neq 0 \,\})}_{\mathsf{restricted\ subterm}\ R} \tag{15b}$$

$$\mathsf{where}\; \{\, n := (a_1 + a_2), \tag{15c}$$

$$a_1 := 200,\; a_2 := 40,\; d := 6 \,\} \tag{15d}$$

The term E, in (16a)–(16c), does not satisfy the restriction:

$$E \equiv \big(n/d \;\mathsf{such\,that}\; \{\, n, d \in \mathbb{Z},\; d \neq 0 \,\}\big) \;\mathsf{where}\; \{ \tag{16a}$$

$$n := (a_1 + a_2), \tag{16b}$$

$$a_1 := 200,\; a_2 := 40,\; d := 0 \,\} \tag{16c}$$

The value $\mathit{Error} \equiv er$ of E, $\mathsf{den}(E) = er$, is algorithmically computed:

$$\mathsf{den}(A) = \mathsf{den}(D) = 40; \qquad \mathsf{den}(E) = er \tag{17a}$$

$$\mathsf{alg}(A) \neq \mathsf{alg}(D) \neq \mathsf{alg}(E) \tag{17b}$$

7 Definite Descriptions

The terms with restricted parameters provide alternatives for semantic representations of definite descriptors, in particular the referential descriptors.

Alternative Operators: such that as an alternative of if ... then Any designated symbol can be used for the constant such that. It's syntactic and semantic roles are determined by the denotational, algorithmic semantics, and reduction calculi. The restricted terms (1e) can be replicated only to certain extend by the traditional terms involving "if ... then", "if ... then ... else ...", especially because $\mathrm{L}^{\lambda}_{\mathrm{ar}}$, and L^{λ}_{rar}, do not include constants (or terms) for the erroneous semantic values er. This is more important, for an extended type-theory of full recursion, by allowing terms that do not obey the Acyclicity Constraint (AC) on page 3.

In this section, as an illustration of the restricted memory variables, we give one of several possible alternatives, for representation of descriptions formed with the definite article "the".

Assume that $q \in \mathsf{RecV}_{\tilde{\mathsf{e}}}$, $p \in \mathsf{RecV}_{(\tilde{\mathsf{e}} \to \tilde{\mathsf{t}})}$, $unique \in \mathsf{Consts}_{(((\tilde{\mathsf{e}} \to \tilde{\mathsf{t}}) \to \mathsf{e}) \to \tilde{\mathsf{t}})}$. Then, we can define:

$$\big[(\mathsf{den}(unique))(g)\big](\bar{p})(y)(s_0) = \begin{cases} 1, & \text{in case (i.e., iff) } y \text{ is} \\ & \text{the unique } y \in \mathbb{T}_{\mathsf{e}}, \\ & \text{ for } \bar{p}(s \mapsto y)(s_0) = 1 \\ \mathsf{er}, & \text{otherwise} \end{cases} \tag{18}$$

We can render the definite article "the" by underspecified term, with the appropriate types of its components:

$$\text{the} \xrightarrow{\mathsf{render}} A_1 \equiv \big(q \text{ s.t. } \{\, unique(p)(q) \,\}\big) \tag{19}$$

The memory variable p is free, i.e., underspecified, in the term A_1, i.e., $p \in \mathsf{FreeV}(A_1)$, in (19). Then, p can be specified when the determiner "the" is combined with some common noun:

$$\text{the dog} \xrightarrow{\mathsf{render}} A_2 \equiv \big(q \text{ s.t. } \{\, unique(p)(q) \,\}\big) \text{ where } \{\, p := dog \,\} \tag{20a}$$

$$\Rightarrow \big(q \text{ s.t. } \{\, U \,\}\big) \text{ where } \{\, U := unique(p)(q),\ p := dog \,\} \tag{20b}$$

Then, a sentence, which has such a definite description, is rendered to a term in (21a)–(21c):

$$\text{the dog runs} \xrightarrow{\mathsf{render}} A_3 \tag{21a}$$

$$A_3 \equiv runs\big[\big(q \text{ s.t. } \{\, unique(p)(q) \,\}\big) \text{ where } \{\, p := dog \,\}\big] \tag{21b}$$

$$\begin{aligned} \Rightarrow runs(Q) \text{ where } \{\, & Q := \big(q \text{ s.t. } \{\, U \,\}\big), \\ & U := unique(p)(q),\ p := dog \,\} \end{aligned} \tag{21c}$$

The term $\mathsf{cf}(A_3)$ in (21c) carries an algorithmic pattern for sentences with a subject NP that is a definite description. The restricted recursion (memory) variable $Q := (q \text{ s.t. } \{\, U \,\})$ is underspecified without a context. Note that $q \in \mathsf{RecV}_{\tilde{\mathsf{e}}}$ is a free memory variable. It can obtain its denotational value, which depends on states c, i.e., on contexts, by variable assignment satisfying the restriction of uniqueness in c.

8 Conclusion and Forthcoming Work

In this paper, we have extended the theory of typed, acyclic recursion $\mathrm{L}^{\lambda}_{\mathrm{ar}}$, by introducing terms with restrictions. The result is a formal language L^{λ}_{rar}, which is a proper extension of the language $\mathrm{L}^{\lambda}_{\mathrm{ar}}$ and its reduction calculus. The same extension applies to the version of the type-theory with full recursion L^{λ}_{r} without the acyclicity. The two subclasses of the terms with restrictions, i.e., the basic restricted memory variables, and restricted memory variables, see Theorem 1, provide parametric algorithmic patterns for semantic representations of memory locations. The memory variables are typed, and thus, can be used to store data of the corresponding types. In addition, the restricted memory variables, introduced in this paper, can be used to store data, which is constrained to satisfy propositional restrictions. The constraints are calculated recursively, by iterative algorithms determined by canonical forms of formal terms of L^{λ}_{rar}. We have introduced a formalization of restricted algorithmic patterns for computational semantics of formal languages, e.g., in programming, and natural languages, by illustrations with mathematical expressions and definite descriptors of natural language, which are typical problems for Natural Language Processing (NLP).

References

1. Loukanova, R.: Gamma-reduction in type theory of acyclic recursion. Fundam. Inform. **170**(4), 367–411 (2019). https://doi.org/10.3233/FI-2019-1867
2. Loukanova, R.: Gamma-star canonical forms in the type-theory of acyclic algorithms. In: van den Herik, J., Rocha, A.P. (eds.) Agents and Artificial Intelligence. ICAART 2018. Lecture Notes in Computer Science, vol. 11352, pp. 383–407. Springer, Cham (2019). https://doi.org/10.1007/978-3-030-05453-3_18
3. Loukanova, R.: Type-theory of acyclic algorithms for models of consecutive binding of functional neuro-receptors. In: Grabowski, A., Loukanova, R., Schwarzweller, C. (eds.) AI Aspects in Reasoning, Languages, and Computation, vol. 889, pp. 1–48. Springer, Cham (2020). https://doi.org/10.1007/978-3-030-41425-2_1
4. Moschovakis, Y.N.: Sense and denotation as algorithm and value. In: Oikkonen, J., Väänänen, J. (eds.) Logic Colloquium '90: ASL Summer Meeting in Helsinki. Lecture Notes in Logic, vol. 2, pp. 210–249. Springer, Berlin (1993)
5. Moschovakis, Y.N.: A logical calculus of meaning and synonymy. Linguist. Philos. **29**(1), 27–89 (2006). https://doi.org/10.1007/s10988-005-6920-7

Author Index

A
Adamatti, Diana Francisca, 221, 231
Agostinho, Nilzair Barreto, 231
Analide, Cesar, 168, 187
Aoki, Toshinori, 206
Arisaka, Ryuta, 41
Arora, Ashish, 258

B
Bartolomé del Canto, Álvaro, 303
Bedia, Manuel González, 295
Bergenti, Federico, 60
Bernal, E. Frutos, 241
Bravo, Narledis Núñez, 81

C
Camacho, Alberto Rivas, 284
Carneiro, João, 314
Castillo, Luis Fernando, 295
Castillo, Luis, 11
Cera, Juan Manuel, 266
Chamoso, Pablo, 258
Chamoso-Santos, Pablo, 303
Chen, Wen-Hui, 102
Conceição, Luís, 314
Corchado, Emilio, 258
Corchado, Juan Manuel, 159, 284
Corchado-Rodríguez, Juan M., 303

D
de la Prieta-Pintado, Fernando, 303
de Luis-Reboredo, Ana, 142
del Rey, A. Martín, 241
Durães, Dalila, 211

F
Fantozzi, Paolo, 113
Fernandes, Bruno, 187
Fonseca, Joaquim, 211
Freitas, A., 274
Fukuyama, Kohei, 159

G
Gil-González, Ana B., 142
Gonçalves, Filipe, 211
González-Briones, Alfonso, 258

H
Hamada, Mitsuru, 159
Hernández-Nieves, Elena, 303
Hutzler, Guillaume, 89

I
Iikura, Riku, 21
Iotti, Eleonora, 60
Isaza, Gustavo, 11
Ito, Takayuki, 41

Y. Dong et al. (Eds.): DCAI 2020, AISC 1237, pp. 333–334, 2021.
https://doi.org/10.1007/978-3-030-53036-5

J
Jimenez, Roberto, 179

K
Kato, Yumiko O., 159
Klaudel, Hanna, 89
Kuo, Han-Yang, 102

L
Laura, Luigi, 113
Lezama, Omar Bonerge Pineda, 81, 134, 152, 179, 198, 251, 266
Lin, Yu-Chen, 102
Llerena, Isabel, 251
Loukanova, Roussanka, 321

M
Machado, José, 211
Marcondes, Francisco S., 211
Marreiros, Goreti, 314
Martins, Vinícius Borges, 221
Mateos, Alberto Martín, 284
Mathieu, Philippe, 1
Matsui, Kenji, 159
Mirto, Ignazio Mauro, 124
Monica, Stefania, 60
Moreno-García, María N., 142
Mori, Naoki, 21, 51
Muñoz, Fabián, 11

N
Nakahara, Tomonori, 159
Nakatoh, Yoshihisa, 159
Nongaillard, Antoine, 1
Novais, Paulo, 187, 211, 314

O
Okada, Makoto, 21
Oliveira, Pedro, 187
Orozco-Alzate, Mauricio, 31
Ospina-Bohórquez, Alejandra, 142

P
Patiño, Yisel Pinillos, 152
Petrosino, Giuseppe, 60

R
Ramalho, A., 274
Rivas, Alberto, 159
Rosa, Luís, 168

S
Sali, Abderrahmane, 89
Sati, Vishwani, 258
Satoh, Ichiro, 71
Serrano, D. Hernández, 241
Shoeibi, Niloufar, 258, 284
Silva, Fábio, 168
Silva, Jesús, 81, 152, 179, 198, 266
Solano, Darwin, 179
Souza, J., 274

T
Terauchi, Akira, 51
Tsai, Cheng-Han, 102

U
Ueno, Miki, 51, 206
Uribe-Hurtado, Ana-Lorena, 31

V
Vara, R. Casado, 241
Varela, Noel, 81, 134, 152, 198
Vargas, Jesús, 134, 266
Vélez, Jairo I., 295
Villegas-Jaramillo, Eduardo-José, 31
Viloria, Amelec, 134, 251

W
Wherhli, Adriano Velasque, 231

Z
Zilberman, Jack, 81, 152